CBS PROBLEMS and SOLUTIONS SERIES

Problems and Solutions in Integrated Electronics

Problems and Solutions in

Integrated Electronics

- This book is suitable for examinations of all Indian universities.
- Solutions are given in most simplified manner.
- Each type of problems is included and the solutions given in detail.
- To explain the mechanism lucidly, computerised figures have been liberally added in the solutions. This volume contains more than 310 line diagrams.
- Problems from the last ten years Indian Engineering Service examinations are added till date at the end of the text as a separate section.

CBS PROBLEMS and SOLUTIONS SERIES

Problems and Solutions in Integrated Electronics

R GOPAL
Engineer

CBS
CBS PUBLISHERS & DISTRIBUTORS PVT. LTD.
New Delhi • Bangalore • Pune • Cochin • Chennai (India)

ISBN : 81-239-1091-6

First Edition : 2004
Reprint : 2006, 2008, 2009, 2011, 2013

Published by Satish Kumar Jain and produced by V.K. Jain for
CBS Publishers & Distributors Pvt. Ltd.,
CBS Plaza, 4819/XI Prahlad Street, 24 Ansari Road, Daryaganj,
New Delhi - 110002, India. • Website: www.cbspd.com
e-mail: delhi@cbspd.com, cbspubs@airtelmail.in
Ph.: 23289259, 23266861, 23266867 • Fax: 011-23243014

Branches:

- ***Bengaluru:*** Seema House, 2975, 17th Cross, K.R. Road,
  Bansankari 2nd Stage, Bengaluru - 560070
  • Ph.: +91-80-26771678/79 • Fax: +91-80-26771680
  • E-mail: cbsbng@gmail.com, bangalore@cbspd.com
- ***Pune:*** Bhuruk Prestige, Sr. No. 52/12/2+1+3/2,
  Narhe, Haveli (Near Katraj-Dehu Road By-pass), Pune - 411041
  • Ph.: +91-20-64704058/59, 32342277 • E-mail: pune@cbspd.com
- ***Kochi:*** 36/14, Kalluvilakam, Lissie Hospital Road,
  Kochi - 682018, Kerala • Ph.: +91-484-4059061-65
  • Fax: +91-484-4059065 • E-mail: cochin@cbspd.com
- ***Chennai:*** 20, West Park Road, Shenoy Nagar, Chennai - 600030
  Ph.: +91-44-26260666, 26208620 • Fax: +91-44-42032115
  • E-mail: chennai@cbspd.com

Printed at :
J.S. Offset Printers, Delhi

Preface

Integrated Electronics is one of the primary subjects for the students of electronics and electrical engineering. The author feels great pleasure in presenting the subject of Integrated Electronics through typical problems and their solutions. The book covers every topic taught under the subject of integrated electronics at the undergraduate level in an engineering college. All the problems are discussed at length, and line diagrams and figures have been given liberally to help the reader in understanding the logic. This volume is suitable for all university examinations. Also questions of the past ten years from various competitive examinations like Indian Engineering Service and Civil Service examinations, till date, have been added in the form of a separate section at the end of the text chapters.

Although every care has been taken to ensure accuracy, yet some errors might have crept in while compiling this voluminous work. The author will be grateful if these are brought to his notice so that these could be rectified in the subsequent printings and editions of the book. Readers are requested to send their suggestions for further improvement and revision at the following e-mail address: cbspubs@del3.vsnl.net.in.

R Gopal

January 2004

Contents

Chapter 1

Transport Phenomenon in Semiconductors

Problem 1.1: Prove that the concentration n of free electrons per cubic meter of a metal is given by

$$n = \frac{dv}{AM} = \frac{A_o dv \times 10^3}{A}$$

where

d = density, kg/m^3

v = valence, free electrons per atom

A = atomic weight

M = weight of atom of unit atomic weight, kg

A_o = Avogadro's number, molecules/mole

Solution:

The weight of an atom is considered to be equal to the atomic mass times the weight of an atom of unit atomic weight

$\therefore$ weight of one atom = (AM)

Concentration of free electrons/m^3

$$= \left(\frac{\text{atoms}}{AM}\right) \times (v) \times (d) = \frac{dv}{AM} \text{ electrons/m}^3 \quad \text{Proved}$$

If η = no. of atoms/molecule, then

An = molecular weight

$$\therefore n = \left(\frac{1}{An}\text{ mole / g}\right) \times \left(A_o \times \frac{\text{molecules}}{\text{mole}}\right)\left(\eta\frac{\text{atoms}}{\text{molecules}}\right) \times$$

$$\left(10^3 \times \frac{\text{g}}{\text{kg}}\left(d\frac{\text{kg}}{\text{m}^3}\right)\right) \times v \text{ (electrons/atom)}$$

$$= \left(\frac{A_o \cdot d \cdot V \times 10^3}{A}\right) \text{ proved}$$

Problem 1.2: The specific density of tungsten is 18.8 g/cm^3, and its atomic weight is 184.0. Assume that there are two free electrons per atom. Calculate the concentration of free electrons.

Solution:

For finding the concentration of electrons,

$$n = \left(6.02 \times 10^{23} \frac{\text{molecules}}{\text{mole}}\right) \times \left(\frac{1 \text{ mole}}{184 \text{ gm}}\right) \times \left(18.8 \frac{\text{gm}}{\text{cm}^3}\right) \times$$

$$(2 \text{ electrons/atom}) \times (1 \text{ atom/mole})$$

$$= 1.23 \times 10^{29} \text{ electrons/m}^3.$$

Problem 1.3: Compute the mobility of the free electrons in aluminum for which the density is 2.70 g/cm^3 and the resistivity is 3.44 × 10^{-6} Ω-cm. Assume that aluminum has three valence electrons per atom.

Solution:

Here for finding the mobility, we use,

$$\sigma = nq\mu$$

$$\therefore \mu = \frac{\sigma}{nq} = \frac{1}{nq\rho}$$

∴ q = change on e$^-$

μ = mobility

n = concentration of free electrons.

σ = conductivity

$$= \frac{1}{\rho} = \frac{1}{\text{resistivity}}$$

Now, $$n = \left(\frac{d \times V}{A \times M}\right) = \frac{2.7 \times 3 \times 10^{-3}}{26.97 \times 1.66 \times 10^{-27}} \times 10^{6}$$

$$= 18 \times 10^{28} \text{ electrons/m}^3$$

Now, $$\mu = \frac{1}{n\rho q}$$

$$= \frac{1}{\left(18 \times 10^{28}\right) \times \left(3.44 \times 10^{-6} \times \frac{1}{100}\right) \times \left(1.6 \times 10^{-19}\right)}$$

$$= 1.009 \times 10^{-3} \text{ m}^2\text{/V-sec}$$

Problem 1.4: The resistance of No.18 copper wire (diameter = 1.03 mm) is 6.51 Ω per 1,000 ft. The concentration of free electrons in copper is 8.4 × 10^{28} electrons/m³. If the current is 2A, find the (a) drift velocity, (b) mobility, (c) conductivity.

Solution:

(a) The cross-sectional area of wire

$$= \frac{\pi}{4} d^2 = 0.835 \times 10^{-6} \text{ m}^2$$

Now, current density, $$J = \frac{I}{A} = \frac{2}{0.835} \times 10^{6} = 2.4 \times 10^{-6} \text{ A/m}^2$$

Now, by eqn, $J = nqV_d$ (V_d = drift velocity)

$$V_d = \frac{J}{nq} = \left(\frac{2.4 \times 10^{6}}{8.4 \times 10^{28} \times 1.6 \times 10^{-19}}\right)$$

$$= 1.73 \times 10^{-4} \text{ m/sec}$$

(b) Also, $[V_d = \mu e]$

Resistance per meter

$$= \left(\frac{6.51}{304.2}\right) = 0.0214 \ \Omega\text{/m}$$

Now, $$e = 0.0214 \times 2 = 0.0428 \text{ V/m}$$

$$\therefore \qquad \mu = \left(\frac{V_d}{e}\right) = 4.42 \times 10^{-3} \text{ m}^2\text{/v-sec}$$

(c)
$$\sigma = nq\mu$$
$$= (8.4 \times 10^{28}) \times (1.6 \times 10^{-19}) \times (4.42 \times 10^{-3})$$
$$= 5.94 \times 10^{7} \ (\Omega\text{–m})^{-1}$$

Problem 1.5: (a) Determine the concentration of free electrons and holes in a sample of germanium at 300°K which has a concentration of donor atoms equal to 2×10^{14} atoms/cm^3 and a concentration of acceptor atoms equal to 3×10^{14} atoms/cm^3. Is this p- or *n*-type germanium? In other words, is the conductivity due primarily to holes or to electrons?

(b) Repeat part a for equal donor and acceptor concentrations of 10^{15} atom/cm^3. Is this *p*-or *n*-type germanium?

(c) Repeat part a for donor concentration of 10^{16} atom/cm^3 and acceptor concentration 10^{14} atoms/cm^3.

Solution:

(a) As we know the equation,

$$np = n_i^2 \qquad \text{...(a)}$$

and
$$N_A + n = p + N_D \qquad \text{...(b)}$$

where n_i = intrinsic concentration; N_D = concentration of donor ions N_A = concentration of acceptor ions,

Solving eq^4 (a) & (b) for n and p

$$n = -\left(\frac{N_A - N_D}{2}\right) + \left(\left(\frac{N_A - N_D}{2}\right)^2 + n_i^2\right)^{1/2}$$

$$p = -\left(\frac{N_D - N_A}{2}\right) + \left(\left(\frac{N_D - N_A}{2}\right)^2 + n_i^2\right)^{1/2}$$

Here at 300°K, n_i (G_e)

$$= 2.5 \times 10^{19}\text{/m}^3$$

$N_a = 3 \times 10^{14}$ atoms/cm^3, and $N_D = 2 \times 10^{14}$ atom/cm^3

$$\therefore \qquad n = 5.8 \times 10^{18}\text{/m}^3$$

and
$$p = 10.53 \times 10^{19}\text{/m}^3$$

Here the concentration of p is more than n and hence the sample is p-type.

(b) Here $N_A = N_D$

$\therefore$ $n = p = n_i$ and the sample of Ge is intrinsic.

(c) For $N_D = 10^{16}/\text{cm}^3$, $N_A = 10^{14}$ atoms/cm³

and $N_D \gg N_A$

$\therefore$ $n = 10^{16}/\text{cm}^3$, and $p = 6.25 \times 10^{10}/\text{cm}^3$

Hence $(n > p)$ the sample is of n-type.

Problem 1.6: (a) Find the concentration of holes and of electrons in p-type germanium at 300°K if the conductivity is 100 $(\Omega\text{-cm})^{-1}$.]
(b) Repeat part a for n-type silicon if the conductivity is 0.1 $(\Omega\text{-cm})^{-1}$

Solution:

(a) $$\sigma = pq\mu_p \quad (\because n_p \ll P_p)$$

$$\therefore \quad P_p = \frac{\sigma}{q\,\mu_p} = \frac{100}{1.6 \times 10^{-19} \times 1300}$$

$(\because$ at 300°C $\mu_p = 1300$ m²/v-sec)

$$\therefore \quad P_p = 3.47 \times 10^{17} \text{ holes/cm}^3$$
$$= 3.47 \times 10^{23} \text{ holes/m}^3$$

(b) Again for n-type,

$$\sigma = n_n\, q\mu_n \; (\therefore n_p \gg p_n)$$

$$\therefore \quad n_n = \frac{\sigma}{q\,\mu_n}$$

$$= \frac{0.1}{1300 \times 1.6 \times 10^{-19}}$$

$$= 4.81 \times 10^{14} \text{ electrons/cm}^3$$
$$= 4.81 \times 10^{20}/\text{m}^3$$

$$\therefore \quad p_n = \frac{n_i^2}{n_n} = \frac{\left(1.5 \times 10^{10}\right)^2}{4.81 \times 10^{14}} = 4.68 \times 10^5 \text{ holes/cm}^3$$
$$= 4.68 \times 10^{11} \text{ holes/m}^3$$

$[\because n_i\,(300\text{K}) = 1.5 \times 10^{10}/\text{cm}^3]$

Problem 1.7: (a) Show that the resistivity of intrinsic germanium at 300°K is 45 Ω-cm.
(b) If a donor-type impurity is added to the extent of 1 atom per 10^8 germanium atoms, prove that the resistivity drops to 3.7 Ω-cm.

Solution:

(a) Formula for finding conductivity

$$\sigma = (n\mu_n = p\mu_p)\, q$$

As the Germanium is intrinsic, $n = p = n_i$

$$\therefore \quad \sigma = n_i (\mu_n + \mu_p)\, q$$

At 300°K, $\mu_n = 3800 \ \text{cm}^2/\text{v-sec}$

$\mu_p = 1800 \ \text{cm}^2/\text{v-sec}$

and $n_i = 2.5 \times 10^{13}$ electrons/cm^3

$$\therefore \quad \sigma = 2.5 \times 10^{13} (3800 + 1800) \times 1.6 \times 10^{-19}$$

$$= 0.0232 \ (\Omega\text{-cm.})^{-1}$$

$$\therefore \quad \rho = \frac{1}{\sigma} = 44.6 \cong 45 \ \Omega$$

(b) $n_{Ge} = 4.41 \times 10^{22}$ atom/cm^3,

$$\therefore \quad N_D = \frac{n_{Ge}}{10^8} = 4.41 \times 10^{14} \ \text{atoms/cm}^3$$

and then
$$p = \frac{n_i^2}{N_D} = \frac{\left(2.5 \times 10^{13}\right)^2}{4.41 \times 10^{14}}$$

$$= 1.42 \times 10^{12} \ \text{holes/cm}^3$$

$$\therefore \quad \sigma = (4.41 \times 10^{14} \times 3800 + 1.42 \times 10^{12} \times 1800) \times 1.6 \times 10^{-19}$$

$$= 0.268 \ (\Omega\text{-cm})^{-1}$$

$$\therefore \text{ Resistivity, } \rho = \frac{1}{\sigma} \cong 3.72 \ (\Omega\text{-cm})$$

Problem 1.8: (a) Find the resistivity of intrinsic silicon at 300°K.
(b) If a donor-type impurity is added to the extent of 1 atom per 10^8 silicon atoms, find the resistivity.

Solution:

(a) For an intrinsic silicon at 300°K,

$$n_i = 1.5 \times 10^{10} \text{ electrons/cm}^3,$$

$$\mu_n = 1300 \text{ cm}^2\text{/v-sec}$$

and $\mu_p = 500 \text{ cm}^2\text{/v-sec [At 300°K]}$

$$\therefore \quad \sigma = n_i (\mu_n + \mu_p)_q$$

$$= 1.5 \times 10^{10} (1300 + 500) \times 1.6 \times 10^{-19}$$

$$= 4.32 \times 10^{-6} (\Omega\text{-cm.})^{-1}$$

$$\therefore \text{ Resistivity, } \rho = \frac{1}{\sigma} = \frac{1}{4.32 \times 10^{-6}} = 231{,}000 \ \Omega\text{-cm}$$

(b)

$$\eta_{si} = 6.02 \times 10^{23} \left(\frac{\text{atoms}}{\text{mole}}\right)$$

$$\times \frac{1}{28.1} \left(\frac{\text{mole}}{\text{g}}\right) \times 2.33 \text{ (g/cm}^3)$$

$$\cong 5 \times 10^{22} \text{ atoms/cm}^3$$

$$\therefore \quad N_D = \frac{5 \times 10^{22}}{10^8}$$

$$= 5 \times 10^{14} \text{ atoms/cm}^3 \text{ and also } N_D = n$$

$$p = \left(\frac{n_i^2}{N_D}\right) = \frac{(1.5 \times 10^{10})^2}{5 \times 10^{14}} = 4.5 \times 10^5\text{/cm}^3$$

Here it is seen that $p \ll n$;

Hence the conductivity is,

$$\sigma = n\mu_n q = 5 \times 10^{14} \times 1300 \times 1.6 \times 10^{-19}$$

$$\cong 1.04 \times 10^{-1} \ (\Omega\text{-cm})^{-1}$$

$$\therefore \text{Resistivity, } \rho = \frac{1}{\sigma} = 9.6 \ \Omega\text{-cm}$$

Problem 1.9: Consider intrinsic germanium at room temperature (300°K). By what percent does the conductivity increase per degree rise in temperature?

Solution: For finding the percentage change in conductivity per degree change of temperature, we take an assumption that μ does not vary with temperature, i.e. it is more or less constant.

Hence the conductivity (σ) varies as n_i.

From eqn of the intrinsic concentration,

$$n_i^2 = A_o T^3 \, e^{\frac{-E_{Go}}{KT}} \qquad \text{...(a)}$$

$$\therefore \qquad n_i = A_o^{1/2} \times T^{3/2} \times e^{\frac{-E_{Go}}{2KT}} ;$$

Now taking ln both sides

$$\ln (ni) = \frac{1}{2} \ln A_o + \frac{3}{2} \ln T - \frac{E_{GO}}{2kT}$$

Now differentiating,

$$\left(\frac{dn_i}{n_i}\right) = \frac{3}{2}\frac{dT}{T} + \frac{E_{GO}}{2kT^2} dT$$

$$= \left(\frac{3}{2} + \frac{E_{GO}}{2KT}\right) \frac{dT}{T} \qquad \text{...(b)}$$

At $\qquad T = 300°\text{K}, KT = 0.026$

For *Ge*, $\dfrac{dn_i}{n_i} = \left[1.5 + \dfrac{0.785}{0.052}\right]\left(\dfrac{1}{300}\right) \times 100$

$= 5.5\%/\text{degrees}$

Problem 1.10: Repeat Prob.1.9 for intrinsic silicon.

Solution:

For intrinsic silicon,

At 300°K, $E_{Go} = 1.21$, $KT = 0.026$

From eqn (b) of prob. (1.9) above,

$$\frac{dn_i}{n_i} = \left(1.5 + \frac{1.21}{0.052}\right)\left(\frac{1}{300}\right) \times 100$$

$= 8.3\%/\text{degree}$

> **Problem 1.11:** Repeat Prob.1.5 (a) for a temperature of 400°K, and show that the sample is essentially intrinsic.

Solution:

From eqn of intrinsic concentration

$$n_i^2 = A_o T^3 \; e^{\frac{-E_{Go}}{KT}}$$

At $T = 300°K$

$$A_o = \frac{n_i^2}{T^3} \; e^{\frac{E_{Go}}{KT}}$$

$$= \frac{6.25 \times 10^{26}}{27 \times 10^6} \times e^{\frac{0.785}{0.026}}$$

$$\left[\begin{array}{l} \because n_i = 2.5 \times 10^{13} \text{ atoms / cm}^3 \\ E_{GO} = 0.785, KT = 0.026 \end{array} \Bigg|_{\text{at } 300°K} \right]$$

$\therefore$ $A_o = 2.91 \times 10^{32}$

At $T = 400°K$,

$$n_i^2 = 2.91 \times 10^{32} \times (400)^3 \times e^{\frac{-0.785}{8.62 \times 10^{-5} \times 400}}$$

$$= 2.57 \times 10^{30}$$

$\therefore$ $n_i \cong 1.60 \times 10^{15}$ atoms/cm^3

Now, $[n \times p = n_i^2]$ and $[N_A + n = p + N_D]$

$N_D = 2 \times 10^{14}$ atoms/cm^3,

$N_A = 3 \times 10^{14}$ atoms/cm^3

Solving for n and p (taking n, $p > 0$)

$$n = -\frac{N_A - N_D}{2} + \left[\left(\frac{N_A - N_D}{2} \right)^2 + n_i^2 \right]^{1/2}$$

$$p = -\frac{N_D - N_A}{2} + \left[\left(\frac{N_D - N_A}{2} \right)^2 + n_i^2 \right]^{1/2}$$

As, $$\left(\frac{N_A - N_D}{2}\right)^2 \ll n_i$$

$\therefore$ $$n = -\frac{N_A - N_D}{2} + n_i$$

and $$p = -\frac{N_D - N_A}{2} + n_i$$

$\therefore$ $$n \cong -0.05 \times 10^{15} + 1.60 \times 10^{15}$$
$$\cong 1.55 \times 10^{15} \text{ atoms/cm}^3$$

and $$p \cong 1.65 \times 10^{15} \text{ atoms/cm}^3$$

Here n and p are more or less same and hence the sample is essentially intrinsic.

Problem 1.12: A sample of germanium is doped to the extent of 10^{14} donor atoms/cm^3 and 7×10^{13} acceptor atoms/cm^3. At the temperature of the sample the resistivity of pure (intrinsic) germanium is 60 Ω-cm. If the applied electric field is 2 V/cm, find the total conduction current density.

Solution:

$$N_D = 10^{14} \text{ atoms/cm}^3,$$
$$N_A = 7 \times 10^{13} \text{ atoms/cm}^3,$$
$$E = 2 \text{ V/cm}$$

Now, given that the resistivity of intrinsic is 60 Ω-cm.

$\therefore$ $$\sigma = \frac{1}{\rho} = n_i (\mu_n + \mu_p) q$$

$\therefore$ $$n_i = \frac{\sigma\left(\frac{1}{\rho}\right)}{(\mu_n + \mu_p)q} \quad \left(\begin{array}{l} \because \mu_n = 3800 \\ \quad \mu_p = 1800 \end{array}\right)$$
$$= 1.86 \times 10^{13}/\text{cm}^3$$

Also, $$Pn = n_i^2 = (1.86 \times 10^{13})^2 \quad \text{...(i)}$$

and $$p - n = N_A - N_D$$
$$7 \times 10^{13} - 10^{14} = -3 \times 10^{13} \quad \text{...(ii)}$$

Solving eqn (i) and (ii), we have,

$n = 3.85 \times 10^{13}$ electrons/cm^3 and

$p = 0.88 \times 10^{13}$ electrons/cm^3

Now, $J = \sigma E$

$= q\,(n\mu_n + p\mu_p)\,E$

$= 1.6 \times 10^{-19}\,(3.88 \times 3800 + 0.88 \times 1800) \times 10^{13} \times 2$

$= 52.3$ mA/cm^2

Problem 1.13: (a) Find the magnitude of the Hall voltage V_H in an n - type germanium bar used in Fig. 1.13 having majority-carrier concentration $N_D = 10^{17}/\text{cm}^3$. Assume

$B_x = 0.1$ Wb/m^2,

$d = 3$ mm,

and $\varepsilon_x = 5$ V/cm.

(b) What happens to V_H if an identical p-type germanium bar having $N_A = 10^{17}/\text{cm}^3$ is used in part a?

Solution:

(a) In the equilibrium state,

$qE = Bqv$

$\therefore \quad E = Bv \quad$ ...(i)

Also, V_H = Hall voltage

$= E \times d = B.v.d$ ...(ii)

Also, $v = \mu E_x$

$\therefore \quad V_H = B\mu\varepsilon_x.d$

$= 0.1 \times 3800 \times 10^{-4} \times 500 \times 3 \times 10^{-3}$

$= 57$ mV

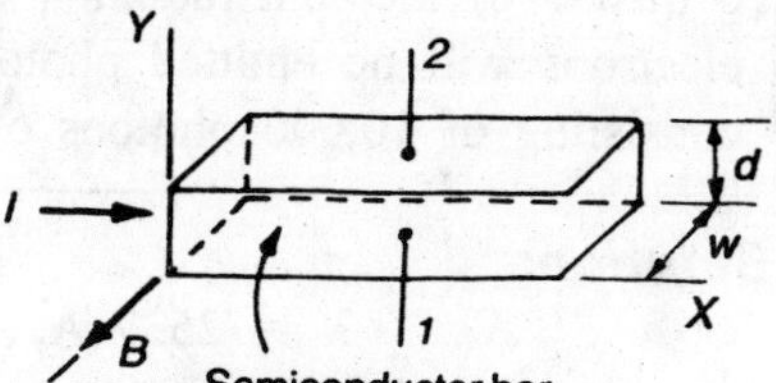

Semiconductor bar

Fig. P. 1.13

(b) In this case also, V_H retains the same value but changes polarity.

Problem 1.14: The Hall effect is used to determine the mobility of holes in a *p-type* silicon bar used in Fig. 1.14. Assume the bar resistivity is 200000 Ω-cm, the magnetic field $B_x = 0.1$ Wb/m^2 and $d = w = 3$ mm. THe measured values of the current and hall voltage are 10 μA and 50 mV find μ_P.

Solution:

From eqn of the hall coefficient,

$$R_H = \frac{V_H w}{BI} \quad \begin{pmatrix} \because V_H = 50 \text{ mV} \\ w = 3 \text{ mm} \\ I = 10\ \mu\text{A} \end{pmatrix}$$

$$= \frac{50 \times 10^{-3} \times 3 \times 10^{-3}}{0.1 \times 10 \times 10^{-6}}$$

$$= 15\ (\text{A-s/m}^3)^{-1}$$

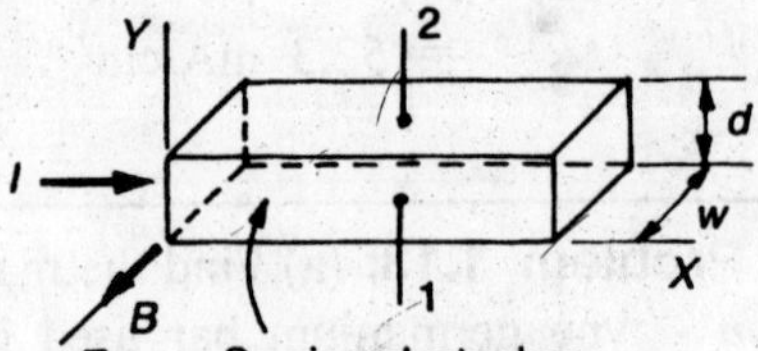

Fig. P. 1.14

Now, The mobility,

$$\mu = \sigma R_H$$

$$= \frac{1}{\rho} R_H = \frac{150}{2000}$$

$$= 75 \times 10^{-3} \text{ m}^2\text{/V-sec.}$$

Problem 1.15: A certain photosurface has a spectral sensitivity of 6 mA/W of incident radiation of wavelength 2,537 Å. How many electrons will be emitted photoelectrically by a pulse of radiation consisting of 10,000 photons of this wavelength?

Solution:

λ = 2537 Å,

no of photons = 10,000

Spectral sensitivity

= 6 mA/W

∴ Energy of one photon

$$E_G = \frac{1.24}{\lambda_G} = \frac{1.24}{0.2537} = 4.89 \text{ ev/photon}$$

∴ Total energy of the photons

$$= 4.89 \times 10^4 \times 1.6 \times 10^{-16}$$

$$= 7.8 \times 10^{-12} \text{ W-sec}$$

∴ Total no. of electrons

$$= \frac{7.8 \times 10^{-12} \times 6 \times 10^{-6}}{1.6 \times 10^{-19}} = 293 \text{ electrons}$$

Problem 1.16: (a) Consider the situation depicted in Fig. 1.16 with the light turned on. Show that the equation of conservation of charge is

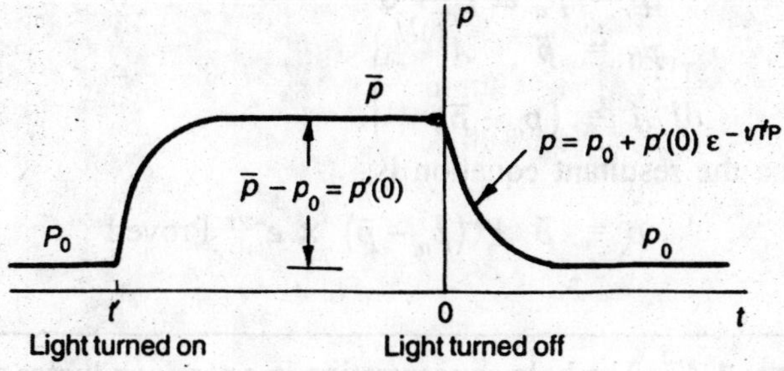

Fig. P. 1.16

$$\frac{dp}{dt} + \frac{p}{\tau} = \frac{\bar{p}}{\tau}$$

Where the time axis in Fig. 1.16 is shifted to t'.

(b) Verify that the concentration is given by the equation

$$p = \bar{p} + (p_o - \bar{p})\,e^{-t/\tau}$$

Solution:

Let $\quad \dfrac{p}{\tau}$ = decrease in No. of holes/sec due to recombination

$\dfrac{p_o}{\tau}$ = g, increase in no. of holes due to thermally generated when light is injected,

$\dfrac{\bar{p}-p_o}{\tau}$ = No. of holes/sec injected due to light (increase in no. of holes)

Hence the eqn of balance takes place,

Total no. of holes generated or decreased/sec,

$$\frac{dp}{dt} = \frac{p_o}{\tau} - \frac{p}{\tau} + \frac{\bar{p}-p_o}{\tau} = \frac{\bar{p}}{\tau} - \frac{p}{\tau}$$

$$\therefore \quad \frac{dp}{dt} + \frac{p}{\tau} = \frac{\bar{p}}{\tau} \quad \text{proved}$$

(b) Solving the differential equation above proved in part (a), this will give exponential solution,

$$p = \bar{p} + A \times e^{-t/\tau}$$

As, $p = \bar{p}$ at $t \to \infty$

and $p = p_o$ at $t \to 0$

i.e., $p_0 = \bar{p} + A$

$\therefore$ $A = (p_o - \bar{p})$

Hence the resultant equation is,

$$p = \bar{p} + (p_o - \bar{p}) \times e^{-t/\tau} \text{ Proved}$$

Problem 1.17: The hole concentration in a semiconductor specimen is shown.

(a) Find an expression for and sketch the hole current density $J_P(x)$ for the case in which there is no externally applied electric field.

(b) Find an expression for and sketch the built-in electric field that must exist if there is to be no net hole current associated with the distribution shown.

(c) Find the value of the potential between the points $x = 0$ and $x = W$ if $p(0)/p_O = 10^3$.

Solution:

(a) From the equation of hole current density

$$J_p(x) = -qD_p \frac{dp}{dx}$$

$$= -qD_p \times \left(\frac{p(0) - p_O}{w}\right)_{[x \ll w]}$$

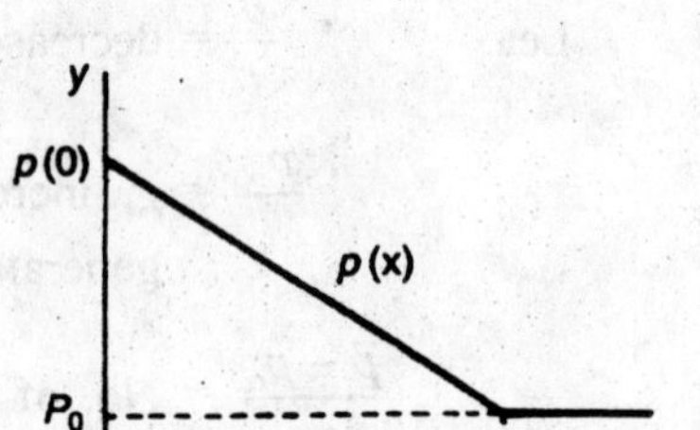

Fig. P. 1.17 (a)

Where

D_p = diffusion constant for holes

(b) When there is no net hole current

$$\therefore \quad J_P = q\mu_p p \times E - q D_p \times \frac{dp}{dx} = 0$$

$$\therefore \quad E = \frac{D_p}{\mu_p \times p(x)} \left(\frac{dp}{dx}\right) = \frac{V_T}{P_{(x)}} \times \frac{dp}{dx}$$

$$= -\frac{V_T}{p(x)} \times \left(\frac{p(0) - p_O}{w}\right) \quad \left(\because \frac{D}{\mu} = \text{constant} = V_T\right)$$

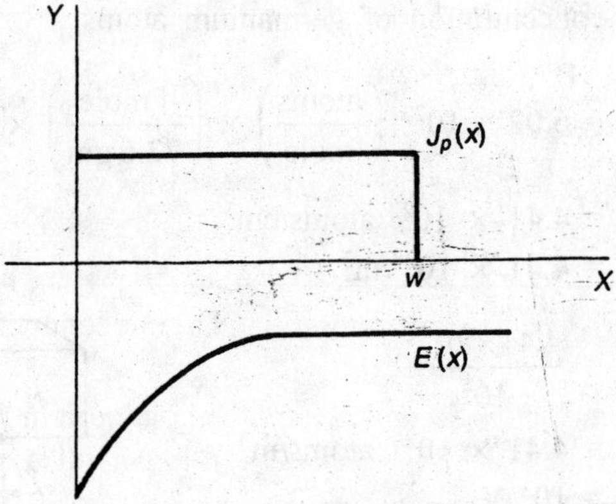

Fig. P. 1.17 (b)

The plot of $J_p(x)$ and $E(x)$ V_s X is given above

(c) $\because$ $$E_x = -\frac{dV}{dx} = \frac{V_T}{p(x)} \times \frac{dp}{dx}$$

$$\Rightarrow \quad \frac{dV}{V_T} = -\frac{dp}{p(x)}$$; Now integrating both sides

$$\int_{V(0)}^{V(w)} \frac{dV}{V_T} = -\int_{p(0)}^{p_0} \frac{dp}{p}$$

$\therefore$ Change in potential, $dV = V(w) - V_{(o)} = V_T \ \ln\left(\frac{p_{(0)}}{p_O}\right)$

Generally we take, for the case assumed, $V_T = 26$ mV

$\therefore \quad dV = 26 \times 3 \times \ln 10 \cong 180$ mV

Problem 1.18: (a) Consider the step-graded germanium semiconductor of Fig.1.18 with $N_D = 10^3 N_A$ and with N_A corresponding to 1 acceptor atom per 10^8 germanium atoms. Calculate the contact *difference of potential* V_o at room temperature.
(b) Repeat part a for a silicon *p-n* junction.

Solution:

For the germanium semiconductor,

$$n_i = 2.5 \times 10^{16}/\text{m}^3$$

Now, the concentration of germanium atoms,

$$n = 6.02 \times 10^{23} \left(\frac{\text{atoms}}{\text{mole}}\right) \times \left(\frac{1\ \text{mole}}{72.6\ \text{gm}}\right) \times 5.32\ (\text{gm/cm}^3)$$

$$= 4.41 \times 10^{22}\ \text{atoms/cm}^3$$

$$= 4.41 \times 10^{28}/\text{m}^3$$

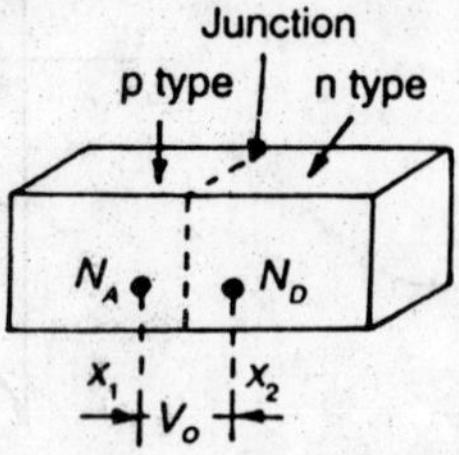

Fig. P. 1.18

$$\therefore \quad N_A = \frac{4.41 \times 10^{28}}{10^8}$$

$$= 4.41 \times 10^{20}\ \text{atoms/m}^3$$

$$N_D = 10^3\ N_A$$

$$= 4.41 \times 10^{23}\ \text{atoms/m}^3$$

Now, from equation of the contact potential difference,

$$V_o = V_T \ln \frac{N_A \times N_D}{n_i^2}$$

$$= 0.026 \ln \left(\frac{4.41 \times 4.41 \times 10^{43}}{6.25 \times 10^{32}}\right) \cong 0.33\text{V}.$$

(b) Now same procedure will repeat in calculation of silicon

$$n = 6.02 \times \frac{1}{28.1} \times 2.33 \times 10^6$$

$$= 5 \times 10^{28}/\text{m}^3 \quad (\text{In Si, density} = 2.33\ \text{gm/cm}^3)$$

$$\therefore \quad N_A = 5 \times 10^{20}\ \text{atoms/m}^3,\ N_D = 5 \times 10^{23}\ \text{atoms/m}^3$$

$$n_i^2\ (\text{Si}) = 2.25 \times 10^{32}\ \frac{\text{atom}^2}{\text{m}^6}$$

$$\therefore \quad V_o = 0.026 \ln \frac{25 \times 10^{43}}{2.25 \times 10^{32}} \cong 0.722\ \text{V}$$

Chapter 2

Junction Diode Characteristics

Problem 2.1: (a) The resistivities of the two sides of a step-graded germanium diode are 2 Ω-cm (*p* side) and 1 Ω-cm (*n* side). Calculate the height E_o of the potential-energy barrier.
(b) Repeat part (a) for a silicon *p-n* junction.

Solution:

(a) The resistivities of two sides, i.e. for *p*-side is 2 Ω-cm and for *n*-side it is 1 Ω-cm.

Now from equation of conductivity,

$$\sigma = \frac{1}{\rho} = Nq\mu$$

$$\therefore \quad \rho_p = \frac{1}{N_A\, q\, \mu_p} = \frac{1}{N_A \times 1.6 \times 10^{-19} \times 1800}$$

(For Ge, $\mu_p = 1800$)

$$\therefore \quad N_A = 1.74 \times 10^{15}/\text{cm}^3$$

Similarly, $$N_D = \frac{1}{\rho_n\, q\, \mu_n}$$

$$= 1.65 \times 10^{15}/\text{cm}^3 \qquad [\because \mu_n = 3800]$$

∴ Value of potential-energy barrier,

$$E_o = 0.026 \ln \frac{1.74 \times 1.55 \times 10^{30}}{6.25 \times 10^{26}} = 0.22 \text{ V}$$

(b) In this case of Si,

$$N_A = \frac{1}{2 \times 1.6 \times 10^{-19} \times 500} = 6.25 \times 10^{15}/\text{cm}^3$$

$$N_D = 4.8 \times 10^{15}\ /\text{cm}^3$$

∴ Value of Energy barrier,

$$E_o = 0.026 \ln \frac{6.25 \times 10^{15} \times 4.8 \times 10^{15}}{2.25 \times 10^{20}}$$

$$= 0.673 \text{ V}$$

Problem 2.2: (a) Prove that the reverse saturation current in a *p-n* diode is given by

$$I_o = Aq \left(\frac{D_p}{L_p N_D} + \frac{D_n}{L_n N_A} \right) n_i^2$$

(b) Starting with the expression for I_O found in part *a,* verify that the reverse saturation current is given by

$$I_O = AV_T \frac{b\sigma_i^2}{(1+b)^2} \left(\frac{1}{L_p \sigma_n} + \frac{1}{L_n \sigma_p} \right)$$

Where σ_n (σ_p) = conductivity of *n* (*p*) side

σ_i = conductivity of intrinsic material

$b = \mu_n / \mu_p$

Solution: Total current in the reversed condition,

$$I = I_{pn}(o) + I_{np}(o)$$

$$= \left(\frac{Aq\, D_p}{L_p} D_{n_o} + \frac{Aq\, D_n}{L_n} n_{p_o} \right) \left[e^{\frac{V}{V_T}} - 1 \right] \qquad \ldots(1)$$

Now equation of current is

$$I = I_o \left[e^{\frac{V}{V_T}} - 1 \right] \qquad \ldots(2)$$

Comparing equations (1) and (2),

$$I_o = Aq\left(\frac{D_p}{L_p}p_{n_o} + \frac{D_n}{L_n}n_{p_o}\right) \quad ...(3)$$

Also, $$p_{n_o} = \frac{n_i^2}{N_D}$$

and $$n_{p_o} = \frac{n_i^2}{N_A} \quad ...(4)$$

Then from equation (3) and (4)

$$I_o = A \,.\, q\left(\frac{D_p}{L_p\, N_D} + \frac{D_n}{L_n\, N_A}\right) n_i^2 \text{ proved}$$

This is the required equation for the reverse saturation current.

(b) Here $D_p = \mu_p V_T$

and $D_n = \mu_n V_T$

$$\therefore \quad I_o = A\left[\frac{q\mu_b V_T}{L_p N_D} + \frac{q\mu_n V_T}{L_n N_A}\right] n_i^2$$

$$= AV_T\left(\frac{q^2\mu_p\mu_n}{L_p \times q\mu_n N_o} + \frac{q^2\mu_n\mu_p V_T}{L_n N_A q \times \mu_p}\right)n_i^2$$

$$= AV_T q^2\mu_p\mu_n\left[\frac{1}{L_p\sigma_n} + \frac{1}{\sigma_p L_n}\right] n_i^2$$

$$[\because \ \sigma = q\mu N]$$

Also this can be written as,

$$I_o = AV_T q^2\ (\mu_p + \mu_n)^2\ n_i^2\ \frac{\mu_p\mu_n}{(\mu_p + \mu_n)^2} \times$$

$$\left(\frac{1}{L_p\sigma_n} + \frac{1}{L_n\sigma_p}\right)$$

$$= AV_T\, \sigma_i^2 \frac{\dfrac{\mu_n}{\mu_p}}{\left(1+\dfrac{\mu_n}{\mu_p}\right)^2} \left(\frac{1}{L_p\sigma_n} + \frac{1}{L_n\sigma_p}\right)$$

$$[\because \sigma_i = q\,(\mu_p + \mu_n)\, n_i]$$

$$\therefore \qquad I_o = AV_T\, \sigma_i^2 \frac{b}{(1+b)^2} \left(\frac{1}{L_p\sigma_n} + \frac{1}{L_n\sigma_p}\right) \quad \text{Proved}$$

Problem 2.3: (a) Using the result of Prob. 2.2 find the reverse saturation current for a germanium *p-n* junction diode at room temperature, 300°K. The cross-sectional area is 4.0 mm^2.

$$\sigma_p = 1.0 \ (\Omega\text{-cm})^{-1}$$
$$\sigma_n = 0.1 \ (\Omega\text{-cm})^{-1}$$
$$L_n = L_p = 0.15 \text{ cm}$$

Other physical constants are given in Table A

(b) Repeat part a for a silicon *p-n* junction diode. Assume

$$L_n = L_p = 0.01\text{cm and } \sigma_n = \sigma_p = 0.01 \ (\Omega\text{-cm})^{-1}$$

Table A

Properties of Germanium and Silicon

Property	Ge	Si
Atomic number	32	14
Atomic weight	72.6	28.1
Density, g/cm^3	5.32	2.33
Dielectric constant (relative)	16	12
Atoms/cm^3	4.4×10^{22}	5.0×10^{22}
E_{Go}, eV, at 0°K	0.785	1.21
E_G, eV, at 300°K	0.72	1.1
n_i at 300°K, cm^{-3}	2.5×10^{13}	1.5×10^{10}
Intrinsic resistivity at 300°K, Ω-cm	4.5	230,000
μ_n, cm^2/V-s at 300°K	3,800	1,300
μ_p, cm^2/V-s at 300°K	1,800	500
D_n, cm^2/s = $\mu_n V_T$	99	34
D_p, cm^2/s = $\mu_p V_T$	47	13

Solution:

(a) For Ge *p-n* junction diode:

$$\sigma_p = 1.0 \ (\Omega\text{-cm})^{-1}$$
$$\sigma_n = 0.1 \ (\Omega\text{-cm})^{-1}$$
$$L_n = L_p = 0.15 \text{ cm}$$
$$\sigma_i = q\,(\mu_p + \mu_p)\, n_i$$
$$= 2.5 \times 10^{13} \times 1.6 \times 10^{-19} \ (3800 + 1800)$$
$$= 2.24 \ (\Omega\text{-cm})^{-1}$$
$$A = 4 \times 10^{-6} \text{ m}^2, \ V_T = 0.026 \text{ V}$$
$$Lp\sigma_n = 0.015 \ (\Omega)^{-1}$$
and $$L_n\sigma_p = 0.15 \ (\Omega)^{-1}$$

Then, I_o will be calculated from above eqn found in the previous problem.

$$I_o = 4 \times 10^{-6} \times 0.026 \times \frac{2.11}{(3.11)^2} \times (2.24)^2 \left(\frac{1}{0.015} + \frac{1}{0.15}\right)$$

$$\cong 8.34 \ \mu A.$$

(b) Similarly for the case of Si diode,

$$A = 4 \times 10^{-2} \text{ cm}^2$$

$$b = \frac{1300}{500}\left(\frac{\mu_n}{\mu_p}\right) = 2.6$$

$$Lp\ \sigma_n = Ln\ \sigma_p = 10^{-4}\ (\Omega^{-1})$$
$$\sigma_i = 1.5 \times 10^{10}\ (1800) \times 1.6 \times 10^{-19}$$
$$= 4.32 \times 10^{-6}$$
$$\therefore \quad I_o \cong 77.8 \text{ pA.}$$

Problem 2.4: (a) For what voltage will the reverse current in a *p-n* junction germanium diode reach 90 percent of its saturation value at room temperature?

(b) What is the ratio of the current for a forward bias of 0.05 V to the current for the same magnitude of reverse bias ?

(c) If the reverse saturation current is 10 μA, calculate the forward currents for voltages of 0.1, 0.2, and 0.3, respectively.

Solution:

(a) $V_T = \dfrac{T}{11{,}600} = 0.026$ V at room temperature.

Then $$I = I_o = \left(e^{\frac{V}{V_T}} - 1\right)$$

$$0.9\ I_o = I_o\left(e^{\frac{V}{V_T}} - 1\right);\quad e^{\frac{V}{0.026}} = 0.1$$

$\therefore$ $$V = 0.060\ \text{V}$$

(b) $$\frac{e^{\frac{0.05}{0.026}} - 1}{e^{\frac{-0.05}{0.026}} - 1} = \frac{I_f}{I_b},$$

Then $$\frac{I_f}{I_b} = \frac{e^{1.92} - 1}{e^{-1.92} - 1}$$

$\therefore$ $$\frac{I_f}{I_b} = \frac{6.82 - 1}{0.14 - 1} \cong 6.767$$

(c) For $$V = 0.1,\ I = 10\left(e^{\frac{0.1}{0.026}} - 1\right) \cong 458\ \mu\text{A}$$

$V = 0.2$, $\quad I = 10\ (e^{200/26} - 1) = 21.7$ mA

$V = 0.3$, $\quad I = 10\ (e^{300/26} - 1) = 1.01$ A

Problem 2.5: (a) Calculate the anticipated factor by which the reverse saturation current of a germanium diode is multiplied when the temperature is increased from 25 to 80°C.
(b) Repeat part *a* for a silicon diode over the range 25 to 150°C.

Solution:

(a) From eqn of the diode current, we find that at every

$$I_o\ (T) = I_{o1} \times 2^{T-T_1/10}$$

10°C rise of temperature the diode current doubles

Hence, I_o (80°C) = I_o (25°C) . $2^{(80-25)/10}$ = $2^{(5.5)}$. I_o (25°C)

Then, I_o (80°C) = 45.25 × I_o (25°C)

Hence the Anticipated factor is 45.25 **Ans**

(b) Same rule of doubling the current of a diode at every 10°C rise of temperature will hold for the case of Si-diode.

$$\text{Then,}\quad I_{o(150°C)} = I_o\,(25°C)\ 2^{12.5}$$
$$= 5792.\ 6 \times I_o\,(25°C)$$

Hence the Ancipated factor here is 5792.6

Problem 2.6: It is predicted that, for germanium, the reverse saturation current should increase by 0.11°C^{-1}. It is found experimentally in a particular diode that at a reverse voltage of 10V, the reverse current is 5 μ A and the temperature dependence is only 0.07°C^{-1}. What is the leakage resistance shunting the diode?

Solution:

From the ckt. shown, we find,

$$I = I_o + I_R \quad ...(1)$$

Since, I_R is independent of T,

Then,

$$\frac{dI}{dT} = \frac{dI_o}{dT} \quad ...(2)$$

[because diode reverse saturation current depends on T]

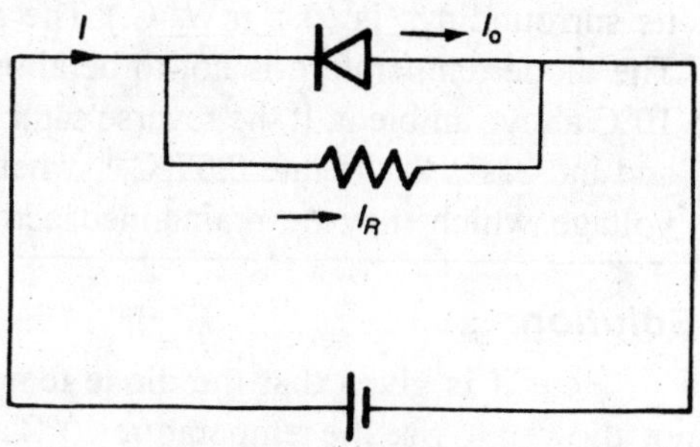

Fig. P. 2.6

Also, From the question, it is given that

$$\frac{1}{I_o} \times \frac{dI_o}{dT} = 0.11$$

$$\text{and}\quad \frac{1}{I}\frac{dI}{dT} = 0.07 \quad ...(3)$$

Now from eqn (2) and (3)

we get, $0.11\ I_o = 0.07I$ ... (4)

multiplying (1) by 0.11 and then applying (1) – (4),

$$0.04I = 0.11\ I_R;$$

or $$I_R = \frac{4}{11}\ I_o$$

Then, $$I_R = \frac{4}{11} \times 5 = \frac{20}{11}\ \mu A$$

[As given $I_o = 5\mu A$]

Hence, $$R = \frac{V}{I_R} = \frac{10 \times 11}{20} \times 10^6 = 5.5\ M\Omega$$

Problem 2.7: A diode is mounted on a chassis in such a manner that, for each degree of temperature rise above ambient, 0.1 mW is thermally transferred from the diode to its surroundings. (The "thermal resistance" of the mechanical contact between the diode and its surroundings is (0.1 mW/°C.) The ambient temperature is 25°C. The diode temperature is not to be allowed to increase by more than 10°C above ambient. It the reverse saturation current is 5.0 μA at 25°C and increases at the rate $0.07°C^{-1}$, what is the maximum reverse-bias voltage which may be maintained across the diode ?

Solution:

Here it is given that the diode temperature must not be 35°C (i.e. not allowed to rise the temperature 10°C above the ambient temperature).

Hence the maximum permissible temperature is 35°C, This is the temperature achieved at which thermal equalibrium takes place i.e. $p_{in} = p_{out}$.

$$\therefore \quad P_{in} = P_{out} = 0.1 \times (35 - 25) = 1\ mW$$

At, 35°C the diode current

$$I_o\,(35) = I_o\,(25) \times 2^{\frac{35-25}{10}} = 2\ I_o\,(25)$$

$$= 2 \times 5 = 10\ \mu A$$

Also, due to this rise in temperature will dissipate heat,

$$\therefore \quad P = VI$$

$$\therefore \quad V = \frac{1\text{ mW}}{10\ \mu\text{A}} = 100 \text{ V}$$

This is the maximum reverse-bias voltage which may be maintained across the diode.

Problem 2.8: An ideal germanium *p-n* junction diode has at a temperature of 125°C a reverse saturation current of 30 μA. At a temperature of 125°C find the dynamic resistance for a 0.2 V bias in (a) the forward direction, (b) the reverse direction.

Solution:

(a)
$$V_T = \frac{T}{11{,}600} \qquad \text{...(1)}$$

At temperature of 125°C

$$V_T = \frac{273+125}{11{,}600}$$
$$= 0.0343 \text{ V}$$

Now,
$$I(T) = I_o\left(e^{\frac{V}{nV_T}} - 1\right) \qquad \text{...(2)}$$

But for Ge–diode, $n = 1$

$$\therefore \quad I = I_o(125)\left(e^{\frac{V}{V_T}} - 1\right)$$

$$= I_o(125) \times e^{\frac{V}{V_T}}$$

$$\left[\text{since } \frac{V}{V_T} \gg 1\right] \qquad \text{...(3)}$$

Now differentiating equation (3) with respect to V,

$$\frac{dI}{dV} = \frac{I_o(125)}{V_T} \times e^{\frac{V}{V_T}}$$

$$\therefore \quad \frac{1}{r} = \frac{dI}{dV} = \frac{30 \times 10^{-6}}{34.3 \times 10^{-3}} \times e^{5.83} = 0.298\ \mho$$

$$\therefore \quad r = 3.36\ \Omega$$

(b) When the reverse direction is taken, the value of V will be negative and then

$$I = I_o (125) \times e^{\frac{-V}{V_T}}$$

Then; $$\left|\frac{dI}{dV}\right| = \frac{I_o(125)}{V_T} \times e^{\frac{-V}{V_T}}$$

or $$\frac{1}{r} = \frac{dI}{dV} = \frac{30 \times 10^{-6}}{34.3 \times 10^{-3}}\ e^{-5.83}$$

$$= 2.73 \times 10^{-6}\ \mho$$

Hence $$r = 0.366\ \text{M}$$

Problem 2.9: Prove that for a an alloy *p-n* junction with ($N_A \ll N_D$), the width W of the depletion layer is given by

$$W = \left(\frac{2\varepsilon\mu_p V_j}{\sigma_p}\right)^{\frac{1}{2}}$$

Where V_i is the junction potential under the condition of an applied diode voltage V_d.

Solution:

As we know from knowledge of our text, the junction barrier potential is related with the depletion layer width by the equation.

$$V_j = \frac{qN_A W^2}{2\varepsilon} \qquad \text{...(1)}$$

But, $$\sigma_p = qN_D\mu_p,$$

then, $$qN_D = \frac{\sigma_p}{\mu_p} \qquad \text{...(2)}$$

From above eqn,

$$V_j = \frac{W^2 \sigma_p}{\mu_p 2\varepsilon}$$

$$\therefore \qquad W = \left(\frac{2\varepsilon\mu_p V_j}{\sigma_p}\right)^{\frac{1}{2}} \qquad \text{Proved}$$

Problem 2.10: (a) Prove that for an alloy silicon *p-n* junction with $N_A \ll N_D$), the depletion-layer capacitance in picofarads per square centimeter is given by

$$C_T = 2.9 \times 10^{-4} \left(\frac{N_A}{V_j}\right)^{\frac{1}{2}}$$

(b) If the resistivity of the *p* material is 3.5 Ω-cm, the barrier height V_o is 0.35 V, the applied reverse voltage is 5 V, and the cross-sectional area is circular of 40 mils diameter, find C_T.

Solution:

(a) For transition capacitance,

$$C_T = \frac{\varepsilon A}{W} \qquad \ldots(1)$$

But W is given by the eqn

$$W = \left(\frac{2\varepsilon V_j}{q N_A}\right)^{1/2} \qquad \ldots(2)$$

Then, from eqn (1) and (2), we have,

$$\frac{C_T}{A} = \frac{\varepsilon}{W} = \frac{\varepsilon}{(2\varepsilon V_j)^{1/2}} \times (qN_A)^{1/2} = \sqrt{\frac{q\varepsilon}{2}} \times \sqrt{\frac{N_A}{V_j}}$$

Now, $\sqrt{\dfrac{q\varepsilon}{2}} = 2.9 \times 10^{-16}$

$$\therefore \quad C_T/\text{cm}^2 = 2.9 \times 10^{-16} \left(\frac{N_A}{V_j}\right)^{1/2} \text{F/cm}^2$$

$$= 2.9 \times 10^{-4} \left(\frac{N_A}{V_j}\right)^{1/2} \text{pF/cm}^2$$

(b) $$A = \frac{\pi D^2}{4} = \frac{\pi}{4}(40 \times 10^{-3} \times 2.54)^2$$

$$= 8.14 \times 10^{-3} \text{ cm}^2$$

Now, $$\rho_p = \frac{1}{\sigma_p} = \frac{1}{q \times \mu_p \times N_A}$$

Then, $$N_A = \frac{1}{3.5 \times 1.6 \times 10^{-19} \times 500}$$

$$= 3.58 \times 10^{15}/\text{cm}^2 \text{ [since } \mu_p = 500 \text{ for Si]}$$

Also, $V_j = 5 + 0.35 = 5.35\text{V}$

Then,

$$C_T = 2.9 \times 10^{-4} \times \left(\frac{3.58 \times 10^{15}}{5.35}\right)^{1/2} \times 8.14 \times 10^{-3} \cong 61 \text{ pF}$$ **Ans**

Problem 2.11: The Zener diode can be used to prevent overloading of sensitive meter movements without affecting meter linearity. The circuit shown represents a dc voltmeter which reads 20 V full scale. The meter resistance is 560 Ω and $R_1 + R_2 = 99.5$ K. If the diode is a 16-V Zener, find R_1 and R_2 so that, when $V_i > 20$V, the Zener diode conducts and the overload current is shunted away from the meter.

Solution:

With $V_i < 16$, the zener diode does not conduct.

Then, $$20 = (R_1 + R_2 + R_m)\, I$$

$$= (99.5 + 0.56) \times 10^3 I$$

or, $I \cong \dfrac{20}{100\,K}$

$= 200\ \mu A$ (Full Scale)

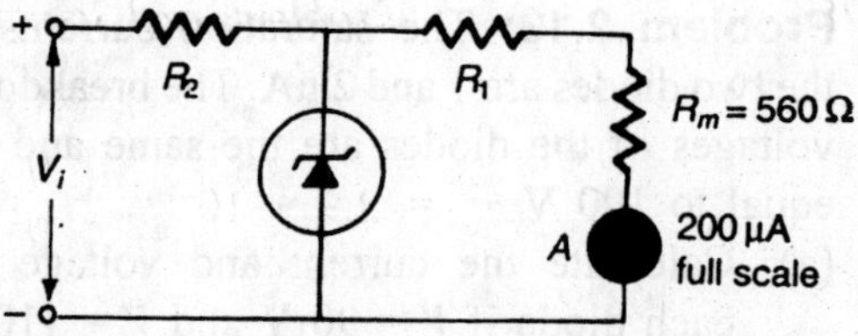

Fig. P. 2.11

With $V_1 > 20V$, the voltage across R_1 and R_m and R_m meter) must be equal to the zener voltage.

Now,

$$V_2 = 16V,$$

$$(R_1 + R_m) = \frac{V_2}{I} = \frac{16}{2 \times 10^{-4}} = 80\ K$$

$\therefore \quad R_1 \cong 79.5\ K$

and $\quad R_2 \cong 20\ K$ Ans.

Problem 2.12: A series combination of a 15-V avalanche diode and a forward-biased silicon diode is to be used to construct a zero-temperature-coefficient voltage reference. The temperature coefficient of the silicon diode is −1.7 mV/°C. Express in percent per degree centigrade the required temperature coefficient of the Zener diode.

Solution:

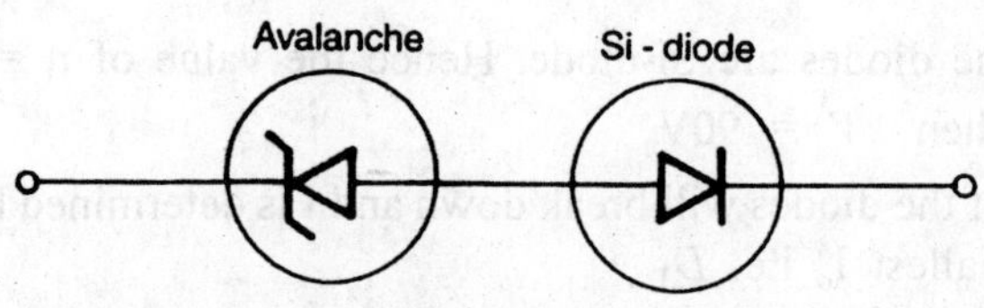

Fig. P. 2.12

The temperature coefficient of Si-diode is −1.7 mV/°C.

The series combination (as made) is to have a zero-temperature coefficient, then the temperature coefficient of the avalanche diode must be at the biasing current + 1.7 mV/°C.

∴ % temperature coefficient

$$= \frac{1.7 \times 10^{-3}}{15} \times 100 = 0.0113\ \%/^\circ C.$$

Problem 2.13: The saturation currents of the two diodes are 1 and 2 μA. The breakdown voltages of the diodes are the same and are equal to 100 V.

(a) Calculate the current and voltage for each diode if $V = 90$ V and $V = 110$ V.

(b) Repeat part a if each diode is shunted by a 10-M resistor.

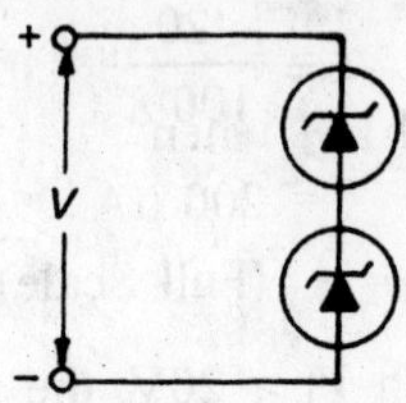

Fig. P. 2.13

Solution:

$$I_{o1} = 1\ \mu A$$

and $$I_{o2} = 2\ \mu A$$

$$V_{Z_1} = V_{Z_2} = 100\text{V}$$

From the equation of diode current,

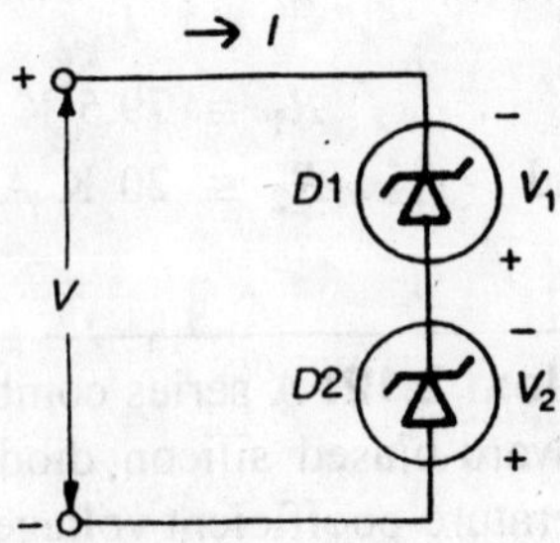

Fig. P. 2.13 (a)

$$I' = I_o\left(e^{\frac{V}{nV_T}} - 1\right)$$

Now,

$$V = n_1 V_T \ln\left(1 + \frac{I'}{I_o}\right) \quad \ldots(1)$$

Here the diodes are Si-diode. Hence the value of $\eta = 2$

(a) (i) when $V = 90$V.

None of the diodes will break down and I is determined by the diode with the smallest I_o i.e. D_1.

Thus, $I = 1\ \mu A$ and For D_2 also

$$I' = -1 = -1\ \mu A$$

Then from equation (1) above,

$$V_2 = 2 \times 0.026 \ln\left(1 - \frac{1}{2}\right) \cong -36 \text{ mV}$$

Then, $V_1 \cong -89.964$ V

(ii) When, $V = 110$ V; As the applied voltage is increased to 110 V, the diode D_1 will breakdown, which D_2 will be reverse biased.

Thus $I = I_{o2} = 2\ \mu A$

$\therefore$ $V_1 = -100V$ and $V_2 = -10V$ Ans.

(b) (i) when $V = 90V$;

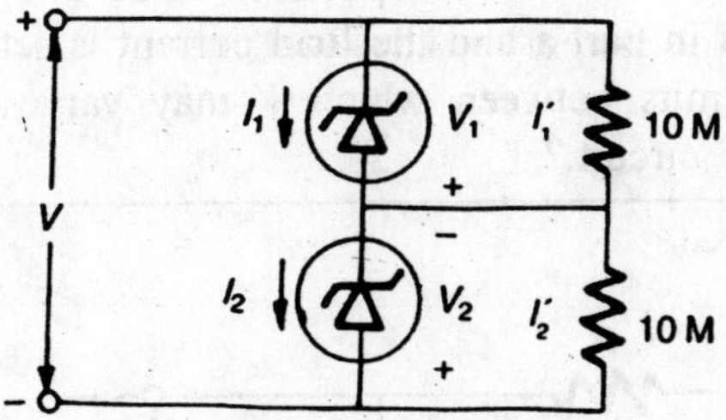

Fig. P. 2.13 (b)

The shunting resistor of value 10M tend to equalize the vottages across the diodes, and hence both diodes are revers biased.

$$I_1 = I_{o1} = 1\mu A$$

and $I_2 = I_{o2} = 2\mu A$

Then from the ckt. above shown,

$$I_1 + I_1' = I_2 + I_2';$$

or, $I_1' = I_2 - I_1 + I_2'$

Hence $I_1' = (1 + I_2')$ ... (2)

Now also, $I_1' = -\dfrac{V_1}{10}$ and $I_2' = \left(\dfrac{V - V_1}{10}\right)$ ...(3)

From equations (2) and (3),

$$\frac{V_1}{10} = 1 + \frac{V - V_1}{10};$$

$$V_1 = -\left(\frac{10 + V}{2}\right)$$

This is valid only for $V < 100$ V

For $V = 90$ V,

$V_1 = -60$ V

and $V_2 = -40$ V **Ans**

(ii) For $V = 110$,

$V_1 = -60$ V

$V_2 = -50$ V **Ans**

Problem 2.14: (a) The avalanche diode regulated at 50 V over a range of diode currents from 5 to 40 mA. The supply voltage $V = 200$ V. Calculate R to allow voltage regulation from a load current $I_L = 0$ up to I_{max}, the maximum possible value of I_L. What is I_{max}? (b) If R is set as in part a and the load current is set at $I_L = 25$ mA, What are the limits between which V may vary without loss of regulation in the circuit.?

Solution:

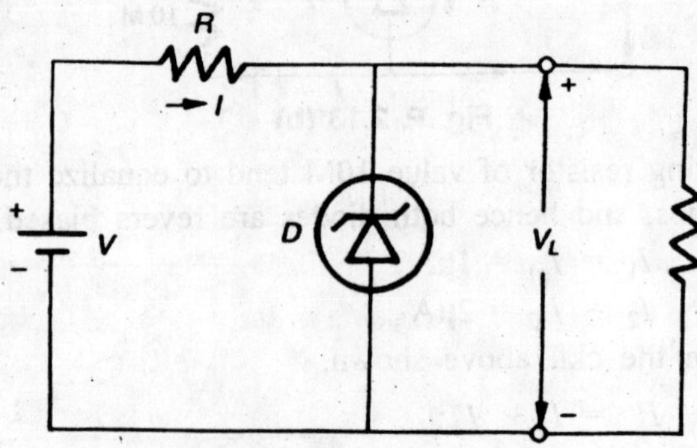

Fig. P. 2.14

The diode current varies from 5 mA to 40 mA for the regulating voltage of 50 V.

$$I_D \leq 40 \text{ mA}$$

i.e. $$I_D = \left(\frac{200-50}{R}\right) \leq 40 \text{ mA}$$

$\therefore$ $$R = \frac{150}{40} \times 10^3 = 3.75 \text{ K}$$

For $\quad I_L = I_{max}\ ;\ I_D \geq 5$ mA

but for $\quad R = 3.75$ K, $I = 40$ mA,

Hence, $\quad I_{max} = 40 - 5 = (35 \text{ mA})$ **Ans**

(b) Minimum current I_D for satisfactory regulation is 5 mA, hence,

$$I = 25 + 5 = 30 \text{ mA}$$

and $$V_{min} = (30 \times 3.75) + 50$$
$$= 162.5 \text{ V}$$

Max[m] allowable current of the diode is 40 mA.

$\therefore \quad I = 65$ mA

and $\quad V_{max} = 65 \times 3.75 + 50 = 293.5$ V

Hence the limits of V is from 162.5 V to 293.5 V.

Problem 2.15: For the diode circuit of Fig. 2.15 predict the currents I_1 and I_2 as functions of the supply voltage V_S, $V_T = 0.7$ V for all diodes (silicon).

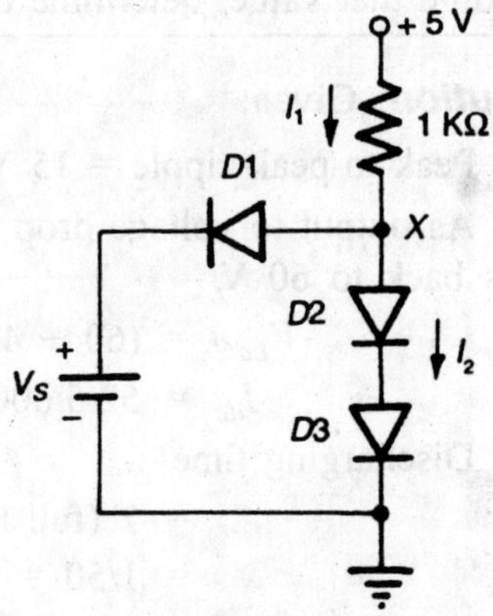

Fig. P. 2.15

Solution:

After replacing all diodes by their circuit models gives the new in Fig. 2.15 (a) it immediately follows from this circuit that the voltage at point X (V_x) cannot exceed 1.4 V because at that voltage, D (equivalent of D_2 and D_3) becomes forward biased and starts conducting. Also V_x cannot exceed $(V_S + 0.7)$ V, otherwise D_1 would become forward biased and conduct.

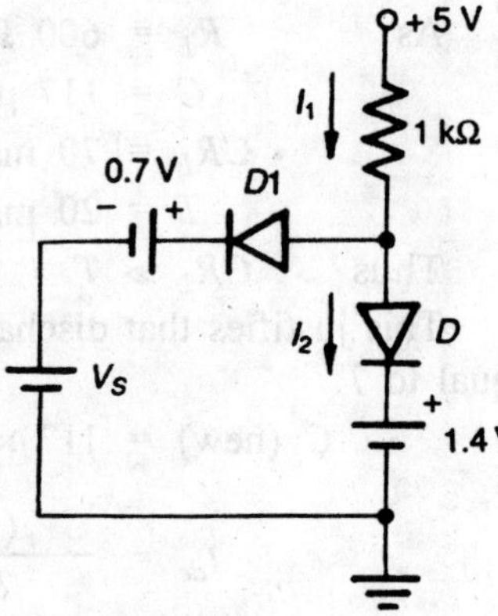

Fig. P. 2.15 (a)

$\therefore$ critical value of V_S is given by

$1.4 = V_S + 0.7$ or $V_S = 0.7$ V

For $0 < V_S < 0.7$ V, D_1 conducts

$\therefore \quad I_1 = (5 - 0.7 - V_S)/1$ mA and $I_2 = 0$

For $\quad V_S > 0.7$ V, D_2, D_3 conduct

$\therefore \quad I_1 = I_2 = (5 - 1.4)/1 = 3.6$ mA

Problem 2.16: (a) A half-wave rectifier operates from a 50 Hz supply and provides a peak output voltage of 60 V across 600 Ω resistance. Estimate the value of the capacitance for the capacitor filter which would limit the peak-to-peak ripple voltage across load to 15 V. (b) It the capacitance as calculated in part (a) is now increased to double that value, determine the ripple voltage and de load current.

Solution: Given:

Peak-to-peak ripple = 15 V

As output to voltage drop from 60 V to 45 V and then cyclically rises back to 60 V.

$$V_{LdC} = (60 + 45)/2 = 52.5 \text{ V}$$

$$I_{dc} = 52.5/600 = 87.5 \text{ mA}$$

Discharging time

$$= T \text{ (full time period)}$$

$$= 1/50 = 0.02 \text{ or } 20 \text{ ms}$$

Considering linear discharge rate

$$C = \frac{I_{dc}T_2}{V_r(p\text{-}p)} = \frac{87.5 \times 20}{15} = 117\ \mu\text{F}$$

As $R_L = 600\ \Omega,$

$$C = 117\ \mu\text{F}$$

$$CR_L = 70 \text{ ms}$$

$$T = 20 \text{ ms}$$

Thus $CR_L \gg T$

This justifies that discharge rate is almost linear over time nearly equal to T.

$$C \text{ (new)} = 117 \times 2 = 234\ \mu\text{F}$$

$$I_{dc} = \frac{CV_r(p\text{-}p)}{T} = \frac{234V_r(p\text{-}p) \times 10^{-3}}{20} \text{A}$$

Also

$$I_{dc} = \frac{60 + (60 - V_r(p\text{-}p))}{2R_L} = \frac{120 - V_r(p\text{-}p)}{2 \times 600} \text{ A}$$

$$= \frac{234\,V_r\,(p\text{-}p)}{20} = \frac{120 - V_r(p\text{-}p)}{1200} \times 1000$$

$$\therefore \quad V_r\,(p\text{-}p) = 8\text{V}$$

$$I_{dc} = (60 - 8/2)/600 = 93.3 \text{ mA}$$

> **Problem 2.17:** A load 1 kΩ is to be supplied 10 mA at 30V dc with ripple voltage of no more than 1% of the dc voltage. Calculate the value of the filter capacitor that would be needed for a half-wave filter circuit. Source voltage = 220 V, 50 Hz. Also calculate transformer turn ratio.

Solution:

We know $$C = \frac{V_{dc}/V_r(p\text{-}p)}{f\,R_L} = \frac{100}{50\times 1000} = 2000\ \mu\text{F}$$

as $V_{dc}/V_r\,(p\text{-}p) = 100$; ripple being 1 % of voltage

Because of capacitor filter

$$V_{dc} \approx V_m - V_r\,(p\text{-}p)/2$$

or $$V_m = 30 + 0.01 \times 30/2 = 30.15\text{V}$$

$$V_2 \text{ (secondary)} = 30.15/\sqrt{2} = 21.32 \text{ V}$$

$\therefore\ N_1 / N_2$ (transformer turn ratio)

$$= 220/21.32 = 10.32$$

The circuit of a full-wave rectifier with capacitor filter is shown in Fig. P. 2.17 (a) and the wave forms $V_L = V_C$, I_{dc} and $V_r\,(p\text{-}p)$, are shown in Fig. P. 2.17 (b). One cycle of charging and discharging of the filter capacitor now takes place in half cycle of the ac source. As per result already established.

$$V_{dc} = V_m - V_r\,(p\text{-}p)/2; \text{ (half-wave and full-wave)}$$

By approximating the output wave form by a triangular wave form, $V_{r(\text{rms})}$ is given by

$$V_r(\text{rms}) = \frac{V_r(p\text{-}p)}{2\sqrt{3}}; \quad \text{(half-wave and full-wave)}$$

Assuming the charging time T_1) to 1/4th of $T/2$), the charging / discharging cycle time in full-wave rectification,

$$\therefore \quad V_r(\text{rms}) = \frac{I_{dc}}{4\sqrt{3}\,fC} \times \frac{V_{dc}}{V_m}; \text{ (full-wave)}$$

At light load $T_1 \approx T/2$ and so the factor (V_{dc}/V_m) disappears.

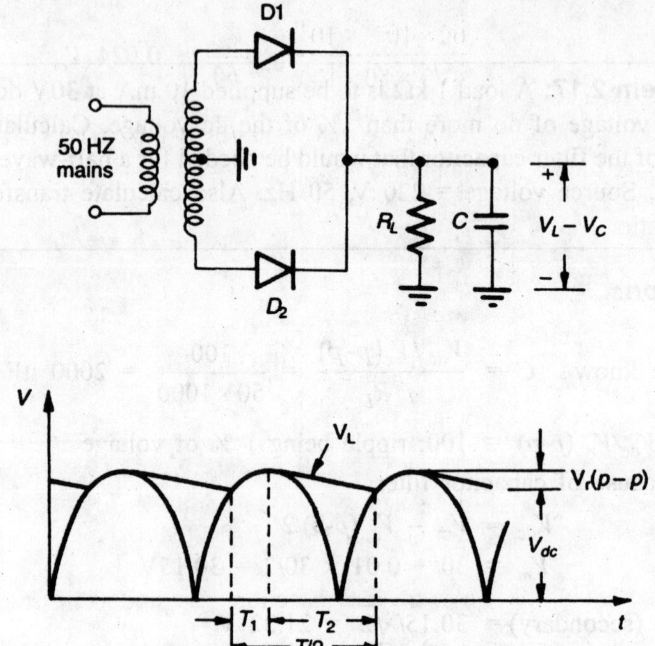

Fig. P. 2.17 (a) & (b)

$$\therefore \quad V_r \text{ (rms)} = \frac{I_{dc}}{4\sqrt{3}\, fC}; \text{ (full wave light load)}$$

Problem 2.18: (a) Calculate the ripple voltage of a full-wave rectifier with a 120 μF capacitor connected to a load of 60 mA. Frequency of ac source = 50 Hz.

(b) If the peak voltage of the rectified wave is 60 V, calculate the dc voltage.

(c) Calculate the ripple factor.

Solution:

(a) we know

$$V_r \text{ (rms)} = \frac{I_{dc}}{4\sqrt{3}\, fC} \times \frac{V_{dc}}{V_m}$$

$$= \frac{60 \times 10^{-3} \times 10^3}{4\sqrt{3} \times 50 \times 120} \times \frac{V_{dc}}{60} = 0.024\ V_{dc}$$

$$V_r\ (p\text{-}p) = 2\sqrt{3}\ V_r\ (\text{rms}) = 2\sqrt{3}\ \times 0.024\ V_{dc}$$

$$= (0.083/2)\ V_{dc}$$

(b) $$V_{dc} = V_m - V_r\ (p\text{-}p)/2$$

$$= 60 - (0.083/2)\ V_{dc};\ V_m = 60\ \text{V}$$

Solving we get

$$V_{dc} = 57.6\ \text{V}$$

(c) Ripple factor = V_r (rms)/V_{dc} = (0.024/57.6) × 100 = 0.042 %

Problem 2.19: Consider the filter circuit of Fig. P. 2.19 with circuit element values as:

$R = 25\ \Omega$, $R_L = 200\ \Omega$, $C = 200\ \mu T$, $L = 20H$

(a) Determine the filter transfer function $\frac{\overline{V}_2}{\overline{V}_1}$ and therefrom the output voltage V_2 (t) for an input voltage of V_1 (t) = ($200\sqrt{2}/\pi$) [1 – (4/3) cos 628 t]

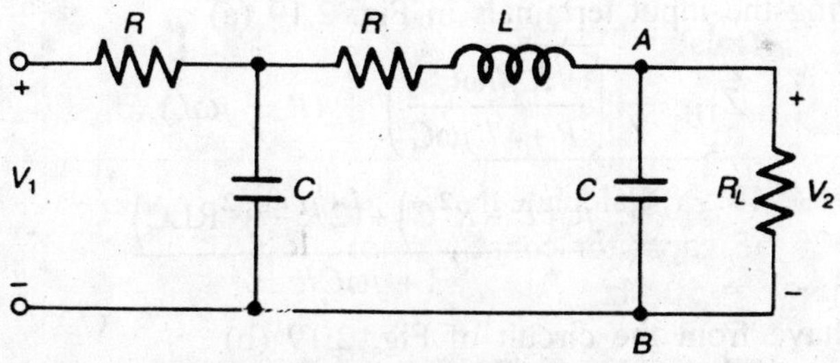

Fig. P. 2.19

wherein harmonics higher than second are ignored.

(b) Find the ripple factor for the output voltage.

Solution:

(a) Consider part to the left of AB. This circuit is drawn in Fig. P. 2.19 (a) Its Thevenin equivalent is shown in Fig. P. 2.19 (b).

We have from the circuit of Fig. P. 2.19 (a)

$$\overline{V}_{TH} = \left(\frac{1/j\omega C}{R + 1/j\omega C}\right)\overline{V}_1 = \frac{\overline{V}_1}{1 + j\omega CR} \quad \text{...(i)}$$

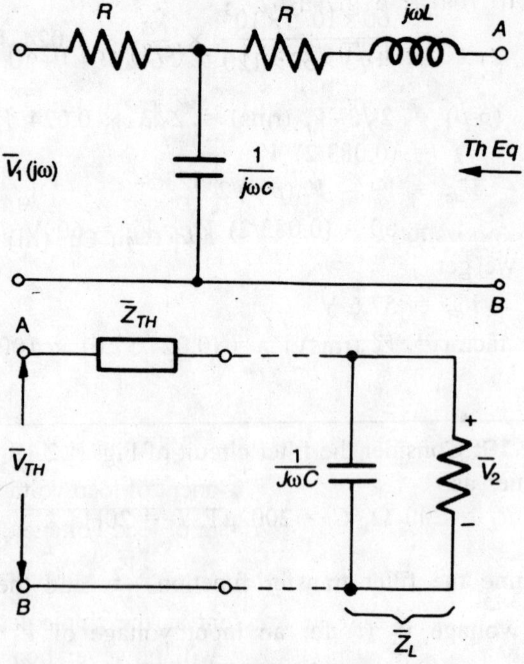

Fig. P. 2.19 (a) & (b)

Sharing the input terminals in Fig. 2.19 (a)

$$\overline{Z}_{\text{TH}} = \left(\frac{RI\ j\omega C}{R + 1/j\omega C}\right) + (R + j\omega L)$$

$$= \frac{j\omega\left(L + R^2C\right) + \left(2R - w^2\text{RLC}\right)}{1 + j\omega CR} \quad \text{...(ii)}$$

We have from the circuit of Fig. 2.19 (b)

$$\overline{V}_2 = \left(\frac{\overline{Z}_L}{\overline{Z}_{TH} + \overline{Z}_L}\right)\overline{V}_1$$

Substituting values we get

$$\overline{V}_2 / \overline{V}_1 = N(\omega)/D(\omega) \quad \text{...(iii)}$$

$$N(\omega) = R_L\ (1 + \omega^2 R^2\ C^2) \quad \text{...(iv)}$$

$$D\ (\omega) = (1 + \omega^2\ RR_LC^2)\ (2R - \omega\ RLC) - \omega^2\ C\ \ (R_L - R)\ (L + R^2C) + R_L\ (1 + \omega^2 R^2\ C^2) + j\omega\ \{RC\ (R_L - R)\ (2 - \omega^2\ LC) + (L + R^2C)\ (C + \omega^2 RR_LC^2)\} \quad \text{...(v)}$$

The input voltage is given by

$$V_1\ (t) = (200\sqrt{2}/\pi)\ [1 - (4/3)\cos 628t] \qquad \ldots\text{(vi)}$$

$$v_2\ (t) = (200\sqrt{2}/\pi)\ \left(\bar{V}_2/\bar{V}_1\right)\Big|_{\omega=0}$$

$$-(800\sqrt{2}/\pi)\ \left(\bar{V}_2/\bar{V}_1\right)\Big|_{\omega=628}\ \cos\left[628t + \angle\bar{V}_2/\bar{V}_1\Big|_{\omega=628}\right]$$

Substituting values of R, R_L, L and C in Eq. (iii), (iv) and (v), we get

$$\left(\bar{V}_2/\bar{V}_1\right)\Big|_{\omega=0} = 0.8;\ \left(\bar{V}_2/\bar{V}_1\right)\Big|_{\omega=628} = 6.43 \times 10^{-4};$$

$$\angle\bar{V}_2/\bar{V}_1\Big|_{\omega=628} = \angle 179.5° = 180°$$

Then $\quad v_2\ (t) = 72.02 + 0.0538\cos 628t$

(b) Ripple factor $= \dfrac{\text{rms of ac component of load voltage}}{\text{dc component of load voltage}}$

$$= 0.0538/(\sqrt{2} \times 72.02) = 5.28 \times 10^{-4}$$

Thus, this filtering is very efficient as the ripple factor is almost zero. Some contribution to ripple factor will, however, be made by higher harmonics of the rectifier output, which have been ignored in the above calculations.

Problem 2.20: For the diode circuit of Fig. 2.20 find the voltage across each diode and the circuit current.

Given:

$V_S = 10\ V$,

$R_L = 100\ k\Omega$;

I_s (for each diode) $= 0.1\ \mu A$.

Take the dioide current as

$I_D = I_s\,(e^{20V_d} - 1)$

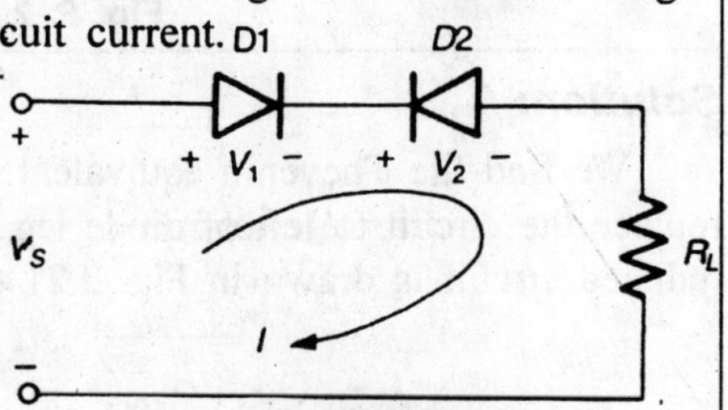

Fig. P. 2.20

Solution:

Applying KVL round the loop is

$$10 = V_1 + V_2 + 100\ I;\ I \text{ in mA} \qquad \ldots\text{(i)}$$

$$I = I_S\left(e^{20V_1} - 1\right) = -I_S\left(e^{-20V_2} - 1\right) \qquad \ldots\text{(ii)}$$

$$\Rightarrow\ e^{20V_1} + e^{-20V_2} = 2 \qquad \ldots\text{(iii)}$$

$$\therefore \quad V_2 = -(1/20)\ \ln\ (2 - e^{20V_1}) \quad \text{...(iv)}$$

Substituting from Eqs. (ii) and (iv) in (i)

$$V_1 - (1/20)\ \ln\ (2 - e^{20V_1}) + 10^{-2}\ (e^{20V_1} - 1) = 10 \quad \text{...(v)}$$

Solving using trial and error

$$V_1 = 0.0346 \text{ V}$$

$$\therefore \quad V_2 = 9.95 \text{ V}$$

Current flowing through the circuit. be

$$I = I_S\ (e^{20V_1} - 1) = 0.0997 \approx 0.1\ \mu\text{A}$$

Problem 2.21: For the diode circuit of Fig. 2.21 find the diode current. The diode parameters are
$V_T = 0.5$ V, $R_F = 5\Omega$, R_r (reverse) = 100 kΩ,
V_Z (breakdown) = 2 V, $R_Z = 100/3\Omega$

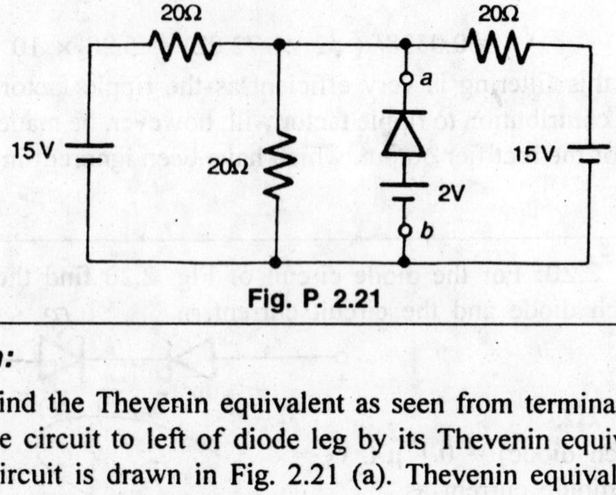

Fig. P. 2.21

Solution:

We find the Thevenin equivalent as seen from terminal *'ab'* First replace the circuit to left of diode leg by its Thevenin equivalent. The reduced circuit is drawn in Fig. 2.21 (a). Thevenin equivalent at *'ab'*

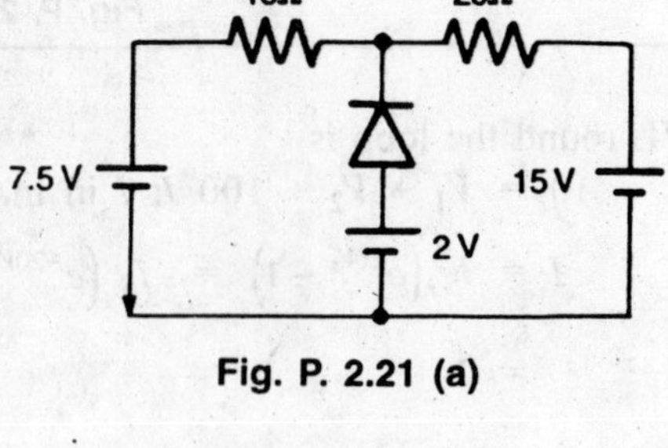

Fig. P. 2.21 (a)

$$V_{ab}\ (OC) = 15 - \left(\frac{15-7.5}{10+20}\right) \times 20 = 10\text{V}$$

$$R_{TH} = 10 \parallel 20 = 20/3\ \Omega$$

The reduced circuit in now drawn in Fig. P. 2.21 (b) The diode is reverse biased by $(10 - 2) = 8\ \text{V} > V_Z = 2$ V. Hence conduction is controlled by $R_Z = 100/3\ \Omega$.

I_D (in direction indicated) = 8 / (100/3 + 20/3) = 0.2A

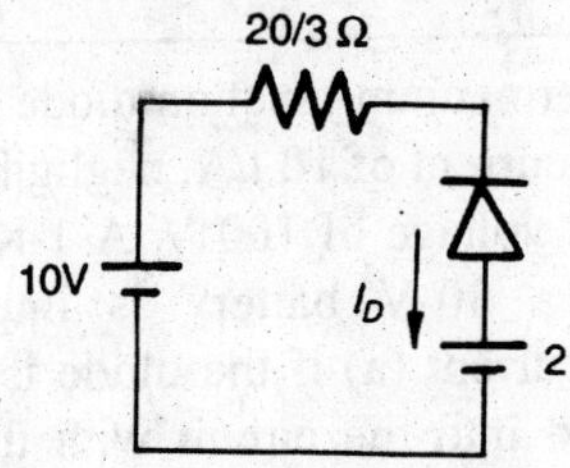

Fig. P. 2.21 (b)

Chapter 3

Diode-Circuits

Problem 3.1: A *p-n* germanium junction diode at room temperature has a reverse saturation current of 10 μA, negligible ohmic resistance, and a Zener breakdown voltage of 100 V. A 1-K resistor is in series with this diode, and a 30-V battery is impressed across this combination. Find the current (a) if the diode is forward-biased, (b) if the battery is inserted into the circuit with the reverse polarity.

Solution: (a) By using the method of successive approximation, with the diode current equal to zero,

$$I = \frac{30}{1\text{K}} = 30 \text{ mA}$$

For this current V is obtained by the equation,

$$30 \times 10^{-3} = 10 \times 10^{-6} (e^{38.4V} - 1)$$

(a) $$e^{38.4} = 3000 \text{ and } V = 0.208\text{V}.$$

$$\therefore \quad I = \frac{30 - 0.208}{1K} \cong 29.8 \text{ mA}$$

For this current

$$V = 0.2 \text{ V}$$

$$\therefore \quad I = 29.8 \text{ mA}$$

(b) The diode voltage is $-30V$ and $I = -I_o = -10$ μA. The voltage drop across the 1K resistor is only 10 mV and this low value may be neglected.

Problem 3.2: Each diode is described by a linearized volt-ampere characteristic, with incremental resistance r and offset voltage V_γ. Diode D_1 is germanium with $V_\gamma = 0.2$ V and $r = 20\ \Omega$, whereas D_2 is silicon with $V_\gamma = 0.6$ V and $r = 15\ \Omega$. Find the diode currents if (a) $R = 10$ K; (b) $R = 1$ K.

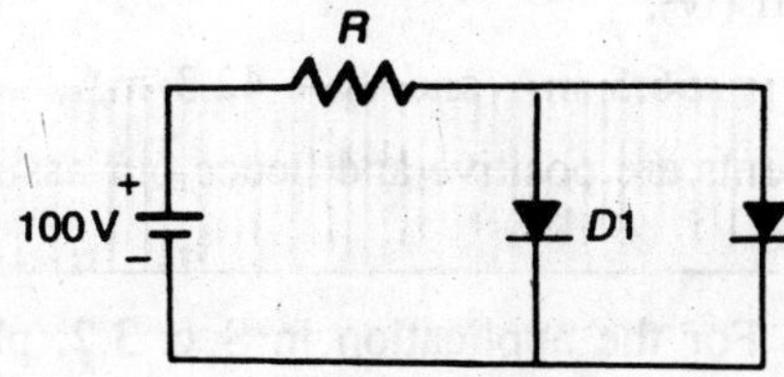

Fig. P. 3.2

Solution:

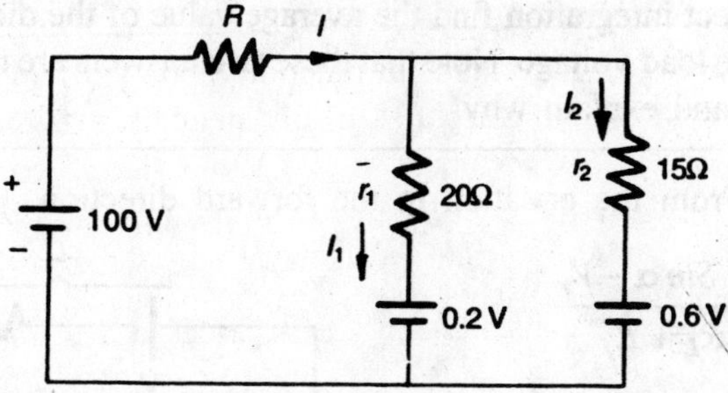

Fig. P. 3.2 (a)

For $D1$, $V_r = 0.2$ V

$r_1 = 20\ \Omega$

For $D2$, $V_r = 0.6$ V

$r_2 = 15\ \Omega$

(a) For $R = 10$ K,

It is assumed that both the diodes are conducting.

$$100 = 10.02\ I_1 + 10\ I_2 + 0.2 \quad \text{...(i)}$$

$$100 = 10.015\ I_2 + 10\ I_1 + 0.6 \quad \text{...(ii)}$$

Solving (i) and (ii), we get $I_2 < 0$.

Thus our assumption of conducting $D1$, is not valid.

Now, assume that $D1$ is on and $D2$ is off.

$$\therefore \quad I_1 = \frac{100 - 0.2}{10.02} \cong 9.97 \text{ mA and } I_2 = 0$$

(b) $R = 1$ K, assume that again in this case both of the diodes are conducting.

$$\therefore \quad 100 = 1.02 + I_2 + 0.2 \quad \text{...(iii)}$$

$$100 = 1.015\ I_2 + I_1 + 0.6 \quad \text{...(iv)}$$

Solving (iii) and (iv),

$$I_1 = 56.5 \text{ mA and } I_2 = 42.3 \text{ mA.}$$

Since both currents are positive and hence our assumption is *valid.*

Problem 3.3: (a) For the application in Sec. 3.2, plot the voltage across the diode for one cycle of the input voltage v_i. Let $V_m = 2.4$ V, $V_y = 0.6$, $R_f = 10\ \Omega$, and $R_L = 100\ \Omega$.
(b) By direct integration find the average value of the diode voltage and the load voltage. Note that these two answers are numerically equal and explain why.

Solution: From the equation in the forward direction

$$I = \frac{V_m \ Sin\,\alpha - V_\gamma}{R_L + R_f}$$

If $V_i = V_m$ Sin $\alpha \geq V_\gamma$;
and $I = 0$ for $V_i = V_\gamma$
The cut in angle ϕ is defined as,

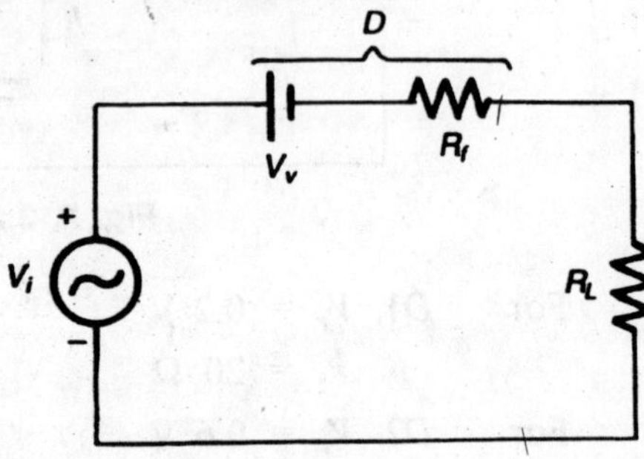

Fig. P. 3.3 (a)

$$\phi = \text{arc sin}\left(\frac{V_\gamma}{V_m}\right)$$

$$= \text{arc sin}\frac{1}{4} = 14.5° \text{ and for } \alpha = \frac{\pi}{2}$$

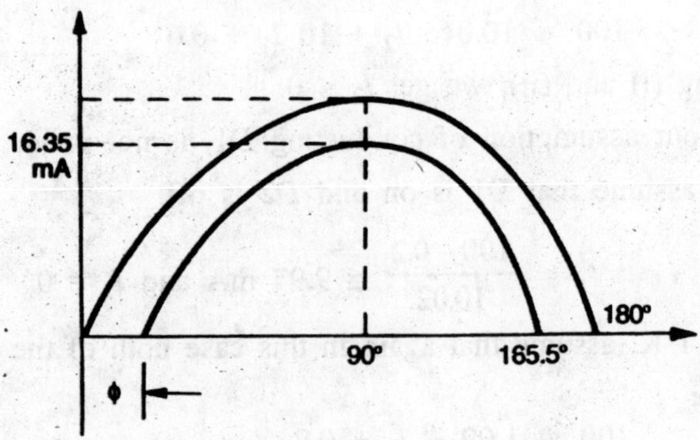

Fig. P. 3.3 (b)

$$I_{max} = \left(\frac{V_m - V_\gamma}{R_i + R_f}\right) = \frac{1.8}{110} \cong 16.35 \text{ mA}$$

$$\therefore \quad V_{Dmax} = 0.6 + 16.35 \times 10^{-3} \times 10 = 0.763 \text{ V}$$

(b)
$$\overline{V}_L = \frac{1}{T}\int_o^T IR_L dt = \frac{1}{2k}\int_0^{2k} IR_L\, d\alpha$$

$$= \frac{1}{2\pi}\left(\int_0^{\phi} IR_L d\alpha + \int_{\phi}^{\pi-\phi} IR_L d\alpha + \int_{\pi-\phi}^{2\pi} IR_L d\alpha\right)$$

$$= \frac{1}{2\pi}\left(0 + \int_{\phi}^{\pi-\phi} \frac{R_L}{R_L + R_f}(V_m \sin\alpha - V_\gamma)d\alpha + 0\right)$$

$$= \frac{1}{2\pi} \times \frac{R_L}{R_L + R_f}\,[2V_m \cos\phi - V_\gamma(\pi - 2\phi)]$$

When the diode is on,

$$V_D = V_\gamma + I R_f = \frac{R_L}{R_L + R_f}\,(V_m \sin\alpha + V_\gamma)$$

When the diode is off,

$$V_D = (V_m \sin\alpha - V_\gamma)$$

Hence,
$$\overline{V}_D = \frac{1}{2\pi} \times$$

$$\int_0^{\pi} (V_m \sin\alpha - V_\gamma)\, d\alpha + \frac{1}{2\pi}\int_{\phi}^{\pi-\phi} \frac{R_L}{R_L + R_f} \times$$

$$(V_m \sin\alpha - V_\gamma)\, d\alpha + \frac{1}{2\pi}\int_{\pi-\phi}^{2\pi} [V_m \sin\alpha - V_\gamma)\, d\alpha$$

$$= -\frac{1}{2\pi}\,\frac{R_L}{R_L + R_f}\,(2V_m \cos\phi - V_\gamma(\pi - 2\phi)]$$

Since, $V_i = V_D + V_L$; for diode on

and $V_i = V_D - V_L = 0$;

for diode off

Hence,
$$0 = \frac{1}{T}\int_0^t V_1\, dt = \frac{1}{T}\int_0^t V_D\, \text{dt} + \frac{1}{T}\int V_L\, dt$$

or,
$$\frac{1}{T}\int V_D\, dt = \overline{V}_D = -\frac{1}{T}\int_0^T V_L\, \text{dt} = -\overline{V}_L$$

Problem 3.4: Calculate the break region over which the dynamic resistance of a diode is multiplied by a factor of 1,000.

Solution: The incremental resistance,

$$r = \frac{dV}{dI} = \frac{1}{g}$$

$$\therefore \quad r = \frac{nV_T}{I_o}\, e^{-V/nV_T}$$

Hence $$\frac{r_1}{r_2} = e^{V_2 - V_1/nV_T}$$

$$\therefore \quad \frac{\Delta V}{nV_T} = \ln\left(\frac{r_1}{r_2}\right)$$

For Si diode, $n = 2$ and $\frac{r_1}{r_2} = 10^3$

and $$\Delta V = V_2 - V_1 = 2 \times 26 \times 3 \times 2.3$$

$$\cong 359 \text{ mV at room temp.}$$

Now, For Ge diode,

$$n = 1$$

$$\therefore \quad \Delta V = 7 \times 26 \times 3 \times 2.3 \cong 180 \text{ mV}$$

Problem 3.5: For the diode clipping circuit of Fig. 3.5 assume that V_R = 10 V, $v_i = 20 \sin \omega t$, and that the diode forward resistance is $R_f = 100\ \Omega$ while $R_r = \infty$ and $V_\gamma = 0$. Neglect all capacitances. Draw to scale the input and output waveforms and label the maximum and minimum values if (a) $R = 100\ \Omega$, (b) $R = 1$ K, and (c) R = 10K.

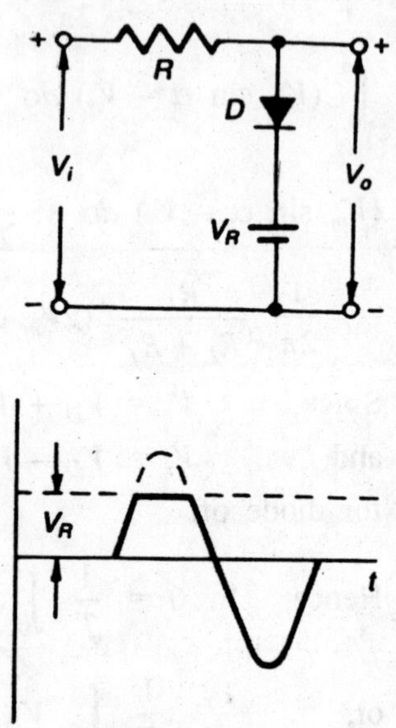

Fig. P. 3.5

Solution: The diode will conduct,
When $V_i = 20 \sin \omega t = 10$

or $\omega t = \sin^{-1}\left(\frac{1}{2}\right) = 30°$

When diode conducts,

$$V_o = (V_i - 10) \times \frac{R_f}{R+R_f} + 10$$

$$V_{omax} = 10 \times \left(\frac{R+2R_f}{R+R_f}\right)$$

(Since, $V_{imax} = 20$)

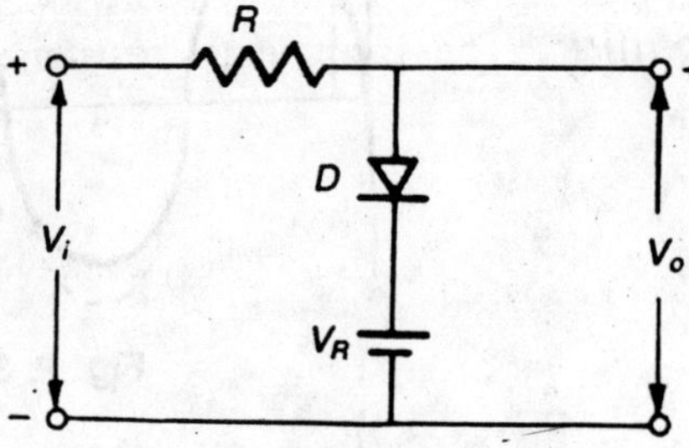

Fig. P. 3.5 (a)

When the diode is non conducting

$$V_o = V_i$$

(a) $R = 100$,

$V_{omax} = 15\ V$

and $V_{omin} = -20$ V

(b) $R = 1$ K,

$V_{omax} = 10.9\ V$

and $V_{omin} = -20$ V

(c) $R = 10$ K,

$V_{omax} = 10.1\ V$

and $V_{omin} = -20$ V

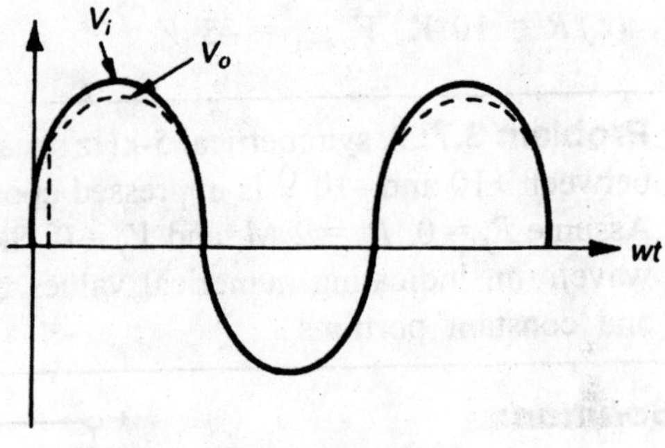

Fig. P. 3.5 (b)

> **Problem 3.6:** Repeat Prob. 3.5 for the case where the reverse resistance is $R_r = 10$ K.

Solution:

From the above ckt shown in (3.5).

When $V_i > 10$ V,

the above equation of problem (3.5) applies well

When $V_i < 10$ V

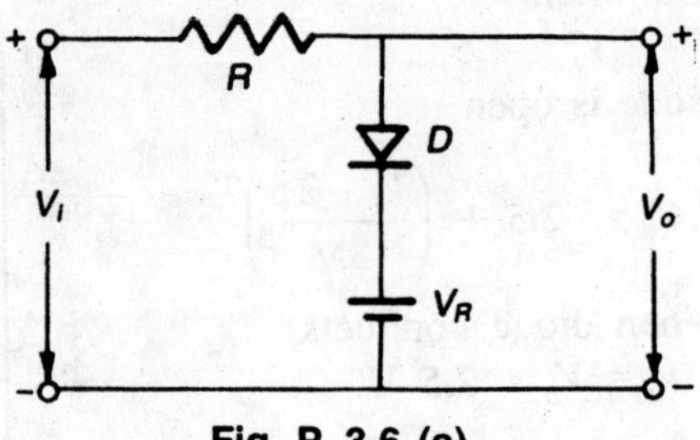

Fig. P. 3.6 (a)

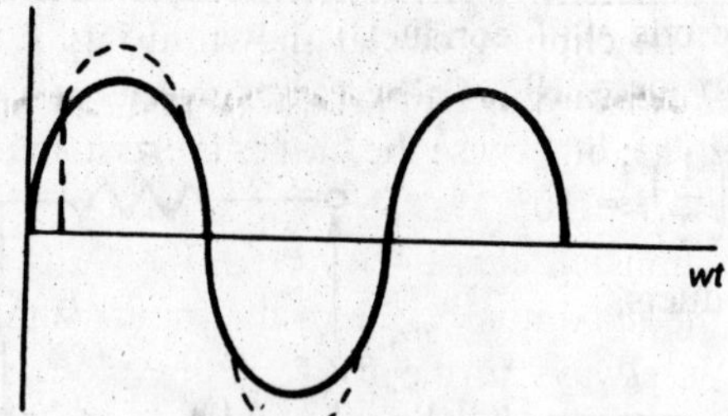

Fig. P. 3.6 (b)

$$V_o = (V_i - 10)\ \frac{R_r}{R + R_r} + 10$$

and $$V_{o\,min} = 10\ \frac{R + 2R_r}{R + R_r}$$

(a) $R = 100,\ V_{o\,min} = -19.7\ V$
(b) $R = 1\ K,\ V_{o\,min} = -17.3\ V$
(c) $R = 10\ K,\ V_{o\,min} = -5\ V$

Problem 3.7: A symmetrical 5-kHz square wave whose output varies between +10 and –10 V is impressed upon the clipping circuit shown. Assume $R_f = 0$, $R_r = 2$ M, and $V_\gamma = 0$. Sketch the steady-state output waveform, indicating numerical values of the maximum, minimum, and constant portions.

Solution:

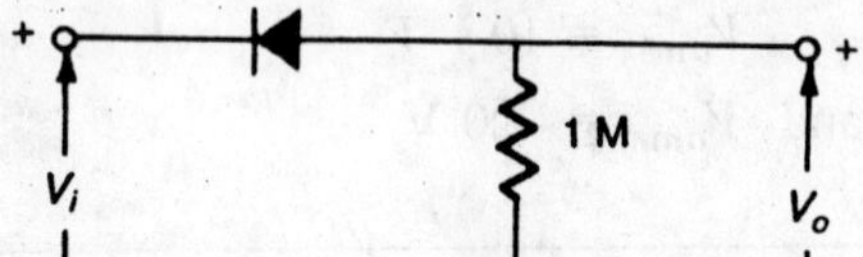

Fig. P. 3.7 (a)

Here $R_f = 0$, $R_r = 2M$
and $V_\gamma = 0$
The diode will conduct only when
$V_i < 2.5$
and when
$V_i > 2.5$,
diode is open

$$\therefore V_o = 2.5 + \left(\frac{V_i - 2.5}{3M}\right)$$

When diode conducts
$V_i = V_o < 2.5$ V

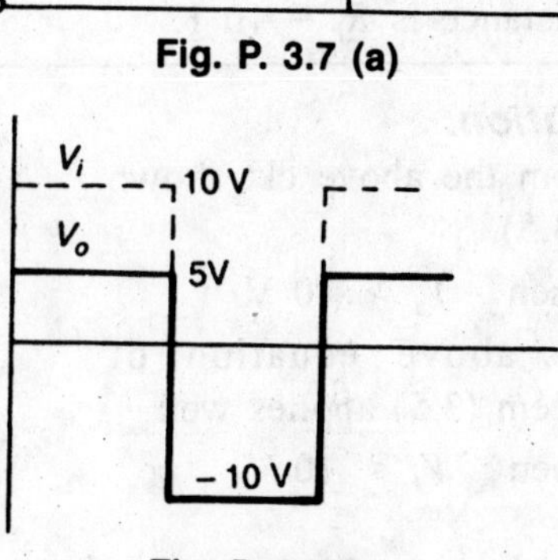

Fig. P. 3.7 (b)

Problem 3.8: The clipping circuit shown in Fig. P. 3.8 (a) employs temperature compensation. The dc voltage source V_γ represents the diode offset voltage; otherwise the diodes are assumed to be ideal with $R_f = 0$ and $R_r = \infty$.

(a) Sketch the transfer curve V_o versus V_i.

(b) Show that the maximum value of the input voltage V_i so that the current in $D2$ is always in the forward direction is

$$V_{i,\ max} = V_R + \frac{R}{R'}(V_R - V_\gamma)$$

(c) What is the temperature dependence of the point on the input waveform at which clipping occurs?

Solution: If $V_1 \le V_R$,

$$V_o = V_R - V_\gamma$$

and if

$$V_1 \ge V_R,\ V_o = V_i - V_R$$

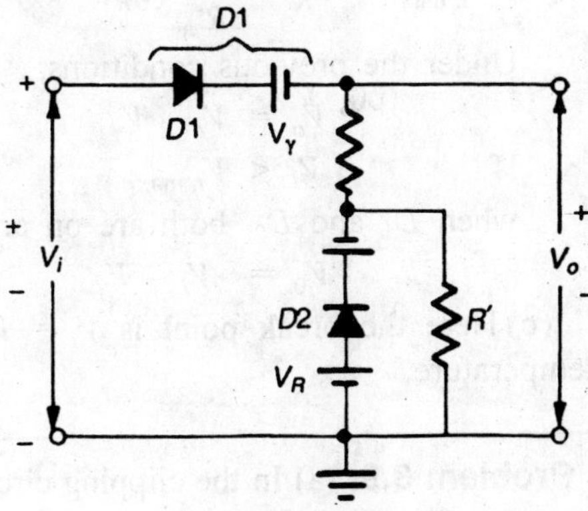

Fig. P. 3.8 (a)

(b) Assuming both the diodes are in on state i.e., conducting;

$$V_i = Ri_1 + V_R + V_\gamma - V_\gamma \quad \text{...(i)}$$

and

$$V_\gamma - V_\gamma = R'i_2 \quad \text{...(ii)}$$

But also,

$$i_1 = i_{D_1} = \frac{V_1 - V_R}{R}$$

(From equation (i))

and

$$i_{D_2} = i_2 - i_1$$

$$\therefore i_{D_2} = i_2 - i_1$$

$$= \frac{V_R - V_\gamma}{R'} - \frac{V_1 - V_R}{R}$$

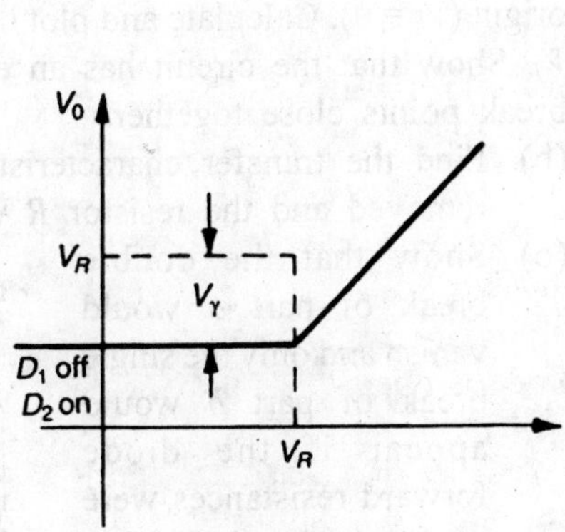

Fig. P. 3.8 (b)

The current i_{D_2} becomes (–)ve when $V_i \geq V_R + \frac{R}{R'}(V_R - V_\gamma)$. Then D_1 is on and D_2 is off. Therefore, the maximum value of the input voltage V_i, so that the current in D_2 is always in the forward direction, is,

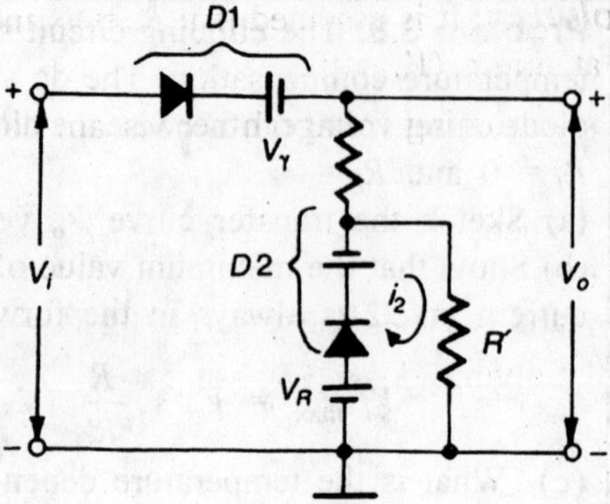

Fig. P. 3.8 (c)

$$V_{i,\,\max} = V_R + \frac{R}{R'}(V_R - V_\gamma)$$

Under the previous conditions,

$$V_o = V_i - V_\gamma$$

If $\quad V_i < V_{i,\,\max}$,

when D_1 and D_2, both are on and

$$V_0 = V_i - V_\gamma$$

(c) Here the break point is $V_i = V_R$ and it is independent of the temperature.

Problem 3.9: (a) In the clipping circuit shown, $D2$ compensates for temperature variations. Assume that the diodes have infinite back resistance, a forward resistance of 50 Ω, and a break point at the origin ($V_\gamma = 0$). Calculate and plot the transfer characteristic V_o against V_i. Show that the circuit has an extended break point, that is, two break points close together.

(b) Find the transfer characteristic that would result if $D2$ were removed and the resistor R were moved to replace $D2$.

(c) Show that the double break of part *a* would vanish and only the single break of part *b* would appear if the diode forward resistances were made vanishingly small in comparison with R.

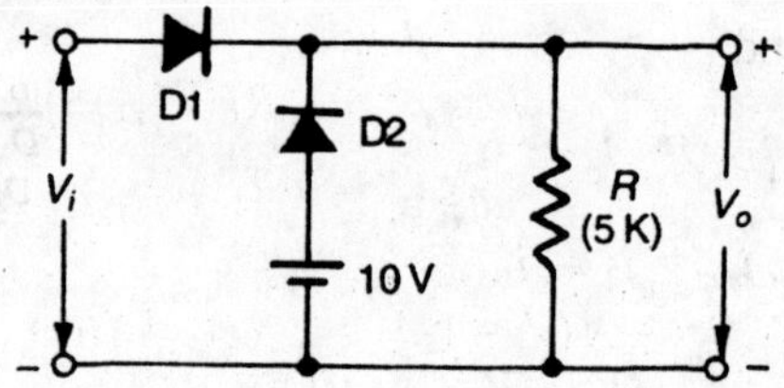

Fig. P. 3.9

Solution: It is assumed that $R_\gamma = \infty$ and $R_f = 50\Omega$ and the break point is at origin ($V_\gamma = 0$).

Assuming that both diodes are conducting.

Then, $$V_1 = 2R_f i_1 - i_2 R_f + 10 + \frac{V_\gamma - V_{\gamma 1}}{O} \quad ...(1)$$

and $$10 - V_\gamma = i_2 (R_f + R) - i_1 R_f \quad ...(2)$$

Solving (1) & (2), we have,

$$i_1 = i_{D_1}$$

$$= \left(\frac{(V_i - 10)(R_f + R) + 10R_f}{R_f^2 + 2RR_f}\right)$$

and

$$i_{D_2} = i_2 - i_1 = \left(\frac{10R_f - R(V_i - 10)}{R_f^2 + 2RR_f}\right)$$

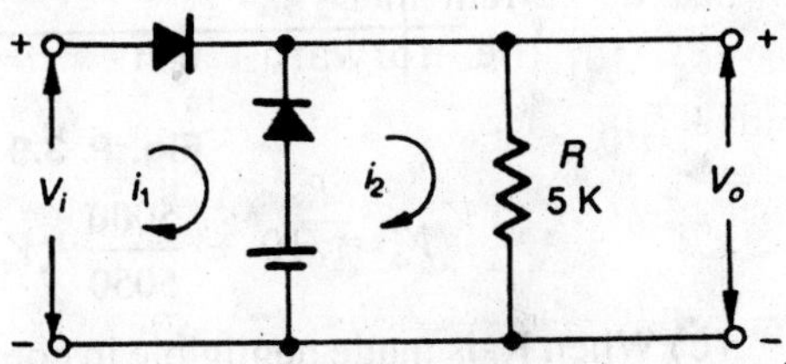

Fig. P. 3.9 (a)

If $i_{D_2} \geq 0$, then $D2$ is on when $V_i \leq \left(10 + \frac{R_f}{R} 10\right)$

If $i_{D_1} \geq 0$, then $D1$ is on when $V_i \geq \left(10 - \frac{R_f}{R + R_F} \times 10\right)$

Hence, break point occurs at $V_i \cong 10.1$ V and another at $V_i = 9.91$ V.

(i) If $V_i < 9.91$ V then $D1$ is off and $D2$ is on and $V_o = 10 \times \frac{5000}{5050}$

(ii) If $V_i > 10.1$ V then $D2$ is off and $D1$ is on, and $V_o = V_i \left(\frac{5000}{5050}\right)$

(iii) If $9.91 < V_i < 10.1$ V, then $D1$ and $D2$, both are on and

$$V_o = i_2 R = \frac{2RR_f \times 10 + R_f (V_i - 10) R}{R_f^{\,2} + 2R_f R} = \frac{20RR_f \times R_f \times R(V_i - 10)}{R_f^{\,2} + 2R_f R}$$

$$= \frac{20R + R(V_i - 10)}{R_f + 2R} = \left(\frac{R(V_i + 10)}{R_f + 2R}\right)$$

The transfer characteristic is as follows :

(b) $D1$ conducts when $V_1 > V_o$. Then,

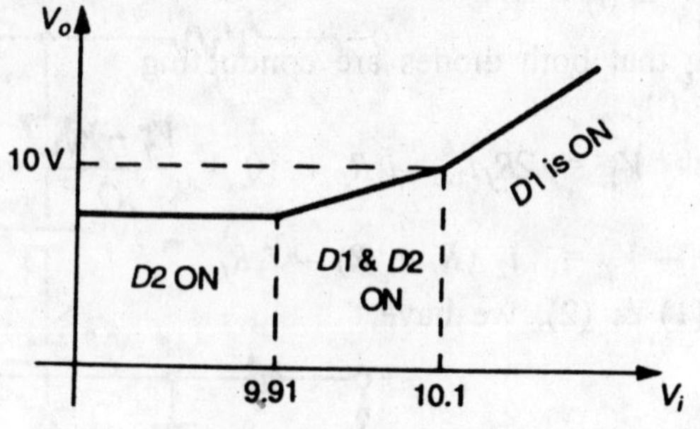

Fig. P. 3.9 (b)

$$V_o = 10 + \frac{5000}{5050}(V_i - 10)$$

(c) When R_f is made negligible in comparison to R then the break point of part a will shift to 10 V, as sketched in part (b).

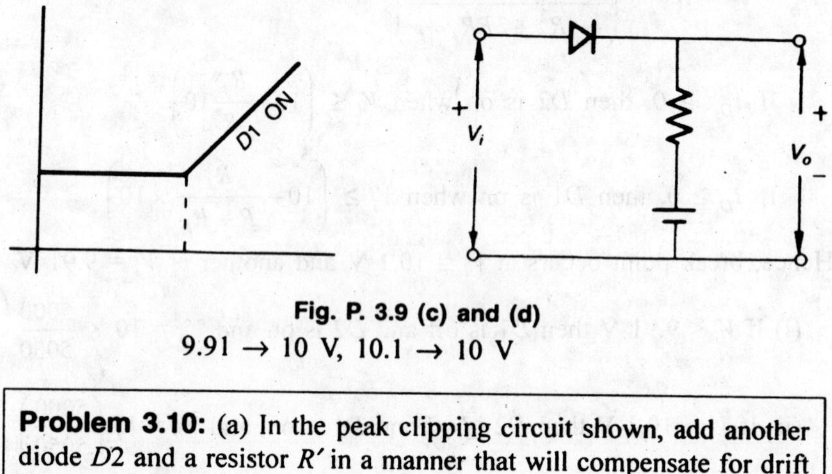

Fig. P. 3.9 (c) and (d)

9.91 → 10 V, 10.1 → 10 V

Problem 3.10: (a) In the peak clipping circuit shown, add another diode $D2$ and a resistor R' in a manner that will compensate for drift with temperature.

(b) Show that the break point of the transmission curve occurs at V_R. Assume $R_r \gg R \gg R_f$.

(c) Show that if $D2$ is always to remain in conduction it is necessary that

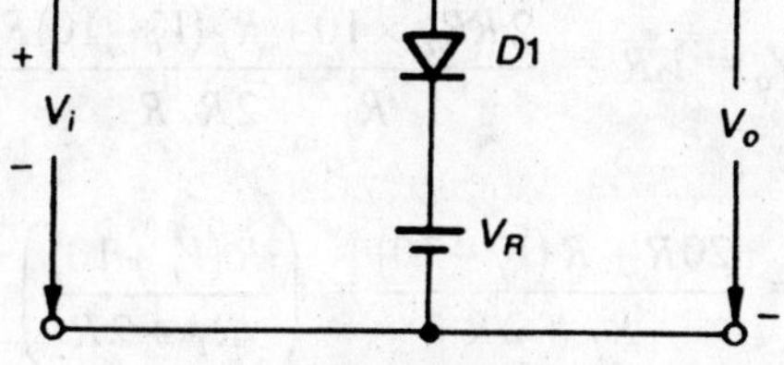

Fig. P. 3.10

$$V_i < V_{i,\max} = V_R + \frac{R}{R'}(V_R - V_\gamma)$$

Solution: (a) To compensate for drift with temperature, the addition of another diode $D2$ and a resister is shown as in Fig. P. 3.10 (a).

(b) Assuming

$R_\gamma \gg R \gg R_f$

If $V_i < V_R + V_\gamma - V_\gamma$

$= V_R$

Then $D1$ is off

and $V_i = V_o$

If $V_i > V_o$

Then $D1$ is on and

$V_0 \approx V_R$

$(\because R_\gamma \gg R \gg R_f)$

If $V_i = V_{i,\, max}$

when the current through $D2$ just drop to zero

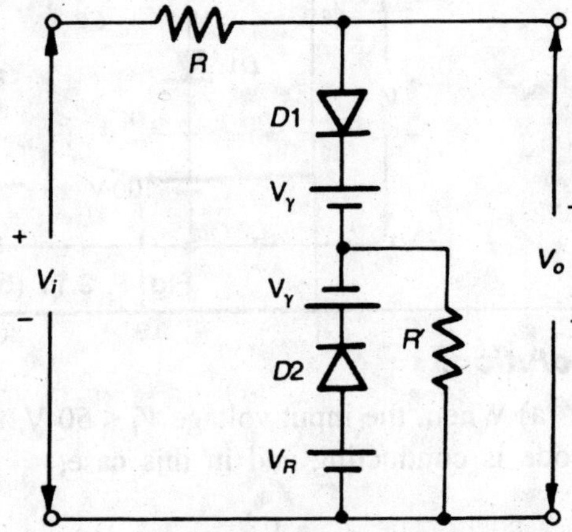

Fig. P. 3.10 (a)

Then $V_{i\,max} = I\,(R + R') + V_\gamma$

But also here it is seen that,

$$IR' = V_R - V_\gamma$$

and finally we get,

$$V_{i\,max} = \frac{V_R - V_\gamma}{R'}\,(R + R') + V_\gamma$$

$$= V_R + \frac{R}{R'}\,(V_R - V_\gamma) \qquad \text{Proved}$$

hence, the condition, to $D2$ remain in conduction always is,

$$V_i < V_{i,max} = V_R + \frac{R}{R'}\,(V_R - V_\gamma)$$

Problem 3.11: (a) The input voltage v_i, to the two-level clipper shown in part *a* of the figure varies linearly from 0 to 150 V. Sketch the output voltage v_o to the same time scale as the input voltage. Assume ideal diodes.

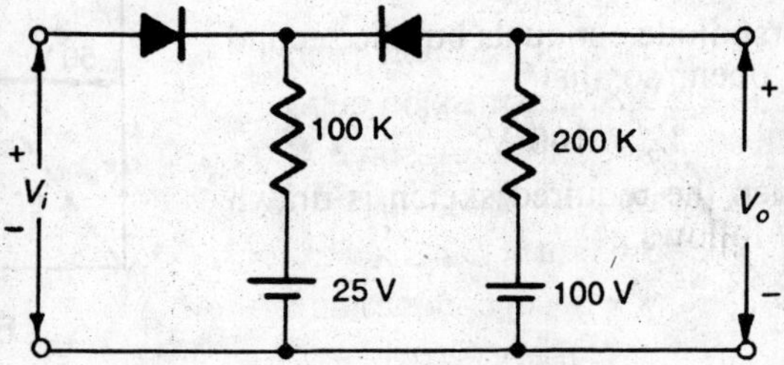

Fig. P. 3.11 (a)

(b) Repeat (a) for the circuit shown in part *b* of the figure.

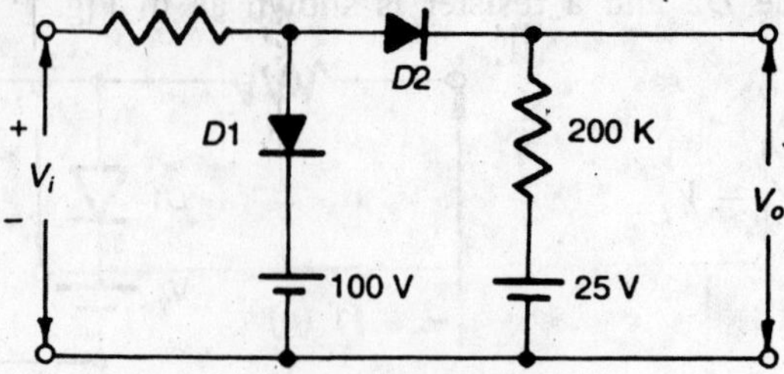

Fig. P. 3.11 (b)

Solution:

(a) When, the input voltage $V_i < 50$ V, first diode is open and second diode is conducting and in this case,

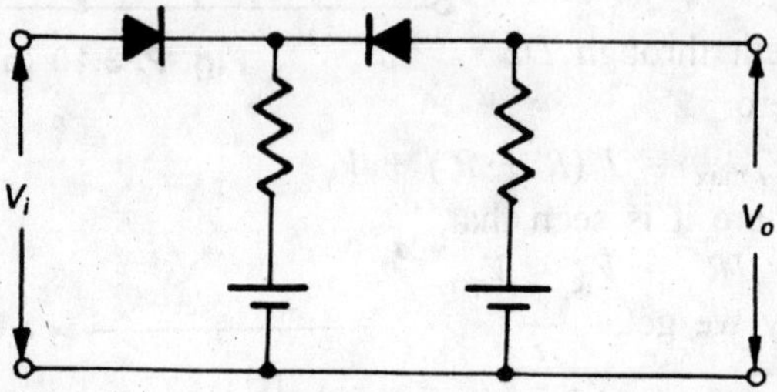

Fig. P. 3.11 (c)

$$V_o = 100 - \frac{2}{3} \times 75 = 50 \text{ V.}$$

When $50 < V_i < 100$ V,

Both the diodes conducts

and $\quad V_o = V_i$

and As $V_i > 100$ V

First diode conducts but the second is open, so that

$$V_o = 100 \text{ V}$$

Then the required sketch is drawn as follows :

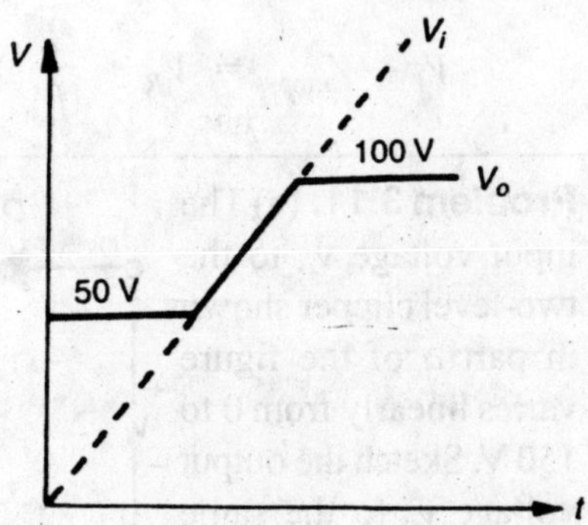

Fig. P. 3.11 (d)

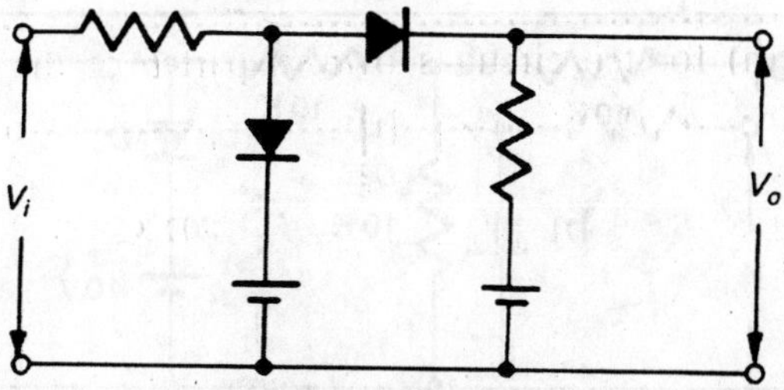

Fig. P. 3.11 (e)

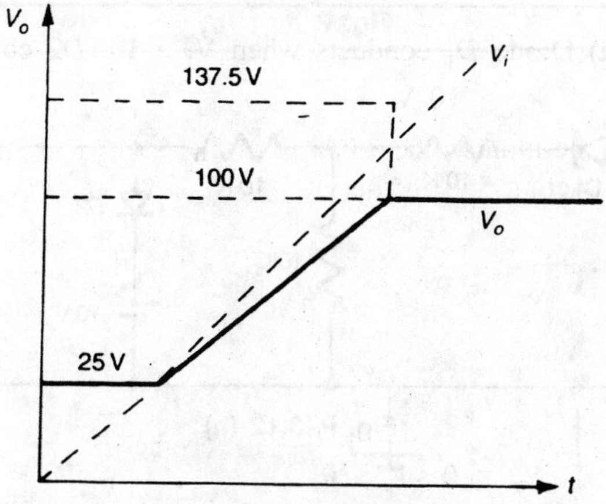

Fig. P. 3.11 (f)

(b) When $V_i < 25$ V, none of the diodes conducts and

$$V_o = 25 \text{ V}$$

When $V_i > 25$ V, upper diode conducts and

$$V_o = (V_i - 25)\frac{2}{3} + 25$$

When V_o reaches 100 V, $V_i = 137.5$ V

For larger V_i, both diodes conduct and then $V_o = 100$ V

Then the required output voltage V_o to the same time scale as the input voltage.

Problem 3.12: (a) The diodes are ideal. Write the transfer characteristic equations (v_o as a function of v_i).

(b) Plot v_o against v_i, indicating all intercepts, slopes, and voltage levels.

(c) Sketch v_o if $V_1 = 40 \sin \omega t$. Indicate all voltage levels.

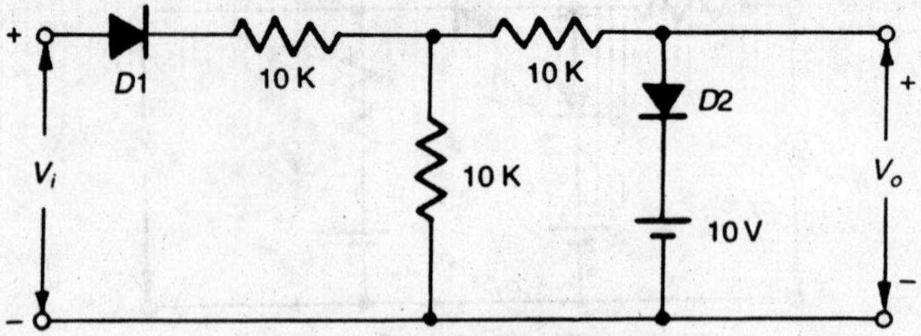

Fig. P. 3.12

Solution: (a) Diode D_1 conducts when $V_i > 10$. $D2$ conducts when $V_o > 10$ V.

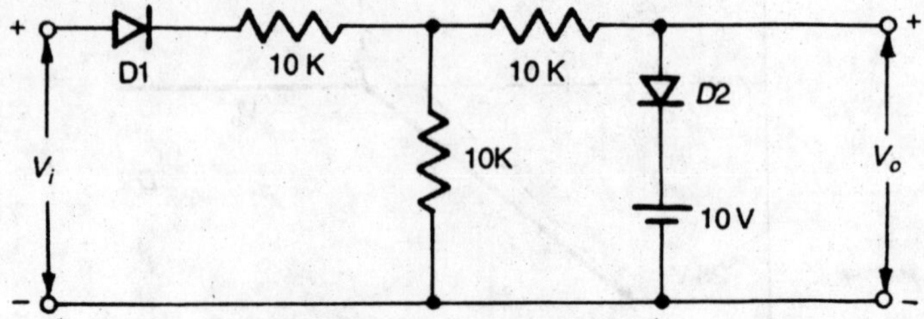

Fig. P. 3.12 (a)

Thus $\quad V_o = 0 \quad V_i \leq 0$

$\qquad V_{1/2} \quad 0 \leq V_i \leq 20$ V

$\qquad 10V \quad 20 \leq V_i$

(b)

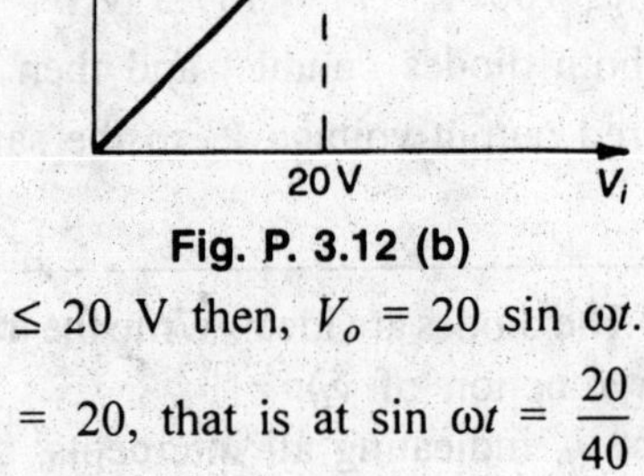

Fig. P. 3.12 (b)

(c) When $0 \leq V_i \leq 20$ V then, $V_o = 20 \sin \omega t$.

For, $\quad V_i = 20$, that is at $\sin \omega t = \dfrac{20}{40} = \dfrac{1}{2}$

$\therefore \quad \omega t = 30°$

$D2$ conducts and $V_o = 10$ V

For $\quad V_i < 0$,

$\qquad V_o = 0$

The sketch of V_o is as follows :

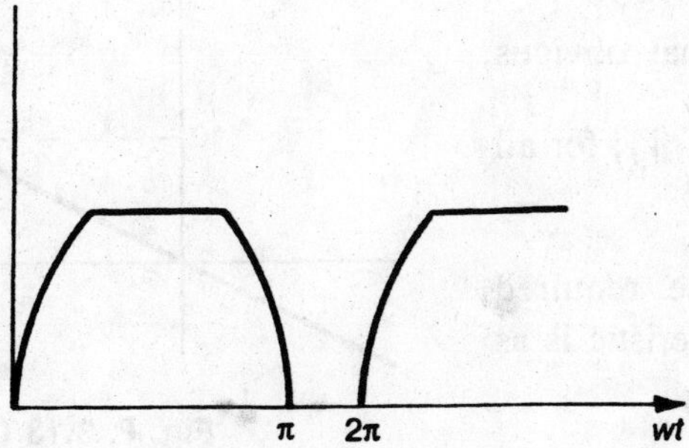

Fig. P. 3.12 (c)

Problem 3.13:
(a) Repeat Prob. 3.12 for the circuit shown.
(b) Repeat for the case where the diodes have an offset voltage $V_\gamma = 1$ V

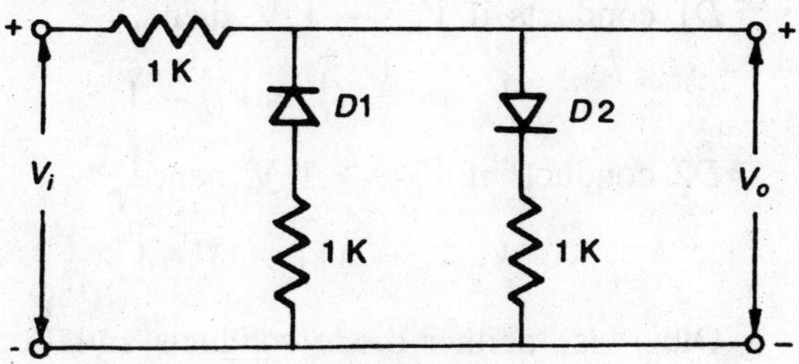

Fig. P. 3.13

Solution: Ckt. for this problem is as,

(a) When $V_i < 0$, i.e., (–)ve, $D1$ conducts and $D2$ remains in off-state

$$\text{then} \quad V_o = \frac{1}{2} V_i$$

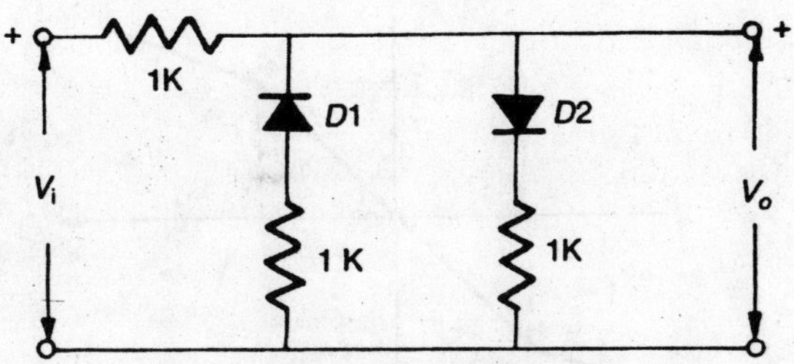

Fig. P. 3.13 (a)

and opposite holds, when $V_i > 0$ i.e. (+)ve, $D2$ conducts but $D2$ will be off. Then in this case also,

$$V_o = \frac{1}{2} V_i$$

Hence it may obvious

that, $\therefore V_o = \frac{1}{2} V_i$; for ail V_i

Hence, the required transfer characteristic is as follows :

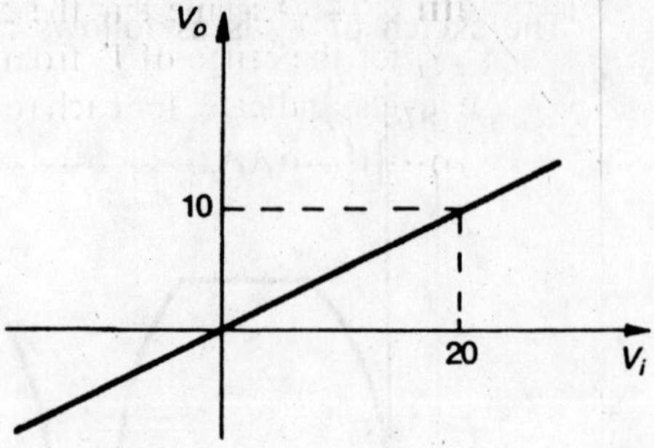

Fig. P. 3.13 (b)

(b) In this case the ckt. connected is same, except the knee voltage $V_\gamma = 1.0$ V introduces. Then,

$D1$ conducts if $V_i < -1$ V, then

$$V_o = \frac{1}{2}(1 + V_1) - 1$$

$D2$ conducts if $V_i = > 1$ V, hence

$$V_o = \frac{1}{2}(V_i - 1) + 1$$

Otherwise neither diode conducts and $V_i = V_o$

$$\text{Hence, } V_o = \begin{cases} \frac{1}{2}(V_i - 1) & ; \quad V_i < -1V \\ V_1 & ; \quad -1 < V_i < 1V \\ \frac{1}{2}(V_i + 1) & ; \quad V_i > 1V \end{cases}$$

The required dynamic characteristic is plotted as follows :

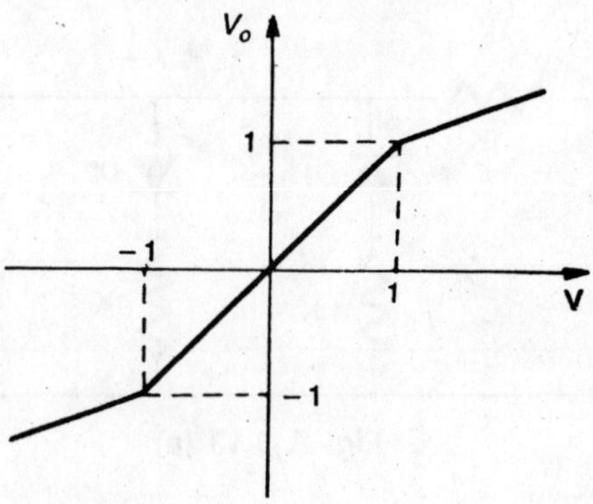

Fig. P. 3.13 (c)

Problem 3.14: Assume that the diodes are ideal. Make a plot of V_o against V_i for the range of V_i from 0 to 50 V. Indicate all slopes and voltage levels. Indicate, for each region, which diodes are conducting.

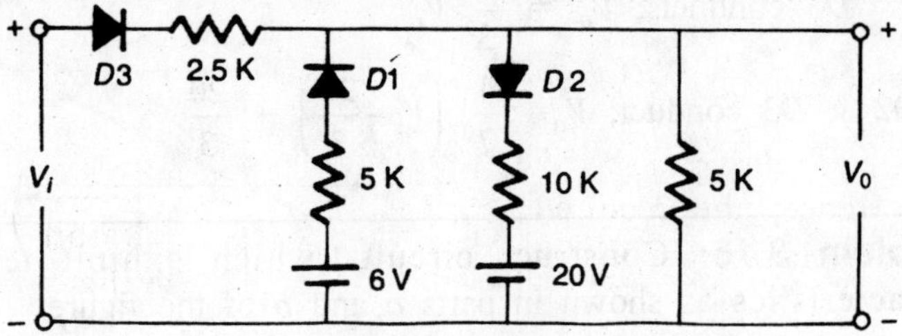

Fig. P. 3.14

Solution: From the above figure (ckt.) one can observe that,

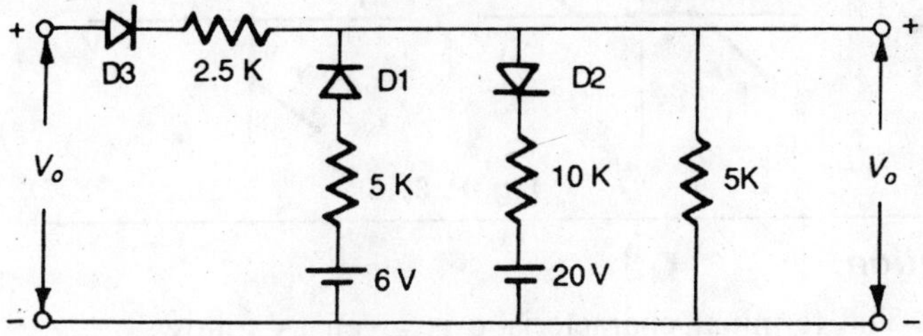

Fig. P. 3.14 (a)

$D1$ conducts when $V_o < 6$ V

$D2$ conducts when, $V_o > 20$ V

and $D3$ conducts when $V_i > V_o$

Now, observations and calculation holds good for the following equation :

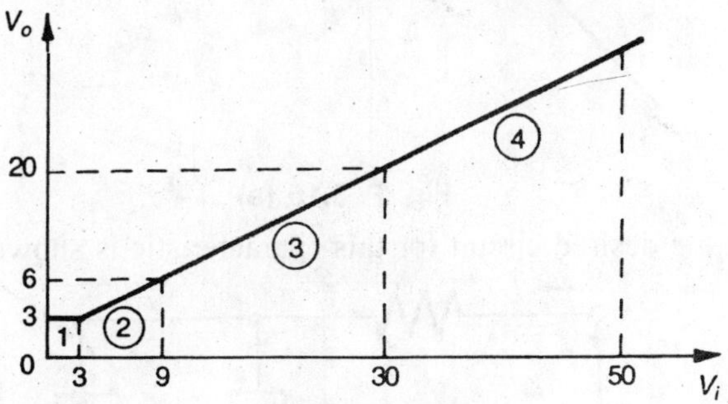

Fig. P. 3.14 (b)

$D1$ will on; $$V_o = \left(\frac{5K}{5K + 5K}\right) \times 6 = 3 \text{ V} \qquad \ldots(1)$$

$D1$ & $D3$ will on; $V_o = \left[\frac{1}{2}(V_i - 3) + 3\right]$ V ...(2)

$D3$ conducts; $V_o = \frac{2}{3} V_i$ ...(3)

$D2$ & $D3$ conduct; $V_o = \frac{4}{7}\left(V_1 - \frac{20}{3}\right) + \frac{20}{3}$ ...(4)

Problem 3.15: Construct circuits which exhibit terminal characteristics as shown in parts *a* and *b* of the figure.

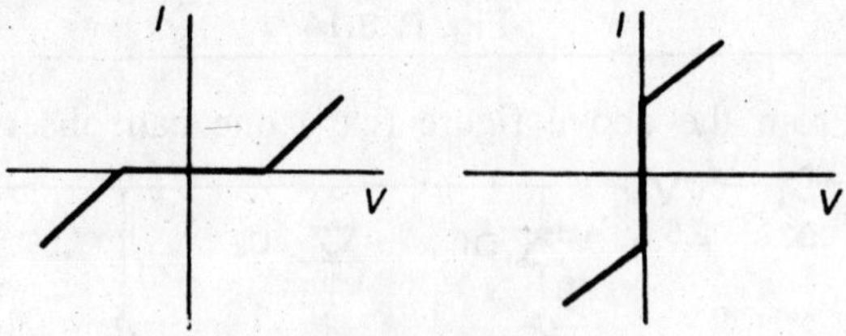

Fig. P. 3.15

Solution:

(a) The terminal characteristic is given as follows :

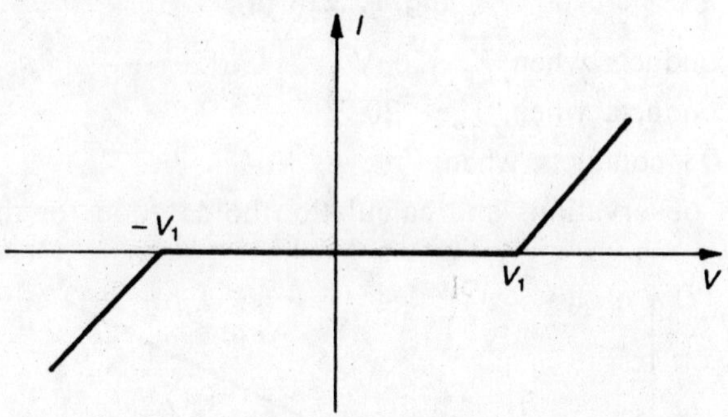

Fig. P. 3.15 (a)

Then the desired circuit for this characteristic is shown as follows:

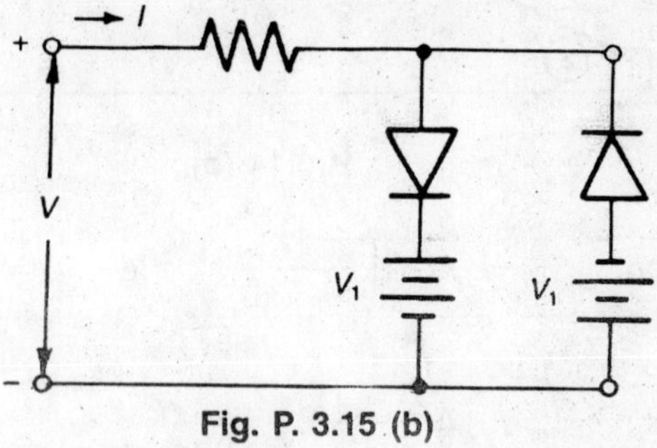

Fig. P. 3.15 (b)

(b) In this case the terminal characteristic is given as

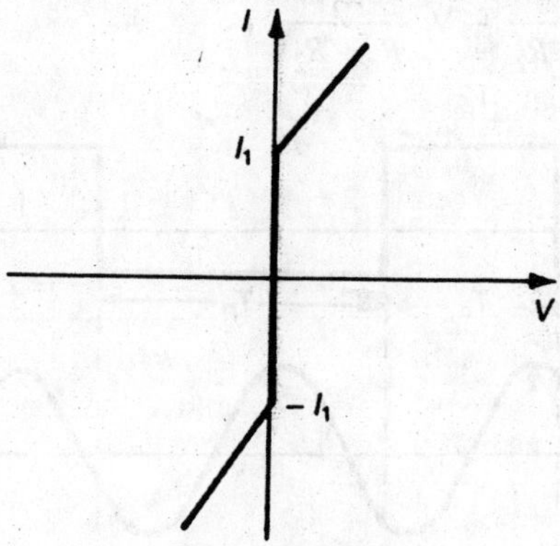

Fig. P. 3.15 (c)

The desired circuit for the characteristic shown above is given as follows :

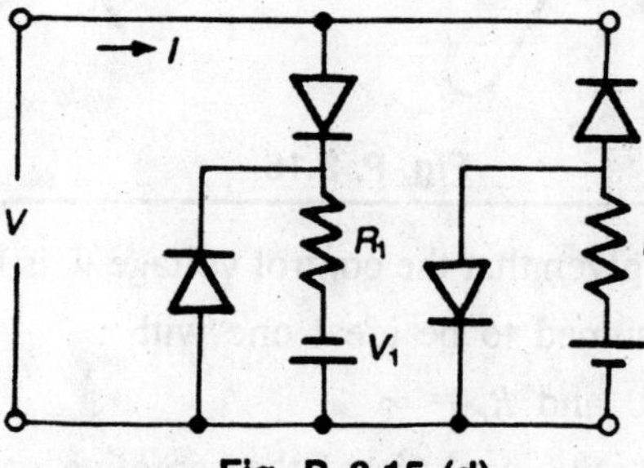

Fig. P. 3.15 (d)

Problem 3.16: (a) Explain qualitatively the operation of the sampling gate shown. The supply voltage V is constant. The control voltage V_c is the square wave of Fig. 3.16. Assume ideal diodes with $V_\gamma = 0$, $R_f = 0$, and $R_r = \infty$.

Verify the following relationship..

(b) $V_{min} = \dfrac{R_c}{R_2} \quad \dfrac{R_1}{R_2 + 2R_L} V_S.$

where

$R_1 = R_c \parallel R_2$

(c) $A = \dfrac{v_o}{v_S} = \dfrac{2R_L}{R_2} \cdot \dfrac{R_1}{R_1 + 2R_L}$

(d) $V_{n,\min} = V_S \dfrac{R_c}{R_c + R_2} - V \dfrac{R_2}{R_c + R_2}$

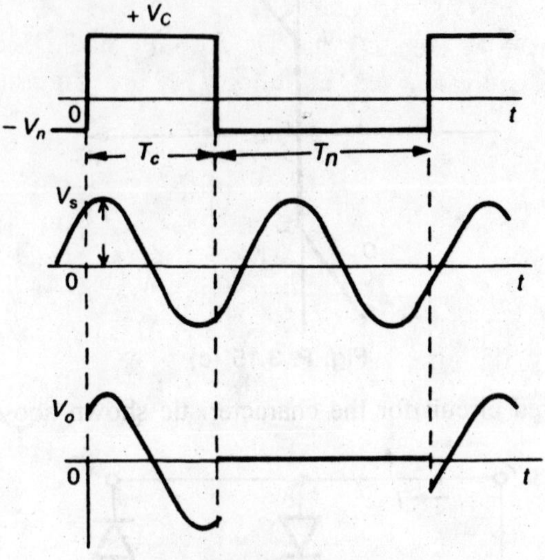

Fig. P. 3.16

Solution: Here it is given that the control voltage v_c is the square wave.

Diodes are assumed to be ideal one with :

$V_\gamma = 0,\ R_f = 0$ and $R_r = \infty$

(a) Now the qualitative analysis of the sampling gate is as :

When $v_c = V_c$ then the diode $D3$ does not conduct and this provided that $V_{s,\min} > -V_c$. Also $D4$ is off. When both the diodes $D3$ and $D4$ are off, $D1$ and $D2$ diodes will be conduct due to the supply end voltages V and V_s will be obtained at the output.

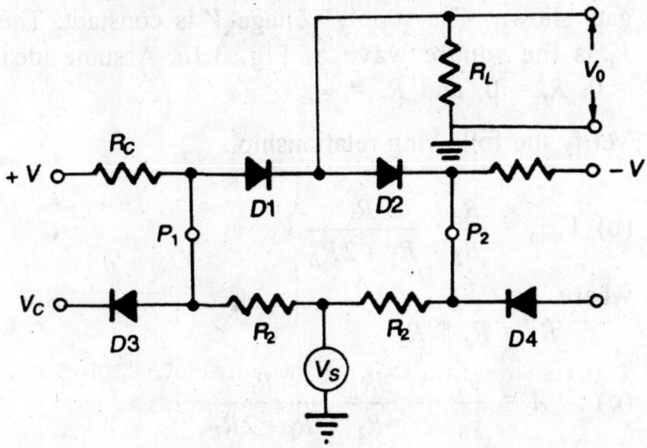

Fig. P. 3.16 (a)

Again when $v_c = -V_n$, $D4$ will be ON and $D3$ will also ON and this will provide that $V_{s,\,min} > -V_n$. When $D4$ is conducting.

Hence in this condition $D1$, $D2$ are reversed biase and the o/p is zero.

(b) When $v_c = V_c$ then $D3$ and $D4$ are OFF and $D1$ and $D2$ are ON. We get the equivalent ckt. as follows, of the sampling gate :

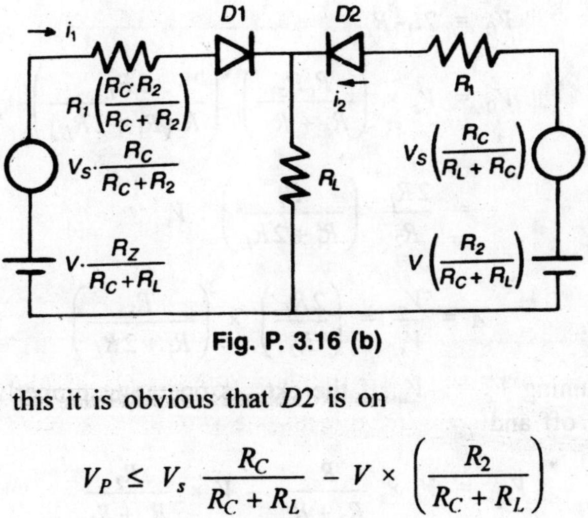

Fig. P. 3.16 (b)

From this it is obvious that $D2$ is on

$$V_P \leq V_s \frac{R_C}{R_C + R_L} - V \times \left(\frac{R_2}{R_C + R_L}\right)$$

$$\text{or, } V\left(\frac{R_2}{R_C + R_L}\right) \times \left(\frac{R_1 + 2R_L}{R_1 + R_L}\right) \leq V_s \left(\frac{R_c}{R_c + R_L}\right) \times \left(\frac{R_1}{R_1 + R_L}\right)$$

Hence the value of V is found to be,

$$V \leq V_s \times \left(\frac{R_c}{R_2}\right) \times \left(\frac{R_1}{R_1 + 2R_L}\right)$$

$$V_{max} = V_s \times \left(\frac{R_c}{R_2}\right) \times \left(\frac{R_1}{R_1 + 2R_L}\right)$$

(c) Also we find that, as the constant supplies in the two loops are of opposite polarities they do not contribute in V_0.

Hence,

$$V_s \left(\frac{R_c}{R_c + R_2} \right) = i_1 \ (R_1 + R_2) + i_2 \ R_L \text{ and}$$

$$V_s \left(\frac{R_c}{R_c + R_2} \right) = i_1 R_L + i_2 \ (R_1 + R_L)$$

Here from symmetry of ckt., $i_1 = i_2$

$$\therefore \qquad i_1 = V_s \times \left(\frac{R_c}{R_c + R_2} \right) \times \frac{1}{R_1 + 2R_L}$$

Then, $\qquad V_0 = 2i_1 \ R_L$

$$V_0 = V_s \times \left(\frac{R_c R_2}{R_c + R_L} \right) \left(\frac{2R_L}{R_2(R_1 + 2R_L)} \right)$$

$$= \frac{2R_L}{R_2} \left(\frac{R_1}{R_1 + 2R_L} \right) . V_s$$

$$\therefore \qquad A = \frac{V_o}{V_s} = \left(\frac{2R_L}{R_2} \right) \times \left(\frac{R_1}{R_1 + 2R_L} \right)$$

(d) Assuming $V_c = -V_n$. If the ckt., is operating properly, $D1$ and $D2$ will be off and

$$V_{P_2} = V_s \times \frac{R_c}{R_2 + R_c} - V \times \frac{R_2}{R_c + R_2}$$

Then for $D4$ to be ON $V_n \geq V_{P_2}$ or

$$V_n = V_s \times \frac{R_c}{R_c + R_2} - V \times \frac{R_2}{R_c + R_2}$$

And it is obvious that $D3$ is ON.
Also, it may possible that, if $v_s = -V_s$ then $D3$ must be conducting

$$-V_n \leq -V_s \frac{R_c}{R_c + R_2} + V \frac{R_2}{R_c + R_2}$$

or, $V_{n,\min} = V_s \left(\dfrac{R_c}{R_c + R_2} \right) - V \left(\dfrac{R_2}{R_c + R_2} \right)$ and this time D_4 will be ON.

Problem 3.17: A diode whose internal resistance is 20 Ω is to supply power to a 1000–Ω load from a 110-V (rms) source of supply. Calculate (a) the peak load current, (b) the dc load current, (c) the ac load current, (d) the dc diode voltage, (e) the total input power to the circuit, (f) the percentage regulation from no load to the given load.

Solution: Here

$$V_m = 110\sqrt{2} \text{ V}$$
$$R_f = 20\Omega,$$
$$R_L = 1000\ \Omega$$

(a)
$$I_m = \frac{V_m}{R_f + R_L}$$
$$= \frac{110\sqrt{2}}{1020} = 152.5 \text{ mA}$$

(b) the dc load current,

$$I_{dc} = \left(\frac{I_m}{\pi}\right) = \left(\frac{152.5}{\pi}\right) \cong 48.5 \text{ mA}$$

(c)
$$I_{rms} = \frac{1}{2} I_m$$
$$= \frac{1}{2} \times 152.5 = 76.25 \text{ mA}$$

(d) the d.c diode voltage,

$$= -\frac{I_m R_L}{\pi} = -I_{dc} \times R_L = -48.5 \times 1 = -48.5 \text{ V}$$

(e) $P_i = I_{rms}^2 (R_f + R_L) = (76.2 \times 10^{-3})^2 \times (1020) \cong 5.92$ W

(f) % Regulation $= \dfrac{V_o - V_L}{V_L} \times 100 = \dfrac{V_{m/\pi} - I_{dc}R_L}{I_{dc} \times R_L}$

$$= 100 \times \frac{49.5 - 48.5}{48.5} \cong 2.06\%$$

Problem 3.18: Show that the maximum dc output power $P_{dc} = V_{dc}I_{dc}$ in a half-wave single-phase circuit occurs when the load resistance equals the diode resistance R_f.

Solution: The d.c output power is,

$$P_{dc} = I_{dc}^2 R_L = \left(\frac{V_m}{\pi(R_f + R_L)}\right)^2 \times R_L$$

For maximum P_{dc}; $\frac{d\,P_{dc}}{d\,R_L} = 0$

or $$\frac{V_m^{\;2}}{\pi^2}\left(\frac{(R_f + R_L)^2 - 2(R_f + R_L)R_L}{(R_f + R_L)^4}\right) = 0$$

or, $R_f + R_L = 2R_L$

$\therefore$ $R_f = R_L$ Proved

Hence, $P_{dc,\,max}$ takes place when $R_f = R_L$

Problem 3.19: The efficiency of ratification n_r is defined as the ratio of the dc output power $P_{dc} = V_{dc}I_{dc}$ to the input power $P_i = (1/2\pi)\int_0^{2\pi} v_i i \; d\alpha$.

(a) Show that, for the half-wave-rectifier circuit,

$$\eta_r = \frac{40.6}{1 + R_f / R_L}\,\%$$

(b) Show that, for the full-wave rectifier, n_r has twice the value given in part *a*.

Solution: From the expression of efficiency

$$\eta = \frac{P_{dc}}{P_i} \times 100$$

$$P_{dc} = I_{dc}^2 \; R_L$$

and $$P_i = I_{rms}^2 \times (R_f + R_L)$$

$$\therefore \quad \eta = \left(\frac{I_{dc}}{I_{rms}}\right)^2 \frac{100\%}{1 + R_f / R_L}$$

(a) For half wave rectifier,

$$I_{dc} = \frac{I_m}{\pi}, \; I_{rms} = \frac{I_m}{2}$$

$$\therefore \quad \eta = \left(\frac{I_m / \pi}{I_m / 2}\right)^2 \times \frac{100}{1 + R_f / R_c} = \left(\frac{40.6}{1 + R_f / R_2}\right)\% \text{ Proved}$$

(b) For full wave rectifier,

$$\eta_r = \left(\frac{I_{dc}}{I_{rms}}\right)^2 \times \frac{100}{1 + R_f / R_L}$$

$$I_{dc} = \frac{2I_m}{\pi} \text{ and } I_{rms} = \frac{I_m}{\sqrt{2}}$$

$$\therefore \quad \eta_r = \left(\frac{81.2}{1 + R_f / R_L}\right)\% \text{ Proved}$$

Problem 3.20: Prove that the regulation of both the half-wave and the full-wave rectifier is given by

$$\% \text{ regulation} = \frac{R_f}{R_L} = 100\%$$

Solution: Taking half wave rectifier into a/c,

$$V_{NL} = \frac{V_m}{\pi}$$

and $$V_{FL} = \frac{V_m}{\pi} \times \left(\frac{R_L}{R_L + R_f}\right)$$

Then, $$\% \text{ regulation} = \frac{V_{NL} - V_{FL}}{V_{FL}} \times 100\%$$

$$= \frac{\frac{V_m}{\pi}\left(1 - \frac{R_L}{R_L + R_f}\right)}{\frac{V_m}{\pi}\left(\frac{R_L}{R_L + R_f}\right)} \times 100\% = \left(\frac{R_f}{R_L}\right) \times 100\%$$

For the case of full wave rectifier, both V_{NL} and V_{FL} are doubled. Hence, the regulation will be of same value.

Problem 3.21: A full-wave single-phase rectifier consists of a double-diode vacuum tube, the internal resistance of each element of which may be considered to be constant and equal to 500 Ω. These feed into a pure resistance load of 2000 Ω. The secondary transformer voltage to center tap is 280 V. Calculate (a) the dc load current, (b) the direct current in each tube, (c) the ac voltage across each diode, (d) the dc output power, (e) the percentage regulation.

Solution:

(a) The dc load current,

$$I_{dc} = \frac{2I_m}{\pi} = \frac{2V_m}{\pi(R_L + R_f)}$$

$$= \frac{2 \times 280\sqrt{2}}{\pi \times 2500}$$

$$\cong 101 \text{ mA}$$

(b) $(I_{dc})_{\text{tube}} = \frac{1}{2} I_{dc} = 50.5 \text{ mA}$

(c) In the centre tap, the voltage impressed across the diode resistance R_f is series with R_L. Hence, the voltage across the conducting tube is Sinusoidal with a peak-value

$$V_m \times \frac{500}{2500} = 0.2 \text{ V}_\text{m}$$

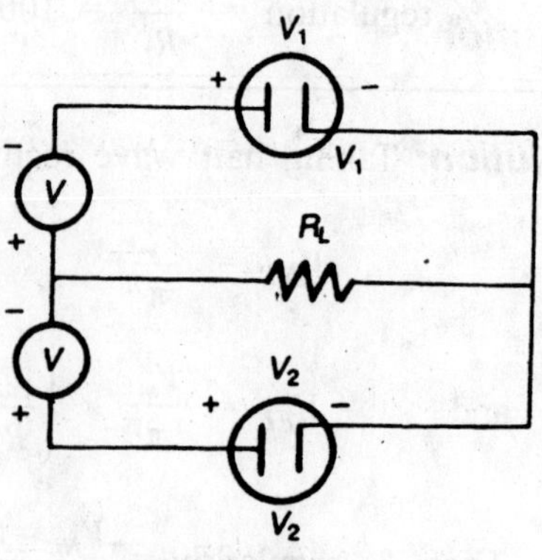

Fig. P. 3.21 (a)

During the non-conducting of V_1, we see that by transversing the outside path of the circuit sketched that $V_i = -2V - V_1$. Since V_2 is now conducting, its peak value is 0.2 V_m and that of V_i is

$-2V_m + 0.2V_m = -1.8\ V_m$

Thus, the voltage V_1 across V_1 is as shown and its ac value is :

$$V_{rms} = \left(\frac{1}{2\pi}\int_0^{\pi}(0.2\,V_m)^2 \sin^2\alpha\, d\alpha + \frac{1}{2\pi}\int_{\pi}^{2\pi}(-1.8\,V_m)^2 \sin^2\alpha\, d\alpha\right)^{1/2}$$

$$= \frac{V_m}{2}\ [(0.2)^2 + (1.8)^2]$$

$$= 0.905\ V_m$$

$$= 0.905 \times 280 \times \sqrt{2}$$

$$= 358 \text{ V}$$

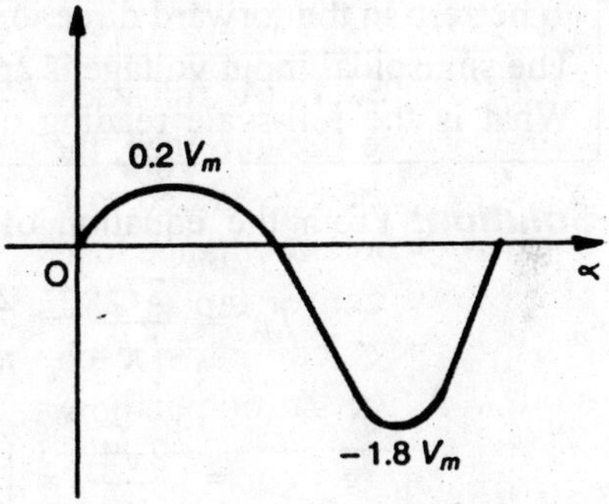

Fig. P. 3.21 (b)

(d) $P_{dc} = I_{dc}^2 \times R_L$

$$= (0.101)^2 \times 2000$$

$$= 20.4 \text{ W.}$$

(e) % Regulation $= \dfrac{R_f}{R_L} \times 100\% = \dfrac{500}{2000} \times 100 = 25\%$

Problem 3.22: In the full-wave single-phase bridge, can the transformer and the load be interchanged? Explain carefully.

Solution:

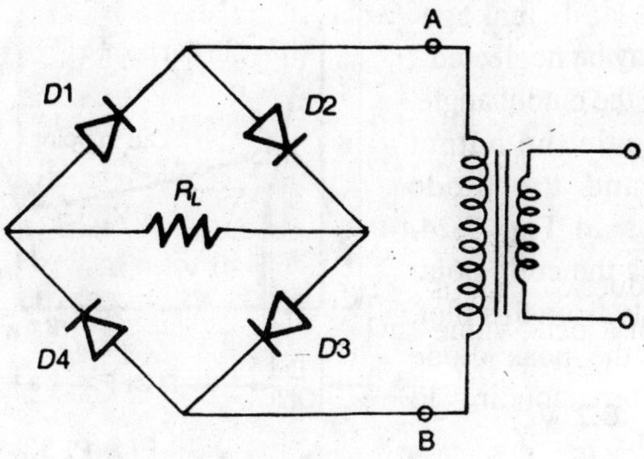

Fig. P. 3.22

The load and the transformer cannot be interchanged. If they were, the circuit shown would result. Note that if A is at positive potential with respect to B, all the four diodes will become reverse biased.

But if B is made positive with respect to A, then all the four diodes conduct and thus short circuiting of transformer will take place.

Problem 3.23: 1-mA dc meter whose resistance is 10 Ω is calibrated to read rms volts when used in a bridge circuit with semiconductor diodes. The effective resistance of each element may be considered to be zero in the forward direction and infinite in the inverse direction. The sinusoidal input voltage is applied in series with a 5-K resistance. What is the full-scale reading of this meter?

Solution: From the equation of I_{dc},

$$I_{dc} = \frac{2I_m}{\pi} = \frac{2}{\pi}\frac{V_m}{R_L}$$

$$= \frac{2\sqrt{2}}{\pi} \times \left(\frac{V_{rms}}{R_L}\right) = \frac{2\sqrt{2}}{\pi} \times \left(\frac{V_{rms}}{5010}\right)$$

Here, $I_{dc} = 1$ mA

$$\therefore \quad V_{rms} = \frac{5010 \times \pi}{2\sqrt{2}} \times 1 \times 10^{-3} \cong 5.56 \text{ V } \textbf{Ans}$$

Problem 3.24: A single-phase full-wave rectifier uses a semiconductor diode. The transformer voltage is 35 V rms to center tap. The load consists of a 40 μF capacitance in parallel with a 250 Ω resistor. The diode and the transformer resistances and leakage reactance may be neglected.

(a) Calculate the cutout angle.

(b) Plot to scale the output voltage and the diode current as in Fig. 3.24. Determine the cutin point graphically from this plot, and find the peak diode current corresponding to this point.

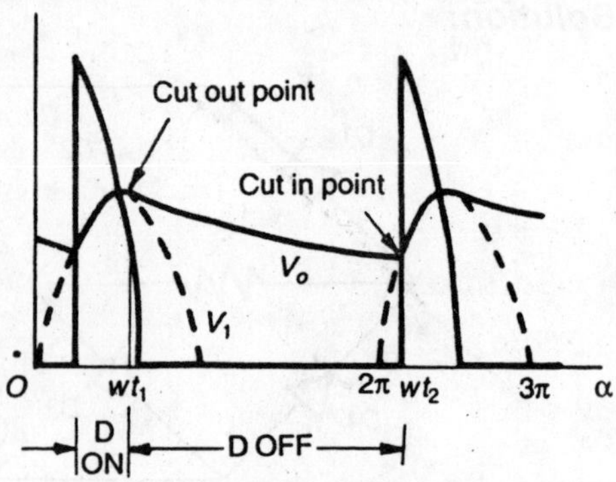

Fig. P. 3.24

Solution:

(a) $\omega t_1 = \alpha_1 = \pi - \phi = 180^\circ - \tan^{-1}(377 \times 250 \times 40 \times 10^{-6})$

$= 180^\circ - 75^\circ = 105^\circ$

(b) $I_m = V_m \left(\sqrt{\omega^2 c^2 + \frac{1}{R_L^2}}\right) \sin(\omega t - \phi)$

$$= 35\sqrt{2} \sqrt{(377 \times 40 \times 10^{-6})^2 + \left(\frac{1}{250}\right)^2} \sin(\alpha + 75^\circ)$$

$\therefore\ i = 772 \sin(\alpha + 75^\circ)$ mA

When Diode conducting

$$V_o = V = 3\sqrt{5} \sin\alpha = 49.5 \sin\alpha$$

When Diode non-conducting,

$$V_o = 49.5 \sin 105^\circ \times e^{-\alpha - \alpha \frac{1}{wR_LC}} = 47.8\, e^{-(\alpha - 105^\circ)/216}$$

Problem 3.25:

For the diode–resistance circuit of Fig. 3.25 the diode cut-in voltage is 0.6 V and voltage drop across a conducting diode is 0.7 V. Calculate V_o and indicate the state of each diode for

(a) $V_1 = 10$ V, $V_2 = 0$ V

(b) $V_1 = 5$ V, $V_2 = 0$ V

(c) $V_1 = 10$ V, $V_2 = 5$ V

(d) $V_1 = V_2 = 5$ V

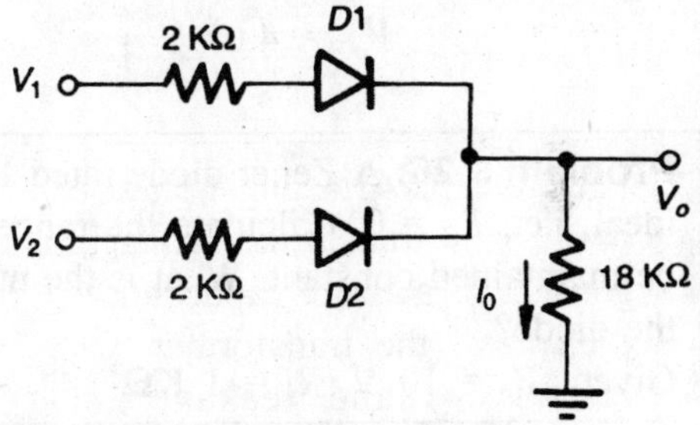

Fig. P. 3.25

Solution:

(a) $V_1 = 10$ V, $V_2 = 0$ V, D_1 would conduct.

$$I_o = \frac{10 - 0.7}{2 + 18} = 0.465 \text{ mA}$$

$$V_o = 0.465 \times 18 = 8.37 \text{ V}$$

(b) $V_1 = 5$ V, $V_2 = 0$ V, D_1 would conduct

$$I_o = \frac{5-0.7}{2+18} = 0.215 \text{ mA}$$

$$V_o = 0.215 \times 18 = 3.87 \text{ V}$$

(c) $V_1 = 10$ V, $V_2 = 5$ V

Assume that D_1 only conducts. Then as per part (a)

$$V_o = 8.37 \text{ V}$$

$$V_2 - V_o = 5 - 8.37 \text{ V} = -3.37 \text{ V}$$

So D_2 does not conduct as assumed. Therefore,

$$V_o = 8.37 \text{ V}$$

(d) $V_1 = V_2 = 5$ V

Considering that D_1 and D_2 both conduct and output voltage is V_o. Then

$$I_{D1} = \frac{5-0.7-V_o}{2} = (4.3 - V_o)/2 = I_{D2}$$

$$I_o = I_{D1} + I_{D2} = (4.3 - V_o) = V_o/18$$

$$\therefore \quad V_o = 4.07 \text{ V}$$

$$V_1 - V_o = V_2 - V_2 = 5 - 4.07 = 0.93 \text{ V} > 0.6 \text{ V}$$

So our assumption is correct.

$$\therefore \quad V_o = 4.07 \text{ V}$$

Problem 3.26: A Zener diode rated 10 V, 32 mA can be considered ideal, i.e., $R_Z = 0$. Calculate the range of R_L (load) and I_L for V_L to be maintained constant. What is the maximum wattage consumed by the diode?

Given: $V_s = 10$ V, $R_s = 1$ KΩ

Solution:

The circuit with Zener diode regulator is drawn below. Fig. 3.26

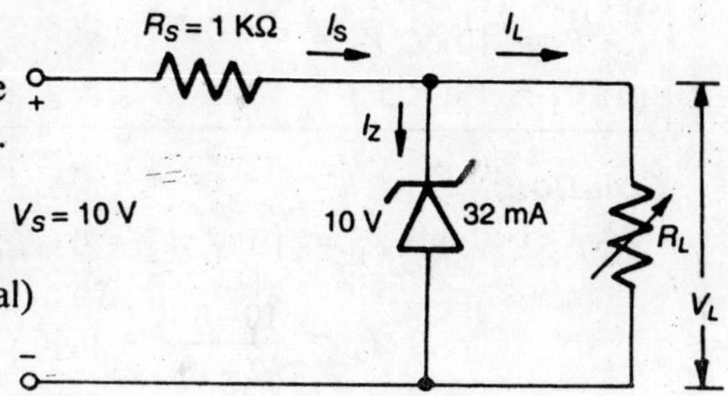

Fig. P. 3.26

When Zener is conducting

$V_L = 10$ V (Zener is ideal)

$$I_s = \frac{50-10}{1}$$

$$= 40 \text{ mA (constant)}$$

(a) R_L is maximum when $I_Z = 0$. Zener is like open circuit.

$$\therefore \quad V_L = \frac{R_L(\text{min})}{1 + R_L(\text{min})} \times 50 = 10 \text{ V}$$

or $\quad R_L \text{ (min)} = 250\ \Omega$

$$I_L = 10/250 = 40 \text{ mA}$$

(b) R_L is maximum when $I_Z = 32$ mA

$$I_s = 40 \text{ mA}$$

$$I_L = 40 - 32 = 8 \text{ mA}$$

$$R_L \text{ (max)} = 10/8 = 1.25 \text{ k}\Omega$$

Range of R_L is 250 Ω to 1.25 kΩ

Max. diode wattage consumed = 10 × 32 = 320 mW

Problem 3.27: For the Zener regulator circuit of Fig. 3.27 determine the range of input voltage (V_s) for the Zener diode to remain in "ON" state.

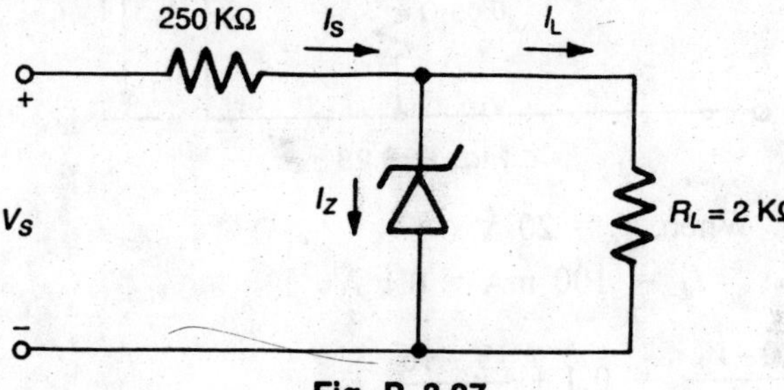

Fig. P. 3.27

Given: Zener diode; $V_Z = 20$ V, I_Z (max) = 50 mA. $R_Z = 0$

Solution: Given

$$R_Z = 0,$$

$$V_L = V_Z = 20 \text{ V (fixed).}$$

$$I_L = 20/2 = 10 \text{ mA}$$

(a) V_s (min), diode is just floating, i.e., $I_Z = 0$

$$V_s \text{ (min)} = (0.25 + 2) \times 10 = 22.5 \text{ V}$$

(b) V_s (max), diode is drawing, i.e., I_Z (max) = 50 mA

$$I_s = 50 + 10 = 60 \text{ mA}$$

$$V_s \text{ (max)} = 20 + 0.25 \times 60 = 35 \text{ V}$$

Range of input voltage = 22.5 to 35 V

Problem 3.28: For the Zener diode regulator, I_L (max) = 100 mA, I_L (min) = 0, V_Z = 16 V, R_Z = 0.05 Ω and V_s varies from 18 V to 20 V; R_s = 10 Ω. Nominal load voltage = 16 V. Determine

(a) Maximum power dissipated by R_s.
(b) Maximum power dissipated by the diode.
(c) Minimum diode current.
(d) Voltage regulation for an input of 20 V
(e) The maximum power that has to be dissipated in R_s, if the output gets accidentally short circuited.

Solution: The Zener regulator with circuit equivalent of Zener diode is drawn in Fig. 3.28

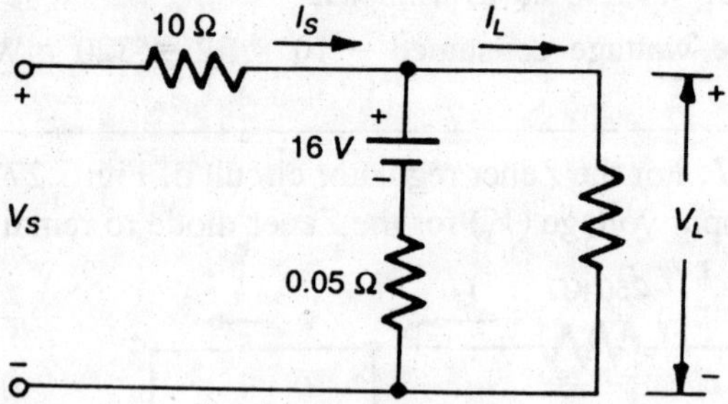

Fig. P. 3.28

Case I : When V_s = 20 V

(i) $$I_L = 100 \text{ mA} = 0.1 \text{ A}$$

$$\therefore \quad \frac{20-V_L}{10} = 0.1 + \frac{V_L-16}{0.05}$$

or $$V_L = 16.015 \text{ V}$$

$$I_Z = \frac{16.015-16}{0.05} = 0.3 \text{ A or } 300 \text{ mA}$$

$$I_s = 100 + 300 = 400 \text{ mA}$$

(ii) $$I_L = 0$$

$$I_s = I_Z = \frac{20-16}{10+0.05} = 0.398 \text{ A or } 398 \text{ mA}$$

$$V_L = 16 + 0.398 \times 0.05 = 16.02 \text{ V}$$

$$\text{Voltage regulation} = \frac{16.02-16.015}{16(\text{nominal})} \times 100 = 0.03\%$$

Case II : $V_s = 18$ V

(i) $I_L = 100$ mA or 0.1A

$$\frac{18 - V_L}{10} = 0.1 + \frac{V_L - 16}{0.05}$$

or $V_L = 16.005$ V

$$I_Z = \frac{16.005 - 16}{0.05} = 0.1 \text{ A or } 100 \text{ mA}$$

$$I_s = 0.1 + 0.1 = 0.2 \text{ A or } 200 \text{ mA}$$

(ii) $I_L = 0$

$$I_s = I_z = \frac{18 - 16}{10 + 0.05} = 0.199\text{A or } 199 \text{ mA}$$

$$V_L = 16 + 0.199 \times 0.05 = 16.01 \text{ V}$$

(a) Maximum power dissipated by $R_s = (0.4)^2 \times 10 = 1.6$ W

(b) Maximum power dissipated by diode = 16.015 × 0.3 = 4.805 W

(c) Minimum diode current = 0.199 A ≈ 200 mA

(d) Voltage regulation (V_s = 20 V) = 0.03% (very small)

(e) Output shorted

Maximum power dissipated by $R_s = (20/10)^2 \times 10 = 40$ W

Problem 3.29: The transformer of a half-wave rectifier has a secondary voltage of 20 V (rms) and a winding resistance of 10 Ω. The diode has a forward resistance of 5 Ω. Find :

(a) No-load dc voltage.

(b) dc output voltage when load is drawing 20 mA

(c) Percentage regulation at this load

(d) Ripple voltage across load and ripple frequency

(e) Power conversion efficiency

(f) PIV of the diode.

Solution:

(a) V_{dc} (no-load) $= (20\sqrt{2})/\pi = 9$ V

(b) $I_L = 20$ mA

$$V_{Ldc} = V_{dc} - I_{dc}\,(R_s + R_F)$$

$$= 9 - 2 \times 10^{-3}\,(10 + 5) = 8.7 \text{ V}$$

(c) Percentage regulation $= [(R_s + R_F)/R_L] \times 100$

$$R_L = V_{Ldc}/I_{dc} = 8.7/(20 \times 20^{-3}) = 435\ \Omega$$

Percentage regulation $= [(10 + 5)/435] \times 100 = 3.45\%$

(d) $$I_{dc} = I_m/\pi$$

$$I_m = \pi \times 20 = 6.28 \text{ mA}$$

$$I_{L\,rms} = I_m/2 = 62.8/2 = 31.4 \text{ mA}$$

$$V_{L\,rms} = 435 \times 31.4 \times 10^{-3} = 13.66 \text{ V}$$

$$[V_{L\,rms}\ (ac)]^2 = [V_{L\,rms}]^2 - [V_{L\,dc}]^2$$

Substituting values

$$V_{L\,rms}\ (ac) = \sqrt{[(13.66)^2 - (8.7)^2]} = 10.53 \text{ V}$$

$\therefore$ Ripple frequency $= 2 \times 50 = 100$ Hz

(e) Power conversion efficiency,

$$\eta = \frac{(2\pi)^2}{1 + (R_S + R_F)/R_L} \times 100$$

$$= \frac{(2\pi)^2}{1 + (10 + 5)/435} \times 100 = 39.2\%$$

(f) $$\text{PIV} = V_m = 20\sqrt{2} = 28.3 \text{ V}$$

Problem 3.30: A diode is connected in series to a 60 V ac source for charging a 24 V battery. What resistance must be connected in series to limit the peak current to 2.5 A. Estimate the time required to deliver a charge of 20 Ah to the battery.

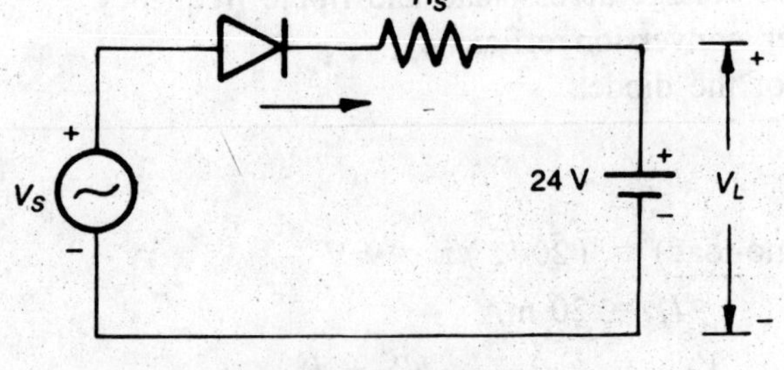

Fig. P. 3.30

Solution: From figure above given,

$$v_s = 60\sqrt{2}\ \sin \omega t$$

$$V_1 = 24\ \text{V}$$

The conduction begins at an angle of

$$\theta_1 = \sin^{-1}(V_L/V_m)$$

$$= \sin^{-1}(24/60\sqrt{2}) = 0.288 \text{ rad or } 16.5^\circ$$

Conduction will cease at an angle $(\pi - 0.288)$ rad.

$$I_{peak} = \frac{60\sqrt{2} - 24}{R_S} = 2.5\ \text{A}$$

$$\therefore \quad R_s = 24.3\ \Omega$$

From Eq. of I_{dc},

$$I_{dc} = \frac{60\sqrt{2}}{\pi \times 24.3} \cos 16.5^\circ - \frac{(\pi - 2 \times 0.288)}{2\pi} \times \frac{24}{24.3}$$

$$= 0.67\ \text{A}$$

Time to deliver 20 Ah = 20/0.67 ≈ 30 h

Problem 3.31: Derive the condition for maximum power output in a half-wave single phase rectifier circuit as the load resistance is varied. Diode forward resistance is R_F.
Does this condition conform to the maximum power theorem in a linear circuit?

Solution: We know

$$I_{dc} = \frac{(V_m/\pi)}{(R_s + R_F + R_L)}$$

Power output,

$$P_{dc} = I_{dc}^2 R_L$$

$$= \frac{(V_m/\pi)^2 R_L}{(R_s + R_F + R_L)^2}$$

$$= \frac{(V_m/\pi)^2}{\left(\frac{R_s + R_F}{\sqrt{R_L}} + \sqrt{R_L}\right)^2}$$

For maximum power output the denominator of this expression should be minimum. This happens when its derivative is zero i.e.

$$-\frac{R_S + R_F}{(R_L)^{3/2}} \quad \frac{1}{2\sqrt{R_L}} = 0$$

or $$R_L = R_S + R_F$$

This is the same condition as the maximum power theorem in a linear circuit.

Problem 3.32: The bridge rectifier of Fig. 3.32 (a) uses diodes with forward resistance of 50 Ω. The bridge output is fed to a dc ammeter of negligible resistance through an external resistance of $R = 25\ \Omega$. The signal voltage is $v_s = 200 \sin \omega t$.

(a) Sketch the wave form of the current through the ammeter. What is its peak value?
(b) Find the value of dc current read by the ammeter.
(c) Sketch the wave form of voltage across D_1. Calculate its peak and average values.
(d) What will be the reading of this ammeter if it is an rms meter?

Solution: (a) The bridge rectifier circuit is redrawn in Fig. 3.32 (a) wherein R_F is shown in series with ideal diodes with signal input to the bridge being

$$v_s = 200 \sin \omega t$$

$$i_{meter}\ (\text{peak}) = i_D\ (\text{peak}) = \frac{200}{2 \times 50 + 25} = 1.6\text{A}$$

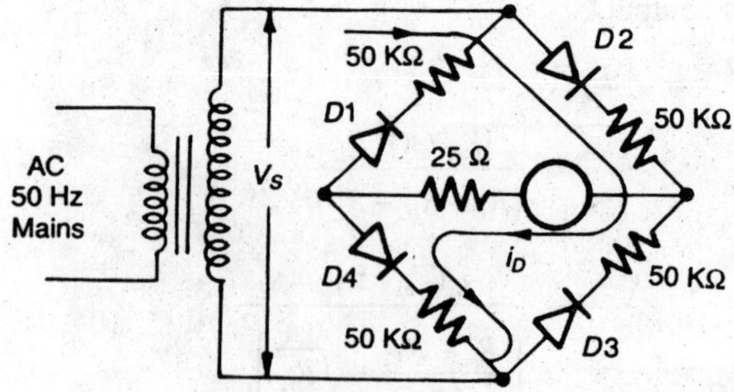

Fig. P. 3.32 (a)

The wave form of the ammeter current is drawn in Fig. 3.32 (b)

(b) $I_{dc} = 2\ i_D\ (\text{peak})/\pi = (2 \times 1.6)\pi = 1.02$ A

(c) $PIV(D_1) = V_m/2$; it is divided equally between D_1 and D_2 when these are not conducting = 100 V

$$PIV(\text{av}) = 100/\pi = 31.8 \text{ V}$$

The wave form is sinusoidal.

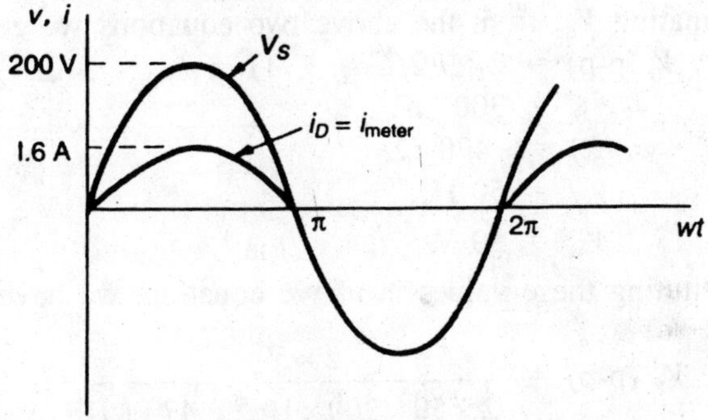

Fig. P. 3.32 (b)

(d) $$I_{rms} = I_m/\sqrt{2} = 1.6\ \sqrt{2} = 1.13\text{A}$$

Problem 3.33: The bridge rectifier connected to a 50 Hz supply is to drive 100 mA load. The maximum and minimum allowable voltages across the load are 25 V and 18 V, respectively. Estimate a suitable value of the capacitor to be connected across the load.

Solution: The capacitor has to discharge from 25 V to 18 V in one half cycle of the ac source.

∴ time of discharge = 1/(2 × 50) = 0.01 s

As $$I_c = C\ (dv/dt)$$

∴ $$C = I_c/(dv/dt) = \frac{100 \times 10^{-3}}{(25-18)/0.01} = 142.86\ \mu\text{F}$$

Problem 3.34: (a) A half-wave rectifier operates from a 50 Hz supply and provides a peak output voltage of 50 V across a 470 Ω resistive load. Estimate the value of the capacitor which limits the peak-to-peak ripple voltage across the load to 10 V.

(b) Given that the capacitance is increased to 300 μF, determine the load voltage ripple and the average load current.

Solution: (a) We have

$$V_{dc} = -V_r\,(\text{p-p})/2 + V_m = 50 - 10/2 = 45 \text{ V}$$

also $V_r\,(\text{p-p}) = V_{dc}/(fCR_L)$

Substituting values $C = \dfrac{45}{10 \times 50 \times 470} = 191.5\ \mu\text{F}$

(b) We have $V_{dc} = V_m - V_r\,(\text{p-p})/2$

$$V_r\,(\text{p-p})/2 = V_{dc}/(fCR_L)$$

Eliminating V_{dc} from the above two equations we get

$$V_r\,(\text{p-p}) = 2V_m/(2fCR_L + 1)$$

Now $C = 300\ \mu\text{F}$,

$R_L = 470\ \Omega$,

$f = 50$ Hz,

$V_m = 50$ V.

Substituting these values in above equation, we have

$$V_r\,(\text{p-p}) = \frac{2 \times 50}{2 \times 50 \times 300 \times 10^{-6} \times 470 + 1} = 6.62 \text{ V}$$

$$V_{dc} = 50 - (6.62/2) = 46.69 \text{ V}$$

$$I_{dc} = V_{dc}/R_L = 46.69/470 = 993.4 \text{ mA}$$

Problem 3.35: The input signal in the circuit of Fig. 3.35 is $v_i(t) = 7.5 \sin 10^4 t$, sketch the output wave form $v_o(t)$ when the diodes are characterized by $V_{ON} = 0.5$ V. Mark the timings on the wave form

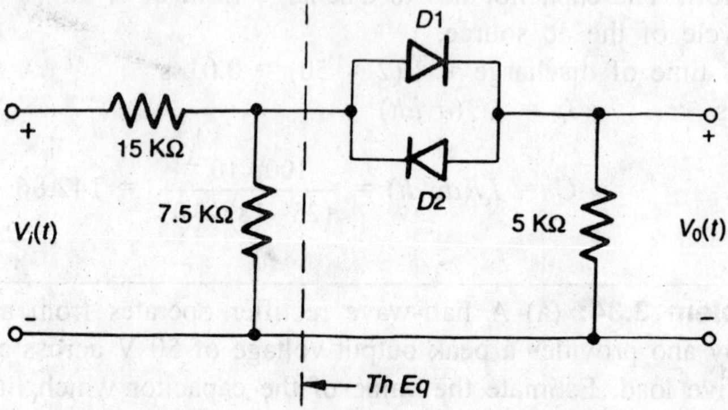

Fig. P. 3.35

Solution: Given input signal is

$$v_i(t) = 7.5 \sin 10^4 t$$

It is sketched in Fig. 3.35 (a)

Using the equivalent of the circuit of Fig. 3.35 (a) to the left of dotted line, we have

$$v_{TH} = v_i \times 7.5/\ (7.5 + 15) = (1/3)v_i = 2.5 \sin 10^4 t$$

$$R_{TH} = (7.5 \times 15)/(7.5 + 15) = 5\ k\Omega$$

The equivalent circuit is drawn in Fig. 3.35 (b)

Only one of the diodes will conduct at a time. The voltage across the conducting diode remains fixed at 0.5 V, while the nonconducting diode acts as an open circuit. Neither diode would conduct when

$$|v_{TH}| < 0.5 \text{ V or } |v_i| < 3 \times 0.5 = 1.5 \text{ V}$$

It then easily follows that on the input wave form D_1 conducts from t_1 to t_2 (during positive half cycle of input) and D_2 conducts from t_3 to t_4 (during negative half cycle of input). These times are found as

$$\text{Time period}, T = 2\pi/10^4 = 0.2\pi \text{ ms}$$

$$0.5 = 2.5 \sin 10^4 t_1; \text{ Fig. 3.35 (c)}$$

which gives $t_1 = 20.14\mu s$

From symmetry

$$t_2 = T/2 - t_1 = 0.1\pi - 0.020 = 0.294 \text{ ms}$$

t_3 and t_4 then follow easily.

These times and the corresponding voltage (± 1.5 V) are indicated in Fig. 3.35 (c).

During nonconduction period, voltage $v_0 = 0$. In conducting period, current in the output resistance is obtained from the circuit of Fig. 3.35 (c) as

$$i_o = \frac{v_{TH} - 0.5}{5+5} = 0.1\ (v_{TH} - 0.5) \text{ mA}$$

So output voltage is

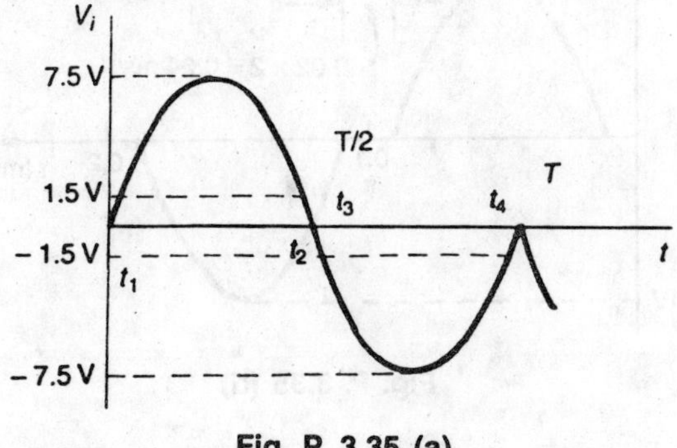

Fig. P. 3.35 (a)

$$v_o = 5i_0 = 0.5(v_{TH} - 0.5) = 1.25 \sin 10^4 t - 0.25;$$

During conducting periods only Peak value of output is

$$|v_o \text{ (peak)}| = 1.25 - 0.25 = 1 \text{ V}$$

The corresponding output wave form is drawn in Fig. 3.35 (d)

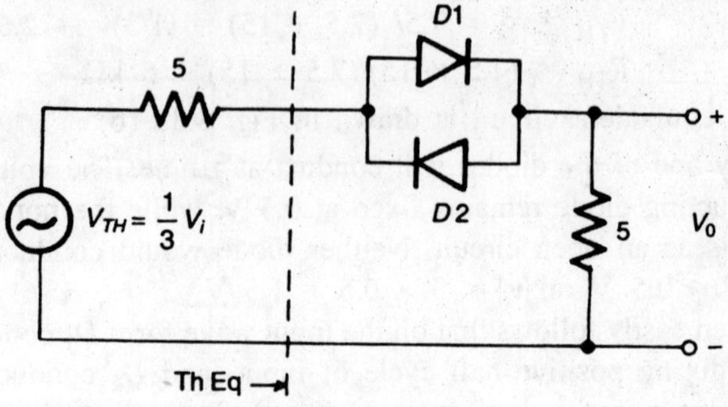

Fig. P. 3.35 (b)

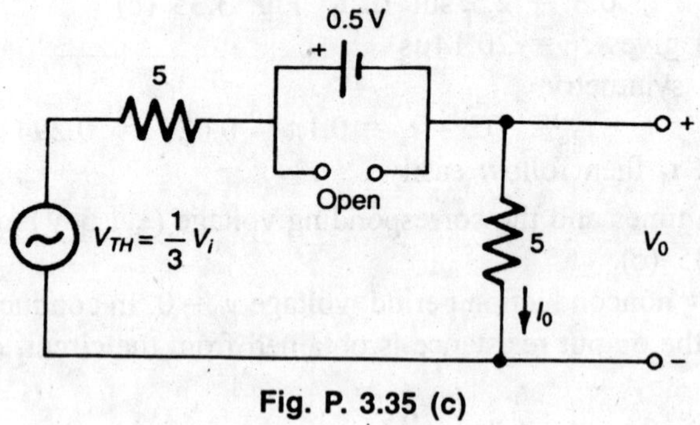

Fig. P. 3.35 (c)

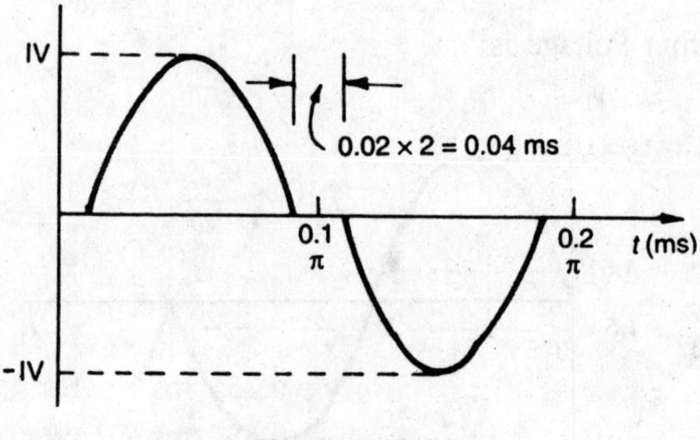

Fig. P. 3.35 (d)

Problem 3.36: For the slicer circuit of Fig. 3.36 (a) the input voltage is $v_1(t) = 20 \sin 10^4 t$. Sketch the output voltage $v_o(t)$ for one cycle assuming diodes to be ideal.
Indicate on the sketch the time corresponding to each diode becoming conducting/nonconducting.

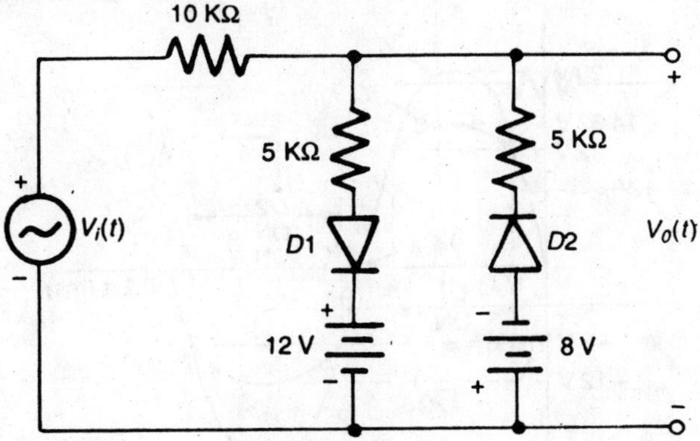

Fig. P. 3.36 (a)

Solution: Considering diodes ideal when

$$-8 < v < 12$$

during which $v_o(t) = v_i(t) = 20 \sin 10^4 t$

For Positive half cycle

$$12 = 20 \sin 10^4 t_1$$

or $t_1 = 0.064$ ms

$$t_2 = 0.1\pi - 0.064 = 0.25 \text{ ms}$$

The output voltage is given as

$$v_o(t) = \left(\frac{20 \sin 10^4 t - 12}{10+5}\right) \times 5 + 12 = 8 + 20/3 \sin 10^4 t$$

(Check : At t_1 = (0.064 ms), v_o = 12 V)

$v_o(t)$ (peak) (at t = 0.05 π ms)

$$= 8 + 20/3 \sin (10 \times 2.05\ \pi) = 14.67 \text{ V}$$

For Negative half cycle

$$8 = 20 \sin 10^4 t_1$$

or $t_1 = 0.041$ ms

$$t_3 = 0.1\pi + 0.041 = 0.355 \text{ ms}$$

$$t_4 = 0.2\pi - 0.041 = 0.587 \text{ ms } D2 \text{ conducts}$$

$$v_o(t) = \left(\frac{20\sin 10^4 t + 8}{10+5}\right) \times 5 - 8 = -16/3 + (20/3)\sin 10^4 t$$

$v_o(t)$ (peak) (at $t = 0.15\pi$ ms) $= -16/3 + (20/3)\sin 10^4 \times 0.15\pi$

The input and output wave forms are sketched in Fig. 3.36 (b)

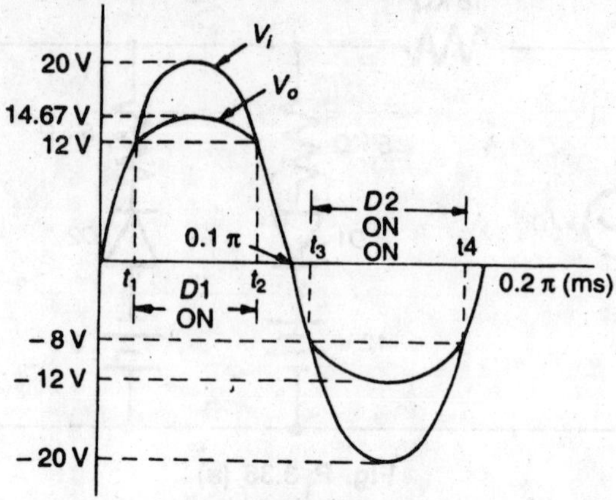

Fig. P. 3.36 (b)

Problem 3.37: For the clamping circuit of Fig. 3.37 with pulse input shown. Sketch the output pulses.

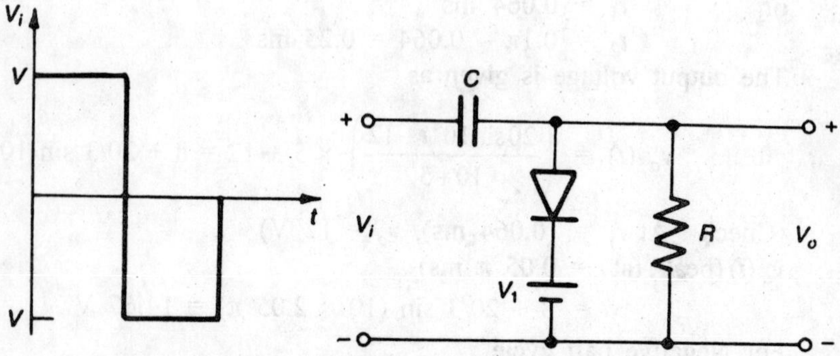

Fig. P. 3.37

Solution: For input $+V$ and $-V$ the circuit behaviour is shown in Fig. 3.37 (a) and (b). The output wave form (pulses) is drawn in Fig. 3.37 (c)

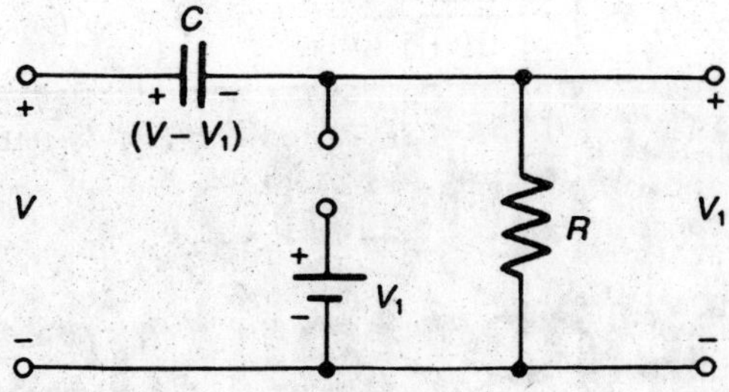

Fig. P. 3.37 (a)

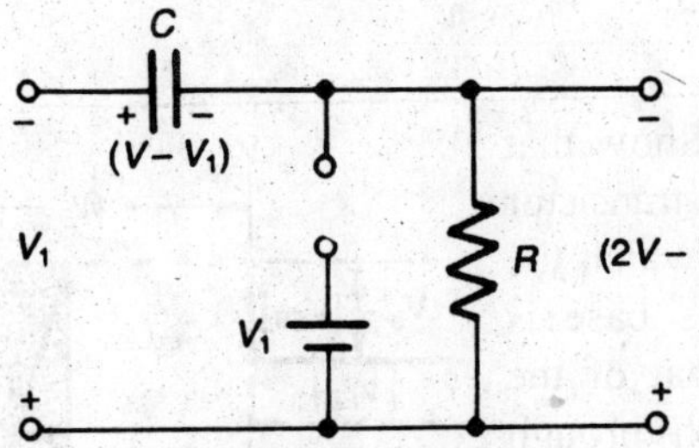

Fig. P. 3.37 (b)

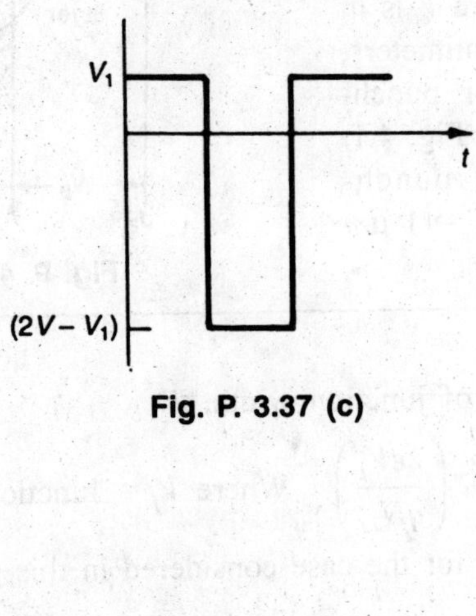

Fig. P. 3.37 (c)

Chapter 4

Transistor Characteristics

Problem 4.1: (a) Show that for an *n-p-n* silicon transistor of the alloy type in which the resistivity ρ_B of the base is much larger than that of the collector, the punch-through voltage V is given by $V = 6.1 \times 10^3\ W\ B^2/\rho_B$, where V is in Volts, ρ_B, in ohm-centimeters, and W in mils. For punch-through, $W = W_B$ in Fig. 4.1. (b) Calculate the punch-through voltage if $W = 1\ \mu m$ and $\rho_B = 0.5\ \Omega$-cm.

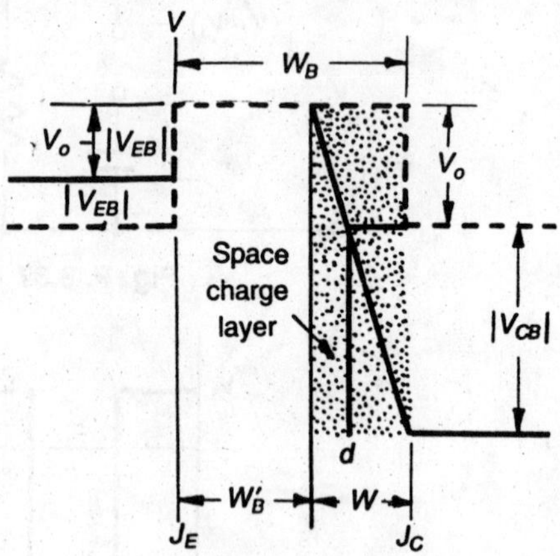

Fig. P. 4.1

Solution:

(a) From equation of junction width,

$$W^2 = \left(\frac{2\varepsilon V_j}{qN_D}\right) \quad \text{Where } V_j = \text{Junction voltage}$$

This equation is valid for the case considered in this problem because

$$\rho_B = \frac{1}{\sigma_B} = \frac{1}{q\mu_p N_A}$$

and $$\rho_E = \frac{1}{\sigma_E} = \frac{1}{q\mu_p N_D}$$

Since, $\rho_B \gg \rho_E$

$\therefore$ $N_D \gg N_A$

Now punch through occurs when $W^2 = W_B^2$

or, $$V_o - V_d = \frac{q\,N_A}{2\varepsilon} W_B^2$$

where V_o = barrier voltage

V_d = externally applied voltage.

or, $$V_o - V_d = \frac{1}{2\varepsilon\mu_p\rho_B} W_B^2$$

Also, $V = -V_d$;

Hence, $$V + V_o = \frac{1}{2\varepsilon\mu_p\rho_B} W_B^2$$

As, V_0 is the order of mV,

$\therefore$ $V \gg V_o$

$$\therefore \quad V = \left(\frac{1}{2\varepsilon\mu_p\rho_B} W_B^2\right)$$

$$V = \frac{(2.54)^2 \times 10^{-6}}{\dfrac{2\times 12}{36\times 10^{11}} \times 500} \times \frac{W_B^2}{\rho_B}$$

$$= 6.1 \times 10^3 \left(\frac{W_B^2}{\rho_B}\right) \text{ Proved}$$

(b) when $W = 1\ \mu = 10^{-4}$ cm $= 3.93 \times 10^{-2}$ mils

$$V = 6.1 \times 10^3 \times \frac{(3.93)^2 \times 10^{-4}}{0.5} = 18.9 \text{ V } \textbf{Ans}$$

Problem 4.2: The transistor of Fig. 4.2 (a) has the characteristics given in Fig. 4.2 (b) and 4.2 (c) Let V_{CC} = 6 V, R_L = 200 Ω, and I_E = 15 mA.

(a) Find I_C and V_{CB}

(b) Find F_{EB} and V_L.

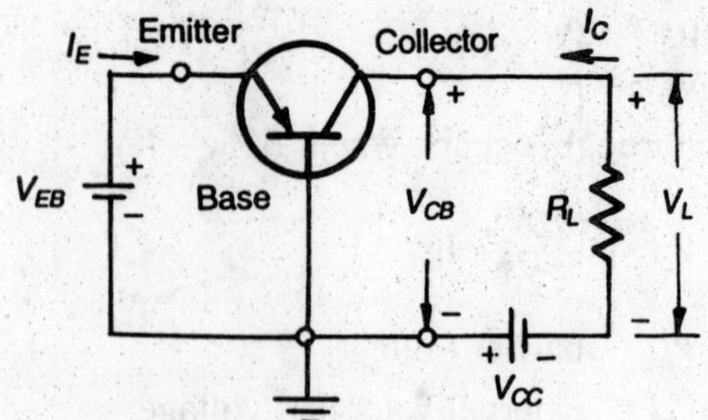

Fig. P. 4.2

Solution:

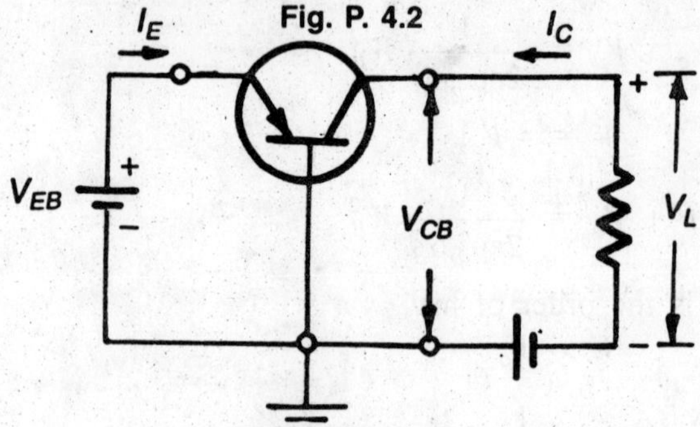

Fig. P. 4.2 (a)

(a) Now from the load line equations,

$$V_{CC} = V_{CB} + I_C R_L,$$

or

$$V_{CB} = -6 - I_C \times 200$$

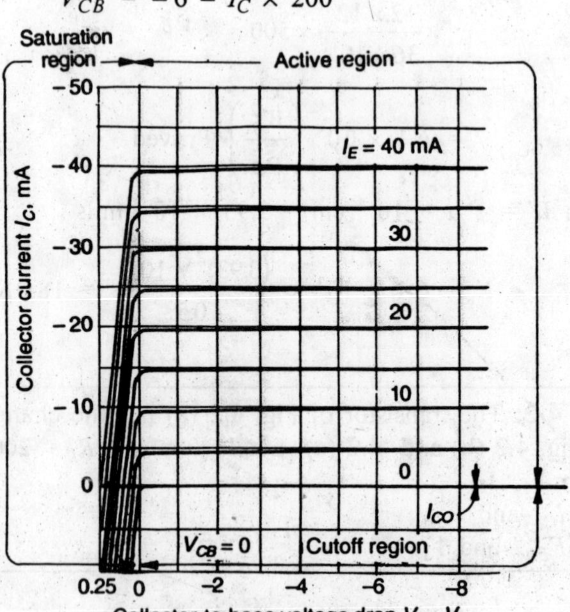

Fig. P. 4.2 (b)

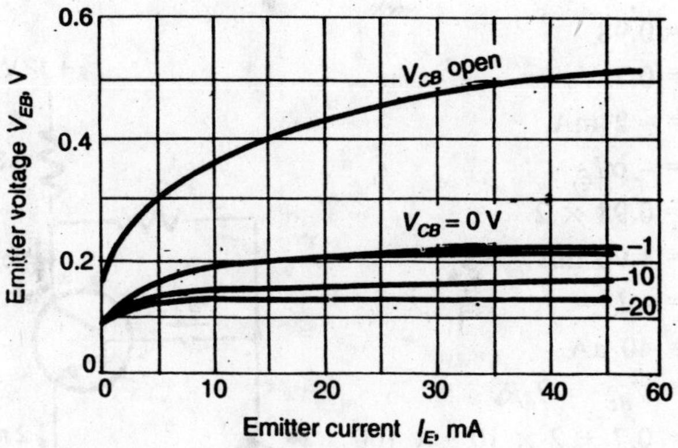

Fig. P. 4.2 (c)

This equation of load line drawn and finding its intersection with the output characteristic having $I_E = 15$ mA

We find the quiscent point as,

$$I_c \cong -15 \text{ mA}$$

$$V_{CB} = -3 \text{ V}$$

(b) $$V_{EB} = f(I_E,\ V_{CB})$$

Now locating $I_E = 15$ mA and $V_{CB} = -3\ V$ in the input characteristic we get,

$$V_{EB} = 0.19 \text{ V},$$

also $$V_L = -I_C R_L = 3\text{V}$$

Problem 4.3: In Fig. 4.3, if $\alpha = 0.98$ and $V_{BE} = 0.7\ V$, find R_1 in the circuit shown for an emitter current.

$$I_E = -2 \text{ mA}.$$

Neglect the reverse saturation current.

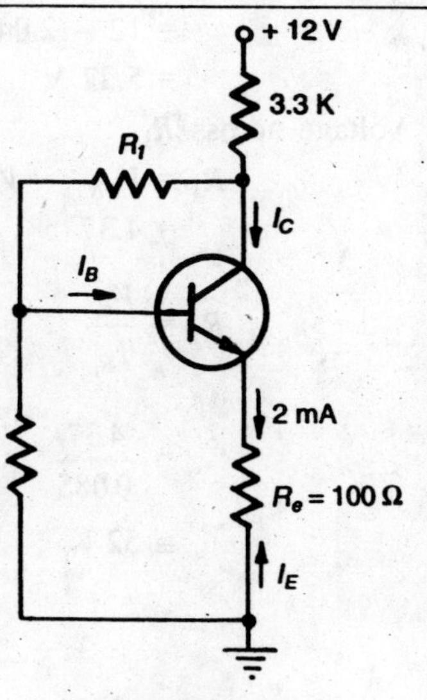

Fig. P. 4.3

Solution:

Given $\alpha = 0.98$

and $V_{BE} = 0.7$ V

$I_E = -2$ mA.

$I_c = -\alpha I_E$

$= 0.98 \times 2$

$= 1.96$ mA

$I_B = -I_C - I_E$

$= 40$ μA

Thus, $V_{BN} = V_{BE} - I_E R_e$

$= 0.7 + 2 \times 10^{-3} \times 100$

$= 0.9$ V

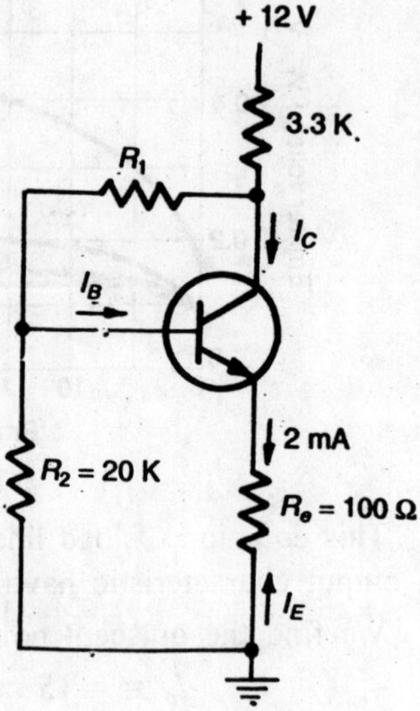

Fig. P. 4.3 (a)

$$\therefore \quad I_{R_2} = \frac{V_{BN}}{R_2} = \frac{0.9}{20}$$

$= 0.045$ mA

$= 45$ μA.

Hence, $I_{R_1} = I_B + I_{R_2}$

$= 40$ μA $+ 45$ μA

$= 85$ μA

Now $V_{CN} = V_{CC} - (I_C + I_{R_1})\, 3.3$

$= 12 - 2.04 \times 3.33$

$= 5.27$ V

∴ Voltage across R_1

$R_1 = V_{CN} - V_{BN}$

$= 4.37$

$$\therefore \quad R_1 = \frac{V_{R_1}}{I_{R_1}}$$

$$= \frac{4.37}{0.085}$$

$\cong 52$ K

Problem 4.4:

(a) Find R_C and R_B in the circuit of Fig 4.4 if $V_{CC} = 10$ V and $V_{BB} = 5$ V, so that $I_c = 10$ mA $V_{CE} = 5$ V. A silicon transistor with $\beta = 100$, $V_{BE} = 0.7$ V, and negligible reverse saturation current is under consideration.

(b) Repeat part a if a 100-Ω emitter resistor is added to the circuit.

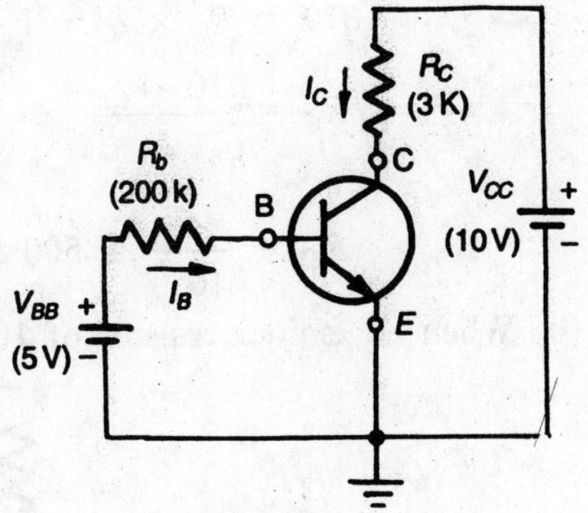

Fig. P. 4.4

Solution:

(a) We have to find R_C and R_b

$V_{CC} = 10$ V

$V_{BB} = 5$ V

$I_C = 10$ mA,

and $V_{CE} = 5$ V

$\beta = 100$

and $V_{BE} = 0.7$ V

(Neglecting reverse saturation current)

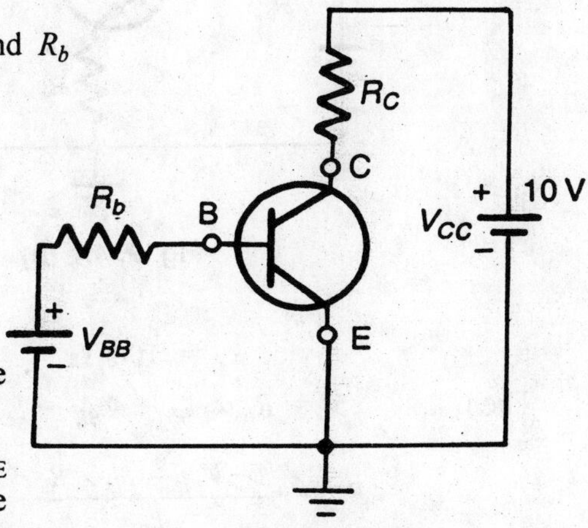

Fig. P. 4.4 (a)

Here it is seen that $V_{CE} = 5$ V $< V_{CC}$, Then the transistor must be in active region and cannot be in cut off or in saturation.

$$\therefore \quad I_C = \beta I_B$$

$$\therefore \quad I_B = \frac{10}{100} = 0.1 \text{ mA}$$

Now from equation of base circuit,

$$5 = R_b \times I_B + V_{BE}; \quad R_B = \frac{5 - 0.7}{I_B}$$

$$\therefore \quad R_B = \frac{4.3}{10^{-4}} = 43 \text{ K}$$

Now from collector ckt.,

$$10 = R_C \times I_C + V_{CE},$$

$$\therefore \quad R_C = \frac{10 - V_{CE}}{I_C}$$

$$\therefore \quad R_C = \frac{5}{10^{-2}} \cong 500\ \Omega$$

(b) When an emitter resistor of 100 Ω is added to the circuit,

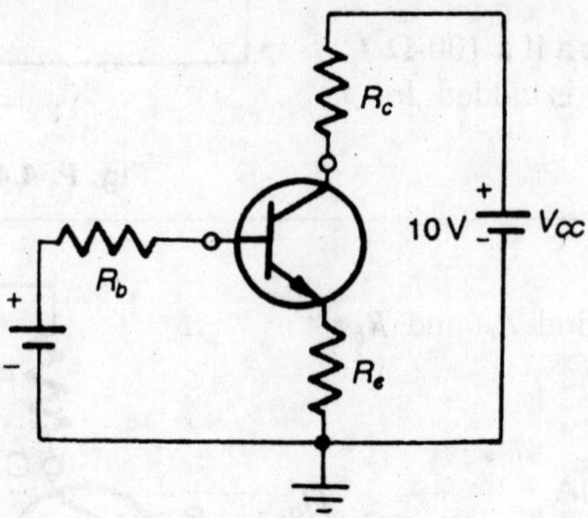

Fig. P. 4.4 (b)

$$R_e = 100\ \Omega$$

Then $$5 = R_b \times I_B + V_{BE} + I_E R_e$$

$$\therefore \quad R_b = \frac{5 - V_{BE} - I_E \times R_e}{I_B}$$

$$= \frac{5 - 0.7 - 1.01}{10^{-4}} \quad (\text{As } I_E = I_B + I_C = 10.1\ \text{mA})$$

$$= 32.9\ \text{k}\Omega$$

Now again viewing into the collector ckt.,

$$10 = 5 + 10.1 \times 10^{-3} \times 100 + 10 \times 10^{-3} \times R_c$$

$$\therefore \quad R_c = \frac{3.99}{10^{-2}} = 399\ \Omega$$

Problem 4.5:
In the circuit shown, V_{cc} = 24V, R_C = 10 K, and R_e = 270 Ω. If a silicon transistor is used with β = 45 and if V_{CE} = 5V, find *R*. Neglect the reverse saturation current.

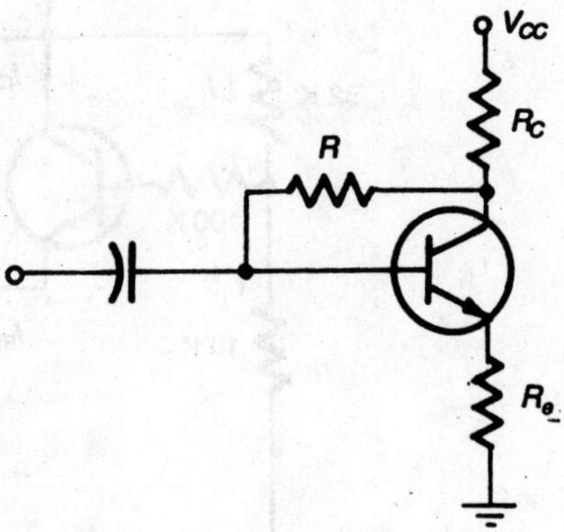

Fig. P. 4.5

Solution: Given

V_{CC} = 24 V
R_C = 10 K
and R_e = 270 Ω
Also, β = 45
and V_{CE} = 5 V
(Neglecting reverse saturation current)

Now from collector emitter ckt,

$$V_{CC} = I_C \times R_C + (I_C + I_B)\, R_e + V_{CE}$$

Here $I_C \gg I_B$

Or, $I_C (R_c + R_e) = V_{CC} - V_{CE}$

Their, $I_C = \dfrac{19}{10.27} \cong 1.85$ mA

∴ $I_B = 41\ \mu A$

Then $I_B \times R = V_{CE} - V_{BE} = 5 - 0.7$

$$\therefore \quad R = \frac{4.3}{41 \times 10^{-3}} \cong 105 \text{ K } \textbf{Ans}$$

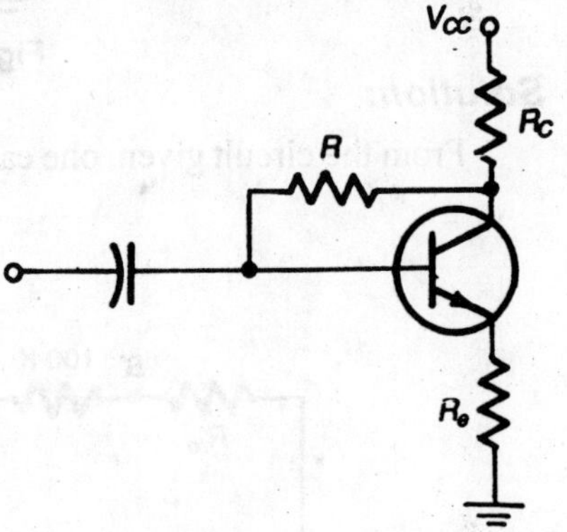

Fig. P. 4.5 (a)

Problem 4.6: For the circuit shown, transistors *Q*1 and *Q*2 operate in the active region with $V_{BE1} = V_{BE2}$ = 0.7 V, β_1 = 100, and β_2 = 50. The reverse saturation currents may be neglected.
(a) Find the currents I_{B2}, I_1, I_2, I_{C2}, I_{B1}, I_{C1}, and I_{E1}
(b) Find the voltages V_{01} and V_{02}.

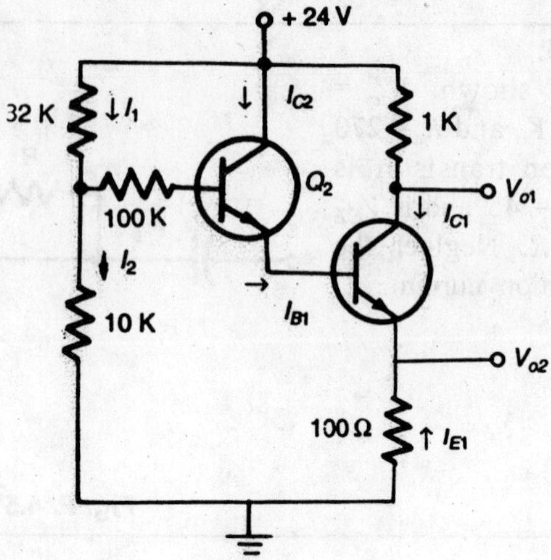

Fig. P. 4.6

Solution:

From the circuit given, one can draw the equivalent circuit as follow:

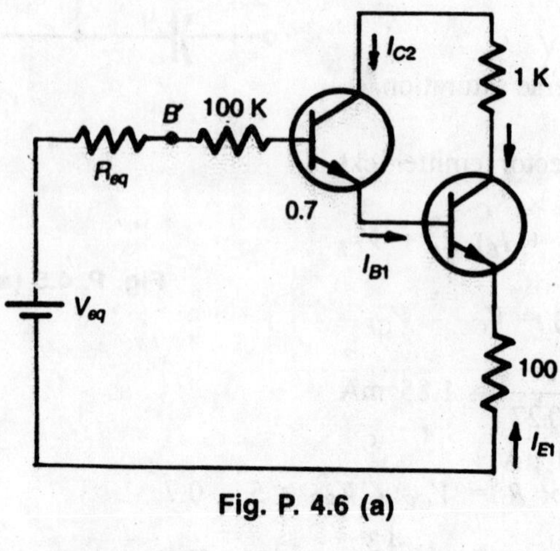

Fig. P. 4.6 (a)

Given $\beta_1 = 100,$

$\beta_2 = 50,$

$V_{BE1} = V_{BE2} = 0.7$ V

(Neglecting the reverse saturation current)

Applying the Thevenin's theorem to left of point B', we abtain the equivalent circuit (as shown above).

$$\therefore \quad V_{eq} = \frac{10}{92} \times 24 = 2.61 \text{ V}$$

and $$R_{eq} = \frac{10 \times 82}{92} = 8.92 \text{ K}$$

$$\therefore\ 2.61 = 108.92 \times I_{B_2} \times 0.1 \times \left(I_{B_1} + I_{C_1}\right) + V_{BE_1} + V_{BE_2} \quad ...(1)$$

$$I_{B_1} = -I_{E_2} = I_{C_2} + I_{B_2} = (\beta_2 + 1) I_{B_2}$$

$$I_{C_1} = \beta_1 I_{B_1} = \beta_1 (\beta_2 + 1) I_{B_2}$$

Hence $I_{B_1} = 51\, I_{B_2}$

and $I_{C_1} = 5100\, I_{B_2}$ ...(2)

Hence from eqn (1) and (2),

$$2.61 = 108.92 \times I_{B_2} + 0.1 \times 5151\ I_{B_2} + 1.4$$

$$\therefore \quad I_{B_2} = 1.94\ \mu\text{A}$$

Then $I_{C_2} = \beta_2 I_{B2} = 97\ \mu\text{A}$

$$I_{B_1} = 51\, I_{B_2} = 99\ \mu\text{A}$$

$$I_{C_1} = 100\, I_{B_1} = 9.9 \text{ mA } \textbf{Ans}$$

$$V_{B'N} = 1.94 \times 10^{-6} \times 100 \times 10^3 + 1.4 + 100 \times 9.99 \times 10^{-3} = 2.59 \text{ V}$$

and $$I_2 = \frac{2.59}{10 \times 10^3} = 259\ \mu\text{A}$$

$$\therefore \quad I_1 = I_{B_2} + I_2 = 261\ \mu\text{A}$$

(b) $$V_{01} = 24 - 9.9 \times 1 = 14.1\text{V}$$

$$V_{02} = 100 \times 9.99 \times 10^{-3} = 0.999 \text{ V}$$

$$\therefore \quad V_{CE_1} = V_{01} - V_{02} = 14.1 - 0.99 = 13.11\text{V}$$

Verification of the operation of transistors Q_1 and Q_2

For $Q_1 \Rightarrow V_{BC} = V_{BN} - V_{CN} = (0.7 + V_{02}) - V_{01} = -12.41 < 0$

Hence Q_1 is in active region.

for Q_2, $V_{BC} = V_{BN} - V_{CN} = 1.4 \times 0.99 - 24 < 0$

Hence it is also in active region.

Problem 4.7:
From the characteristic curves for the type 2N404 transistor given in Fig. 4.7 find the voltages V_{BE}, V_{CE}, and V_{EC} for the circuit shown.

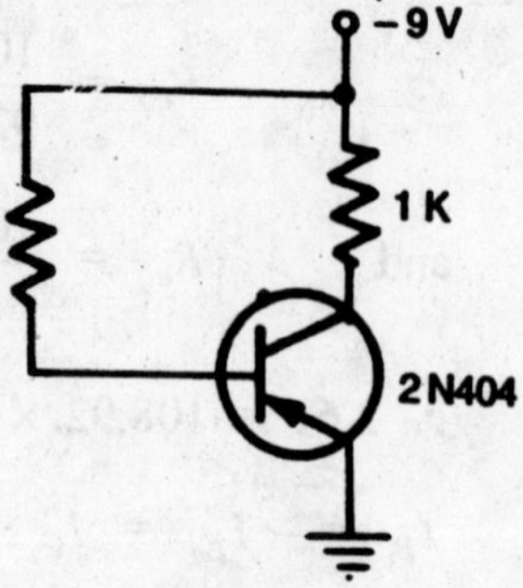

Fig. P. 4.7

Solution:

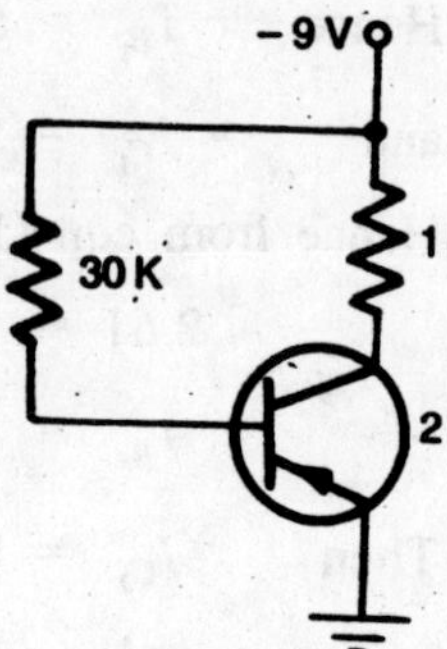

Fig. P. 4.7 (a)

Here $V_{CC} = -9\text{V}$

$R_B = 30$ K and $R_C = 1\text{K}$

First approximation:

$$\therefore \quad I_B = \frac{-9\text{V}}{30\text{K}}$$

$$= -0.3 \text{ mA}$$

$$I_C = -9 \text{ mA}$$

For the values of I_B and I_C found above me obtain

$$V_{CE} = -0.07 \text{ V}$$

Now *Second approximation:*

$$V_{CE} = 0.07 \text{ V}, \qquad V_{BE} = 0 \text{ then}$$

$$I_C = \frac{-9+0.07}{1}$$

$$= 8.93 \text{ mA}$$

and $\quad I_B = -0.3 \text{ mA}$

Now from previously found values,

$$V_{BE} = -0.26 \text{ V}$$

Third approximation:

$$V_{CE} = -0.07 \text{ V},$$

$$V_{BE} = -0.26 \text{ V then } \textbf{Ans}$$

$$I_C = 8.93$$

and $\quad I_B = 0.29$

Then, $\quad V_{BC} = -0.26 - (-0.07)$

$$= -0.19 \text{ V } \textbf{Ans}$$

Problem 4.8:

A silicon transistor with $V_{BE_{sat}} = 0.8$ V, $\beta = h_{FE} = 100$, $V_{CE_{sat}} = 0.2$ V is used in the circuit shown. Find the minimum value of R_C for which the transistor remains in saturation.

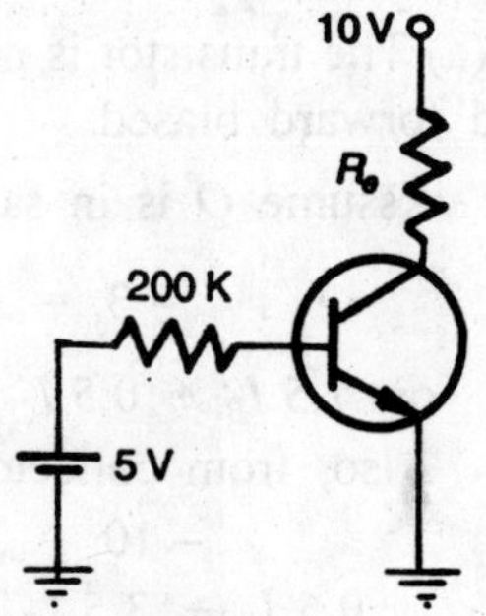

Fig. P. 4.8

Solution:

Given,
$$V_{BE(sat)} = 0.8 \text{ V}$$
$$\beta h_{FE(sat)} = 100$$
$$V_{CE(sat)} = 0.2 \text{ V}$$

For transistor remains in saturation.

$$I_B = \frac{5-0.8}{200\times10^3} = 21\ \mu\text{A}$$

and $$I_C = \beta I_B = 2.1 \text{ mA}$$

and hence $$I_C \le 2.1 \text{ mA}$$

$$I_C = \frac{10-0.2}{R_C} \le 2.1 \text{ mA}$$

$$\therefore \quad R_C \ge \frac{9.8}{2.1} = 4.66 \text{ K}$$

$\therefore$ Minimum value of $R_C = 4.66$ K

Problem 4.9: For the circuit shown, assume $\beta = h_{FE} = 100$

(a) Find if the silicon transistor is in cutoff, saturation, or in the active region.

(b) Find V_0.

(c) Find the minimum value for the emitter resistor R_e for which the transistor operates in the active region.

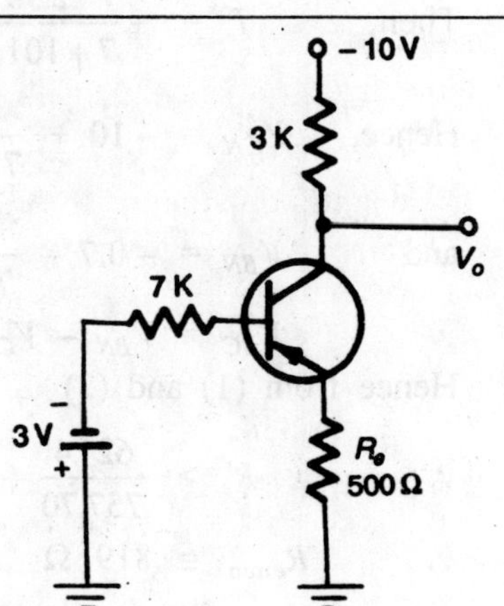

Fig. P. 4.9

Solution:

(a) The transistor is not in cut off because the emitter to bas junction and forward biased.

Assume Q is in saturation, Then,

$$-3 = I_B \times 7.5 + 0.5 \times I_C - 0.8 \ \left(V_{BE,\text{sat}} = 0.8 \text{ V}\right)$$

or, $7.5\, I_B + 0.5\, I_C = -2.2$ ...(1)

Also, from collector to emitter act,

$$-10 = 3.5\, I_C + 0.5\, I_B - 0.2 \ (\text{As } V_{CE,\text{sat}} = 0.2 \text{ V})$$

or, $0.5\, I_B + 3.5\, I_C = -9.8$ ...(2)

Solving (1) and (2), we get,

$$I_B = -0.107 \text{ mA}$$

and $I_C = -2.90$ mA

Also, $$|\, I_B, \min \,| = \frac{|I_C|}{h_{FE}} = \frac{2.9}{100} = 29\ \mu\text{A}$$

Hence the transistor is in saturation as $I_B > |\, I_{B,\min} \,|$

(b) $V_0 = V_{CE,\text{sat}} + I_E \times R_e = -0.2 + (-2.910 \times 0.5) \cong -1.65$ V

(c) Taking transistor into active region

Then $V_{BC} \geq -0.5$, where 0.5 V is the cut in voltage of the junction of collector.

Now, $-3 = I_B \times (7 + R_e) + R_e R_C - 0.7$

and, $I_C = \beta I_B$

$$\therefore \quad I_B = -\frac{2.30}{7 + 101\, R_e} \text{ mA}$$

$$\text{Then,} \quad I_e = -\frac{2.30}{7 + 101\, R_e}$$

$$\text{Hence,} \quad V_{CN} = -10 + \frac{3 \times 230}{7 + 101\, R_e}$$

and $$V_{BN} = -0.7 - \frac{232.30}{7 + 101\, R_e} \times R_e \quad \text{...(1)}$$

$$\therefore \quad V_{BC} = V_{BN} - V_{CN} \geq 0.5 \quad \text{...(2)}$$

Hence from (1) and (2)

$$\therefore \quad R_e > \frac{621.4}{757.70}$$

$\therefore$ $R_{e\min} \cong 819\ \Omega$ **Ans**

Problem 4.10: If the silicon transistor used in the circuit shown has a minimum value of $\beta = h_{FE}$ of 30 and if I_{CBO} = 10 nA at 25°C.

(a) Find V_0 for V_1 = 12 V and show that Q is in saturation.

(b) Find the minimum value of R_1 for which the transistor in part *a* is in the active region.

(c) If R_1 = 15 K and V_i = 1 V, find V_0 and show that Q is at cutoff.

(d) Find the maximum temperature at which the transistor in part c remains at cutoff.

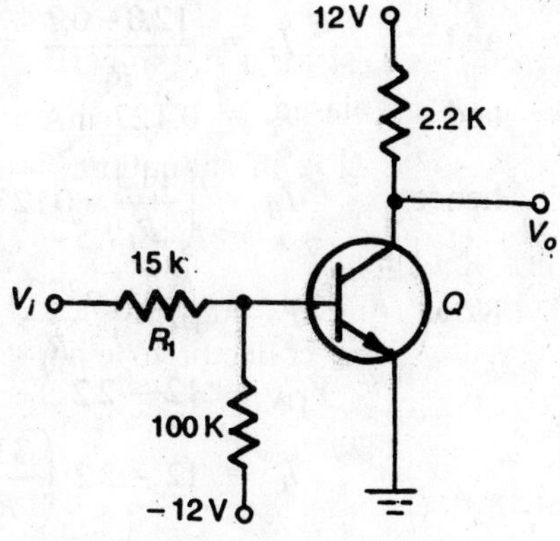

Fig. P. 4.10

Solution: Given:

$$\beta = h_{FE} = 30$$

The reverse saturation current,

$$I_{CBO} = 10 \text{ mA at } 25°\text{C}$$

$$I_1 = \frac{12 - 0.8}{15} \cong 0.747 \text{ mA}$$

$$I_2 = \frac{0.8 + 12.0}{100} = 0.128 \text{ mA}$$

Hence I_B = 0.619 mA

Also

$$I_c = \frac{12.0 - 0.2}{2.2} = 5.36 \text{ mA}$$

But $I_{B,\,\min} = \dfrac{5.36}{30} = 0.179 < I_B$

Hence Q is in saturation.

Hence obviously in this case,

$$V_0 = 0.2 \text{ V}$$

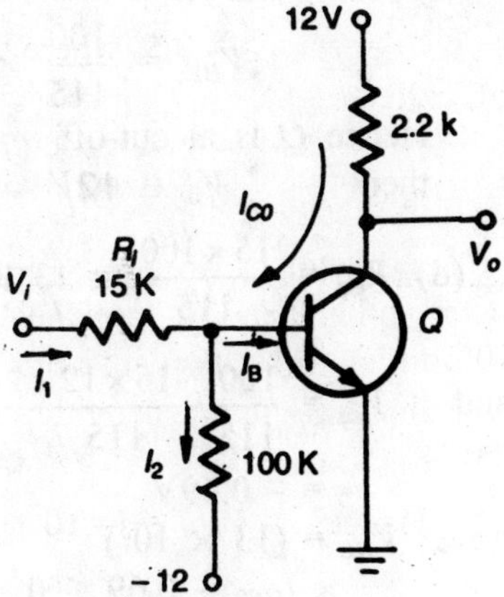

Fig. P. 4.10 (a)

(b) Assuming that Q is in the active region

$$V_{BE} = 0.7$$

and $\quad I_1 = \frac{12.0 - 0.7}{R_1} = \frac{11.3}{R_1}$

and $\quad I_2 = 0.127$ mA

hence $\quad I_B = \left(\frac{11.3}{R_1} - 0.127\right)$

Now, $\quad I_C = \beta I_B = \frac{339}{R_1} - 3.8$ (Neglecting I_{C0})

$\therefore \quad V_{CN} = 12 - 2.2$

$$I_C = 12 - 2.2\left(\frac{339}{R_1} - 3.8\right) = \left(20.35 - \frac{746}{R_1}\right)$$

But In order to keep the transistor in active region,

$$V_{BC} = V_{BN} - V_{CN} \leq 0.5;\ 0.7 - 20.35 + \frac{746}{R_1} \leq 0.5$$

Then, $\quad R_1 \geq \frac{746}{20.25} = 36.9$ K

$\therefore \quad R_{min} = 36.9$ K

(c) For $\quad V_i = 1$ V

$$V_{BE} = \frac{100}{115} \times 1 - \frac{15}{115} \times 12 = -\frac{80}{115} < 0$$

Hence Q is at cut-off

then $\quad V_0 = 12V$

(d) $R_{eq} = \frac{15 \times 100}{115} \cong 13$ K

and $\quad V_{eq} = \frac{100}{115} - \frac{15 \times 12}{115}$

$= -0.69$V

now, $V_{BE} = (13 \times 10^3) \times I_{CB0} - 0.69 \leq 0$

$I_{CO} \leq 53\ \mu$A

and

$I_{CBO,\max} = 53\ \mu$A

from eqn. of current,

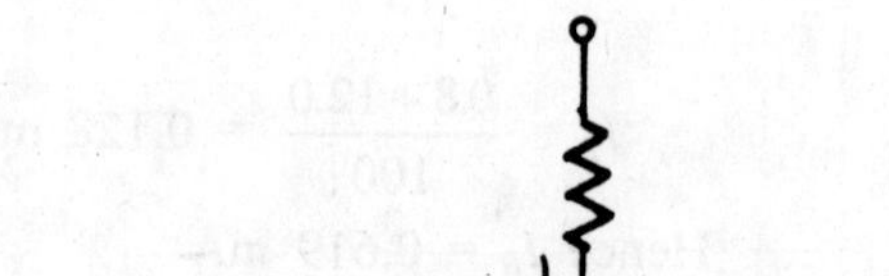

Fig. P. 4.10 (b)

$$I_{CBO,\max} = I_{CBO}\ (25°C)\ (2)^{\frac{\Delta T}{10}}$$

$$\frac{53 \times 10^{-6}}{10 \times 10^{-9}} = 5.3 \times 10^3 = 2^{\frac{\Delta T}{10}}$$

$$\therefore \quad \Delta T = 10\ (3.32 \times 3 + 2.2) = 121.6$$

$$\therefore \quad T_{max} = 121.6 + 25 = 146.6°C \textbf{ Ans}$$

Problem 4.11:

Silicon transistors with $h_{FE} = 100$ are used in the circuit shown. Neglect the reverse saturation current

(a) Find V_o when $V_i = 0$ V. Assume Q_1 is OFF and justify the assumption.

(b) Find V_o when $V_i = 6$V. Assume Q_2 is OFF and justify this assumption.

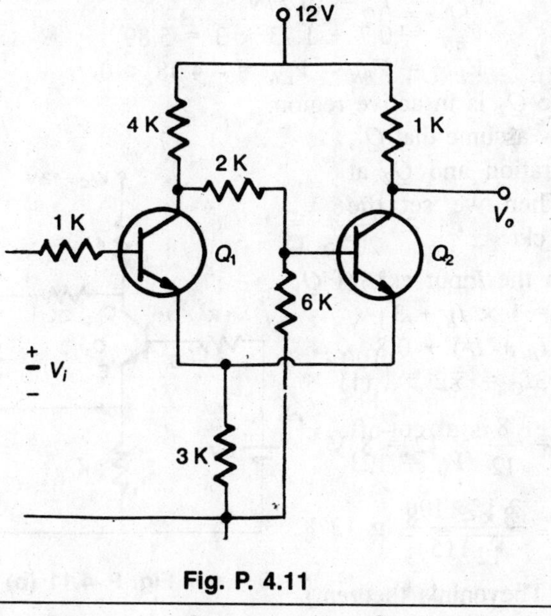

Fig. P. 4.11

Solution:

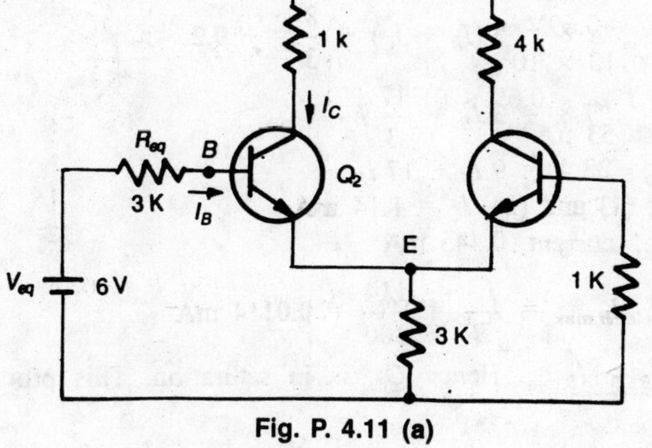

Fig. P. 4.11 (a)

Drawing the equivalent ckt

(a) Applying ckt law,

$6 = 3 \times I_B + 3\,(I_B + I_C) + 0.7$

But $\quad I_C = 100\, I_B$

Then, $\quad I_B = \dfrac{6-0.7}{3+101\times 3} = 1.73 \times 10^{-5} = 17.3\ \mu A$

Then, $\quad I_C = 1.73$ mA

Hence $\quad V_o = -1.73 \times 1 + 12 = 10.27V$

Now we have to verify that Q_2 is in active region

$V_{CN} = V_o = 10.27V$

Also $\quad V_{BN} = 0.7 + 1.73 \times 3 = 5.89$

Also $\quad V_{BN} = V_{BN} - V_{CN} = -4.38 < 0$

Hence Q_2 is in active region.

(b) Now assume that Q_1, is in saturation and Q_2 at cut-off. Then we set the following ckt.

Now from the input ckt, of Q_1

$6 = 1 \times I_B + 3\,(I_B + I_C) + 0.8$

or, $4I_B + 3I_C = 5.2 \quad \ldots(1)$

also, $V_{CN} = \dfrac{8}{12} \times 12 = 8$ V

and $R_{CN} = \dfrac{8\times 4}{12} = \dfrac{8}{3}$ K

(by using Thevenin's theorem)

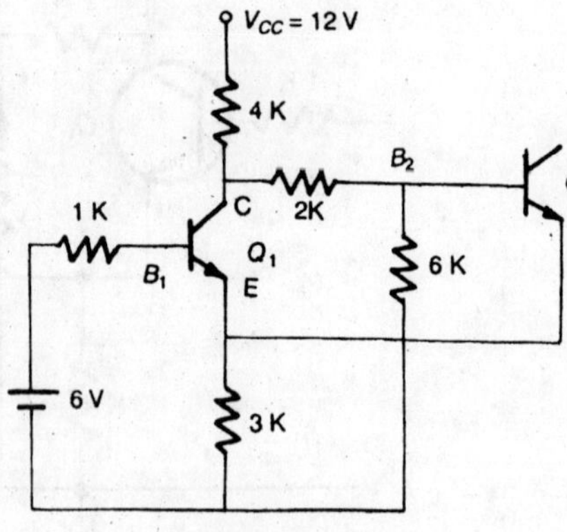

Fig. P. 4.11 (b)

Now from the collector ckt, of θ_1,

$$8 = 3\,(I_B + I_C) + \frac{8}{3} I_C + 0.2$$

or, $\quad 7.8 = 3\, I_B + \dfrac{17}{3} I_C$

or, $\quad 23.4 = 9\, I_B + 17\, I_C \quad \ldots(2)$

Solving (1) and (2), $I_C = 1.14$ mA

and $\quad I_B = 0.445$ mA

since $\quad I_{B,\min} = \dfrac{I_C}{B} = \dfrac{1.14}{100} = 0.0114$ mA

Here $I_B > I_{B,\min}$, Hence Q_1, is in saturation. This proves the assumption.

Problem 4.12: For the circuit shown, $\alpha_1 = 0.98$, $\alpha_2 = 0.96$, $V_{CC} = 24\ V$, $R_C = 120\ \Omega$, and $I_E = -\ 100$ mA. Neglecting the reverse saturation currents, determine (a) the currents I_{C1}, I_{B1}, I_{C2}, I_{B2}, and I_c; (b) V_{CE}; (c) I_C/I_B, I_C/I_E

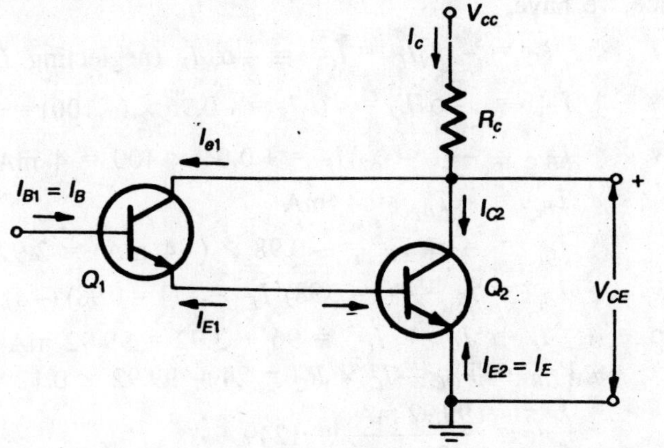

Fig. P. 4.12

Solution: Given

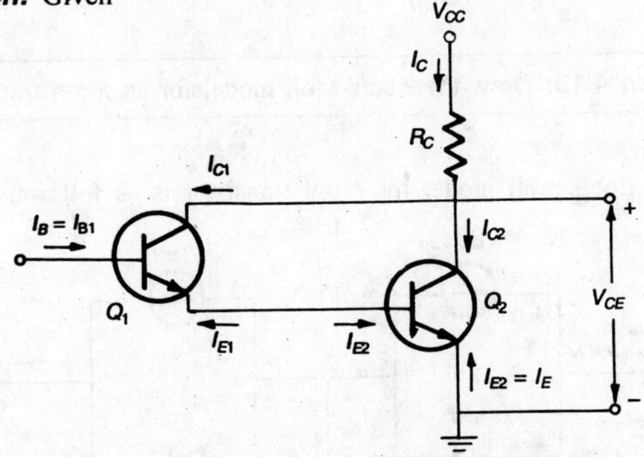

Fig. P. 4.12 (a)

$$V_{CE} = 24\ \text{V}$$
$$\alpha_1 = 0.98$$
$$\alpha_1 = 0.96$$
$$R_C = 120\ \Omega$$

Also, $I_E = -\ 100$ mA

Assuming that the transistor Q_2 is in the active region then $V_c < 0$ and $/V_c/ \gg V_T$

Then $e^{\frac{V_C}{V_T}} \cong 0$

Hence we have,

$$I_C = -\alpha_N I_E + I_{C_0} \cong -\alpha_N I_E \text{ (neglecting } I_{Co})$$

Thus $I_{C_2} = -\alpha I E_2 = -\alpha_2 I_E = -0.96 \times (-100) = +96 \text{ mA}$

Now $I_{B_2} \cong -(1-\alpha_2)I_E = +0.04 \times 100 = 4 \text{ mA}$

$$I_{B_1} = -I_{B_2} = -4\text{mA}$$

$$I_{C_1} = -\alpha_1 I_{E_1} = -0.98 \times (-4 \text{ mA}) = 3.92 \text{ mA}$$

$$I_{B_1} = I_B = -(1-0.98)\, I_{E_1} = -(1-0.98)(-4) = 80 \,\mu\text{A}$$

Then $I_c = I_{C_1} + I_{C_2} = 96 + 3.92 = 99.92 \text{ mA}$

(b) $V_{CE} = V_{CE} - I_C \times R_C = 24 - 99.92 \times 0.12 \cong 12 \text{ V}$

(c) $\frac{I_C}{I_B} = \frac{99.92 \text{ mA}}{80\,\mu\text{A}} = 1249$

and $\frac{I_C}{I_B} = \frac{99.92}{100} = 0.9992$

Problem 4.13: Draw the Ebers-Moll model for an n-p-n transistor.

Solution:

The Ebers-moll model for n-p-n transistor is as follows.

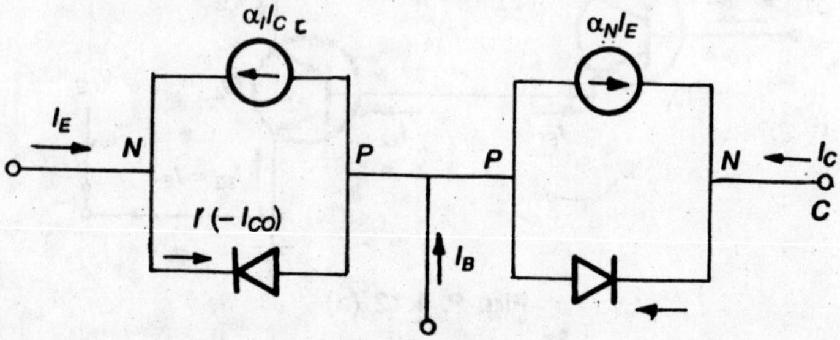

Fig. P. 4.13

Problem 4.14: (a) Show that the exact expression for the CE **output** characteristics of a p-n-p transistor is

$$V_{CE} = V_T \ln \frac{\alpha_I}{\alpha_N} + V_T \ln \frac{I_{CO} + \alpha_N I_B - I_C(1-\alpha_N)}{I_{EO} + I_B + I_C(1-\alpha_I)}$$

Solution:

$$\because \qquad V_{CE} = -V_T \ln \frac{\dfrac{1}{\alpha_I} + \dfrac{1}{\beta_I} \times \left(\dfrac{I_C}{I_B}\right)}{1 - \dfrac{1}{\beta}\left(\dfrac{I_C}{I_B}\right)}$$

Where $\qquad \beta_I = \dfrac{\alpha_I}{1-\alpha_I}$

and $\qquad \beta_N = \beta \equiv \left(\dfrac{\alpha_N}{1-\alpha_N}\right)$

When I_{CO} and I_{EO} are taken into account

$$V_{Co} = -V_T \ln \frac{\dfrac{\dfrac{I_B}{\alpha_I} + \left(\dfrac{1-\alpha_I}{\alpha_I}\right) I_C + \dfrac{I_{EO}}{\alpha_I}}{\dfrac{I_{EO}}{\alpha_I}}}{\dfrac{\dfrac{I_{CO}}{\alpha_N} - I_C \times \left(\dfrac{1-\alpha_N}{\alpha_N}\right) + I_B}{\dfrac{I_{CO}}{\alpha_N}}}$$

Now from equation, $\alpha_I I_{CO} = \alpha_N I_{EO}$

$$V_{CE} = +V_T \ln \frac{\dfrac{1}{\alpha_N}\left(I_{CO} - (1-\alpha_N) I_C + \alpha_N I_B\right)}{\dfrac{1}{\alpha_I}\left(I_B + (1-\alpha_I) I_C + I_{EO}\right)}$$

$$\therefore \qquad V_{CE} = +V_T \ln \left(\frac{\alpha_I}{\alpha_N}\right) + V_T.$$

$$\ln \left(\frac{\alpha_N I_B + I_{CO} - (1-\alpha_N) I_C}{I_{EO} + I_B + (1-\alpha_I) I_C}\right) \quad \text{Proved}$$

Problem 4.15: (a) A transistor is operating in the cutoff region with both the emitter and collector junctions reverse-biased by at least a few tenths of a volt. Prove that the currents are given by

$$I_E = \frac{I_{EO}(1-\alpha_N)}{1-\alpha_N\alpha_I}$$

$$I_C = \frac{I_{CO}(1-\alpha_I)}{1-\alpha_N\alpha_I}$$

(b) Prove that the emitter-junction voltage required just to produce cutoff (I_E = 0 and the collector back-biased) is

$$V_E = V_T \ln (1 - \alpha_N)$$

Solution:

(a) If $| V_C | \gg V_T$

then $e^{\frac{V_C}{V_T}} \cong 0$

and also If $| V_E | \gg V_T$

and $e^{\frac{V_E}{V_T}} \cong 0$

Then, $$I_E = \frac{I_{EO}}{1-\alpha_N\alpha_I} - \frac{\alpha_1 I_{CO}}{1-\alpha_N\alpha_I}$$

$$= \frac{I_{EO} - I_{EO}\alpha_N}{1-\alpha_N\alpha_I} \quad (\because \alpha_I I_{CO} = I_{EO} \times \alpha_N)$$

$$= \left(\frac{I_{EO}(1-\alpha_N)}{1-\alpha_N \times \alpha_I}\right)$$

$$I_C = \frac{I_{CO}}{1-\alpha_N\alpha_I} - \frac{\alpha_N I_{EO}}{1-\alpha_N\alpha_I} = \frac{I_{CO}(1-\alpha_I)}{(1-\alpha_N \times \alpha_I)}$$

(b) When it is given that, $I_E = 0$

and $e^{\frac{V_C}{V_T}} \approx 0$

then $I_e = I_{CO}$

also, $0 = -\alpha_I I_{CO} + I_{EO} (I - e^{\frac{V_E}{V_T}})$

$$\therefore \quad \frac{I_{CO}}{I_{EO}} \alpha_I = \left(1 - e^{\frac{V_E}{V_T}}\right) \qquad ...(1)$$

Also, $\alpha_1 I_{CO} = \alpha_N I_{EO}$ ...(2)

from (i) and (ii)

$$\alpha_N = \left(1 - e^{\frac{V_E}{V_T}}\right)$$

Hence, $e^{\frac{V_E}{V_T}} = (1 - \alpha_N)$

Then $V_E = V_T \ln (1 - \alpha_N)$ Proved

> **Problem 4.16:** (a) Find the collector current for a transistor when both emitter and collector for junction are reverse-baised. Assume I_{CO} = 5 μA, I_{EO} = 3.57 μA, and α_N = 0.98.
> (b) Find the emitter current I_E under the same conditions as in part (a).

Solution:

(a) As found in last problem,

$$I_C = \left(\frac{I_{CO}(1 - \alpha_I)}{1 - \alpha_I \times \alpha_N}\right)$$

$$\alpha_I = \frac{\alpha_N I_{EO}}{I_{CO}} = \frac{0.98 \times 3.57}{5} = 0.7$$

Then $$I_C = \frac{5(1 - 0.7)}{1 - 0.7 \times 0.98} \cong 4.73 \text{ μA}$$

(b) $$I_E = \frac{I_{EO}(1 - \alpha_N)}{1 - \alpha_N \times \alpha_I} = \frac{3.57(1 - 0.98)}{1 - 0.98 \times 0.7} = 0.227 \text{ μA}$$

> **Problem 4.17:** Show that the emitter volt-ampere characteristics of a transistor in the active region is given by

$$I_E \approx I_S \varepsilon^{V_E/V_T}$$

where $I_s = -I_{EO}/(1 - \alpha_N \alpha_I)$. Note that this characteristic is that of a p-n junction diode.

Solution:

When the transistor is operating into active region,

$$V_C < 0$$

and $\quad V_E > 0$

and from here,

$$e^{\frac{V_C}{V_T}} \approx 0$$

and $\quad e^{\frac{V_E}{V_T}} \gg 1$

Hence the equation of I_C and I_E becomes

$$I_C = -\alpha_N I_E + I_{CO}$$

and $\quad I_E = -\alpha_I I_C - I_{EO}\left(e^{\frac{V_E}{V_T}} - 1\right) \equiv -\alpha_I I_C - I_{EO} \times e^{\frac{V_E}{V_T}}$

Again $I_C + \alpha_N I_E = I_{CO}$

and $\alpha_I I_C + I_E = I_{EO}\, e^{\frac{V_E}{V_T}}$

Solving these two equations for I_E

$$I_E = -\frac{\alpha_I I_{CO} + I_{EO} e^{\frac{V_E}{V_T}}}{1 - \alpha_N \alpha_I} \cong = \frac{-I_{EO}}{1 - \alpha_N \alpha_I} \; e^{\frac{V_E}{V_T}}$$

(Neglecting $\alpha_I\, I_{CO}$, as $e^{\frac{V_E}{V_T}} \gg 1$)

$\therefore \quad I_E = I_S \times e^{\frac{V_E}{V_T}}$ Proved

Where $\quad I_S = \left(\frac{-I_{EO}}{1 - \alpha_N \times_{I_I}}\right)$

Problem 4.18: (a) Given an n-p-n transistor for which (at room temperature) $\alpha_N = 0.98$, $I_{CO} = 2\ \mu A$, and $I_{EO} = 1.6\ \mu A$. A common-emitter connection is used, and $Vcc = 12\ V$ and $R_L = 4.0$ K. What is the minimum base current required in order that the transistor enter its saturation region ?

(b) Under the conditions in part a, find the voltages across each junction between each pair of terminals if the base-spreading resistance $\gamma_{bb'}$ is neglected.

(c) Repeat part b if the base current is 200 μA

(d) How are the above results modified if $\gamma_{bb'} = 250\ \Omega$?

Solution:

(a) $$IC_{(sat)} \cong \frac{12}{4K} = 3 \text{ mA}$$

and neglecting $V_{C(sat)}$

Then, $$(I_B)_{min} = \frac{I_{C(sat)}}{h_{FE}}$$

$$h_{FE} = \frac{0.98}{1-0.98} = 49$$

∴ $$(I_B)_{min} = \frac{3}{49} \cong 61.3\ \mu A$$

(b) Now, $$V_E = V_T \ln\left(1 - \frac{I_E + \alpha_I I_C}{I_{Eo}}\right)$$

where $$\alpha_I = \frac{0.98 \times 1.6}{2} = 0.784$$

$$I_E = -3.06 \text{ mA}$$

Now, $$V_E = 26 \ln\left(1 - \frac{-3.06 + 0.784 \times 3}{1.06 \times 10^{-3}}\right) = 159 \text{ mV}$$

Also, $$V_C = V_T \ln\left(1 - \frac{I_C + \alpha_N I_E}{I_{CO}}\right)$$

$$= 26 \ln\left(1 - \frac{3 - 0.98 \times 3.06}{2 \times 10^{-3}}\right) \cong 0$$

Hence $V_{CB} \cong V_C = 0$

$V_{CE} = -V_C + V_E = 159$ mV

(c) Here $I_B = 0.2$ mA

$I_C = 3$ mA.

and $I_E = -3.2$ mA

also, $$V_E = 26 \ln\left(1 - \frac{3.2 + 0.784 \times 3}{1.6 \times 10^{-3}}\right) = 167 \text{ mV}$$

and $$V_C = 26 \ln\left(1 - \frac{3 + 0.98(-3.2)}{2 \times 10^{-3}}\right) = 110 \text{ mV}$$

Thus $V_{CB} = -V_C = -110$ mV

$V_{EB} = -V_E = -167$ mV

(d) The values of V_C, V_E and V_{CE} are unchanged

$$V_{CB} = -V_C - I_B \gamma_{bb'}$$
$$= -0.061 \times 150 = -15.3 \text{ mV}$$
$$V_{EB} = -V_E - I_B\gamma_{bb'} = -159 - 0.061 \times 250 = 174 \text{ mV}$$

Problem 4.19: Plot the emitter current vs. emitter-to-base voltage for a transistor for which $\alpha_N = 0.98$, $I_{CO} = 2$ μA, and $I_{EO} = 1.6$ μA if (a) $V_C = 0$, (b) V_C is back-biased by more than a few tenths of a volt. Neglect the base-spreading resistance.

Solution:

Since, $$I_E = \frac{\alpha_I I_{CO}}{1 - \alpha_N \times \alpha_I}\left(e^{\frac{V_C}{V_T}} - 1\right) - \frac{I_{EO}}{1 - \alpha_N \alpha_I}\left(e^{\frac{V_E}{V_T}} - 1\right)$$

As, $V_C = 0$

then $$I_E = -\frac{I_{EO}}{1 - \alpha_N \times \alpha_I}\left(e^{\frac{V_E}{V_T}} - 1\right) = -\frac{1.6 \times 10^{-3}}{1 - 0.98 \times 0.784}$$

$$= 6.89 \times 10^{-3} \left(e^{\frac{V_E}{V_T}} - 1 \right) \text{ mA}$$

$$\therefore \quad -I_C = 6.89 \times 10^{-3} \left(e^{\frac{V_E}{V_T}} - 1 \right) \text{ mA}$$

V_E/V_T	0	2	4	5	6	6.5	7.0
$-I_E$	0	0.044	0.37	1.2	2.75	4.5	7.6

Problem 4.20: Show that

$$I_{CES} = \frac{I_{CO}}{1-\alpha_N \alpha_I}$$

$$I_{CEO} = \frac{I_{CO}}{1-\alpha_N}$$

Solution:

When the base is short circuited, we have, $V_E = 0$

Then from equation,

$$I_E = -\alpha_I I_C$$

Since the collector is heavily reverse biased, then,

$$e^{\frac{V_C}{V_T}} \cong 0 \text{ we have}$$

$$I_C = -\alpha_N I_E + I_{CO}$$

or,

$$I_C = -\alpha_N \alpha_1 I_C + I_{CO}$$

or

$$I_C = I_{CEB} = \left(\frac{I_{CO}}{1-\alpha_N \alpha_I} \right)$$

When the base is open circuited, i.e. floating,

$$I_B = 0$$

$$\therefore \quad I_C = -I_E,$$

and with $V_C < 0$

then $e^{\frac{V_C}{V_T}} \cong 0$

and $$I_C = -\alpha_N I_E + I_{CO}$$

or, $$(1 - \alpha_N) I_C = I_{CO};$$

or, $$I_C = I_{CEO} = \left(\frac{I_{CO}}{1-\alpha_N}\right)$$

Problem 4.21: A common method of calculating α_N and α_I is by measurement of I_{CO}, I_{CEO}, and I_{CES}. Show that

(a) $$\alpha_N = \frac{I_{CEO} - I_{CO}}{I_{CEO}}$$

(b) $$\alpha_I = \frac{1 - I_{CO} / I_{CES}}{1 - I_{CO} / I_{CEO}}$$

Solution:

As proved in last case (Prob. 4.20)

$$I_{CEO} = \frac{I_{CO}}{1-\alpha_N};$$

$$\therefore \quad \alpha_N = \left(\frac{I_{CEO} - I_{CO}}{I_{CEO}}\right)$$

(b) $$I_{CES} = \frac{I_{CO}}{1 - \alpha_N \alpha_I}$$

Then, $$1 - \left[\frac{(1 - I_{CO})}{I_{CEO}}\right] \alpha_I = \frac{I_{CO}}{I_{CES}}$$

or, $$\alpha_I = \left(\frac{1 - I_{CO} / I_{CES}}{1 - I_{CO} / I_{CEO}}\right)$$

Problem 4.22: A transistor is operated at a forward emitter current of 2 mA and with the collector open-circuited. Find (a) the jucntion voltages V_C and V_E, (b) the collector-to-emitter voltage V_{CE}. Assume $I_{CO} = 2\ \mu A$, $I_{EO} = 1.6\ \mu A$, $\alpha_N = 0.98$. Is the transistor operating in saturation, at cutoff, or in the active region ?

Solution:

As $\qquad I_E = 2$ mA

and with $\quad I_C = 0$

Then from eqn

$$V_E = V_T \ln\left(1 - \frac{I_E}{I_{EO}}\right)$$

[Here for p-n-p transistor I_{Eo} is negative i.e., $I_{EO} = -1.6\ \mu A$]

Thus, $\quad V_E = 26 \ln\left(1 + \frac{2}{1.6 \times 10^{-3}}\right) = 26 \ln 1250$

$= 185.3$ mV (forward biased)

$$V_C = V_T \ln\left(1 - \frac{\alpha_N I_E}{I_{CO}}\right)$$

$$= 26 \ln\left(1 - \frac{0.98 \times 2}{2 \times 10^{-3}}\right) = 26 \ln 9.81$$

$= 179$ mV (forward biased)

Hence the transistor is in saturation

(b) $\qquad V_{CE} = V_C - V_E = 179 - 185.3 = -6.3$ mV

Problem 4.23: The idealized characteristics of an *npn* transistor are drawn in Fig. 4.23. Draw a set of typical output characteristics and

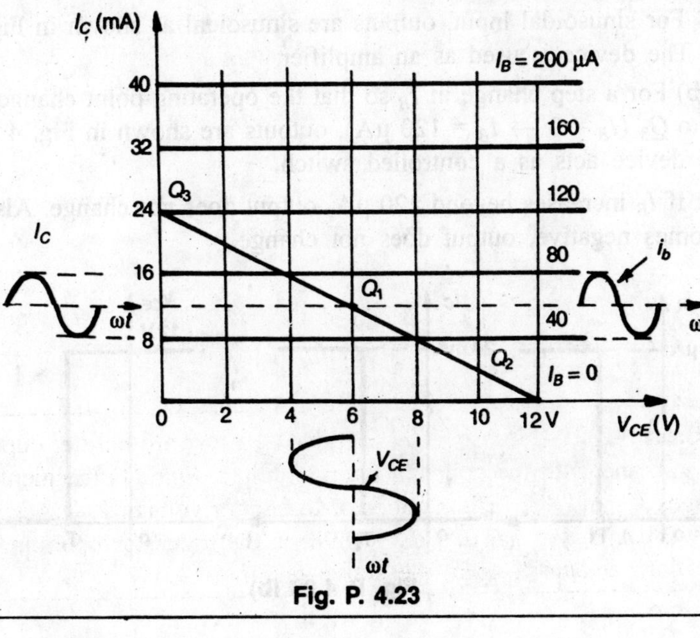

Fig. P. 4.23

then explain how the transistor acts as a controlled switch or as an amplifier. There possible operating points (Q_1, Q_2, Q_3) are indicated on the figure for this purpose.
What happens in case of an actual (non-ideal) transistor? If I_B = 140 μA, what is the value of β?

Solution:

(a) Transistor is biased at Q_1 (I_{BQ} = 60 μA, V_{CEQ} = 6 V, I_{CQ} = 12 mA.

A small changes in I_B are linearly reflected in I_C and V_{CE}.

$$\therefore \quad \beta = \frac{I_C}{I_B} = \frac{12}{60 \times 10^{-3}} = 200$$

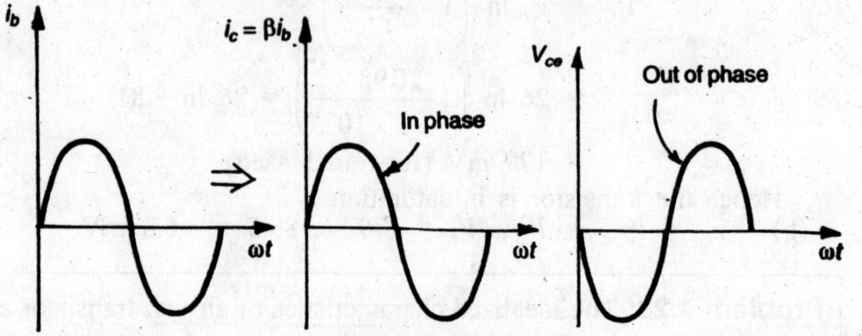

Fig. P. 4.23 (a)

For sinusoidal input, outputs are sinusoidal as shown in Fig. 4.23 (a). The device is used as an amplifier

(b) For a step change in I_B so that the operating point changes from Q_2 to Q_3 ($I_B = 0 \rightarrow I_B$ = 120 μA), outputs are shown in Fig. 4.23 (b). The device acts as a controlled switch.

If I_B increases beyond 120 μA, output does not change. Also if I_B becomes negative, output does not change.

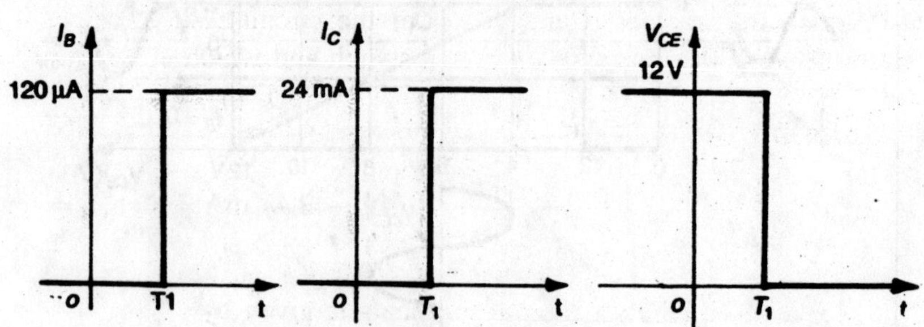

Fig. P. 4.23 (b)

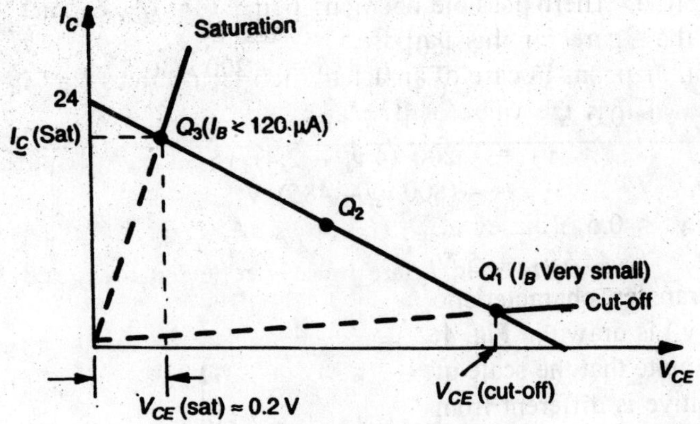

Fig. P. 4.23 (c)

Problem 4.24: A nonideal current-controlled source and its input, output circuits are shown in Fig. 4.24. The cut-in (V_T) and V_{ON} voltages of the diode are 0.6 V and R_F = 50Ω.

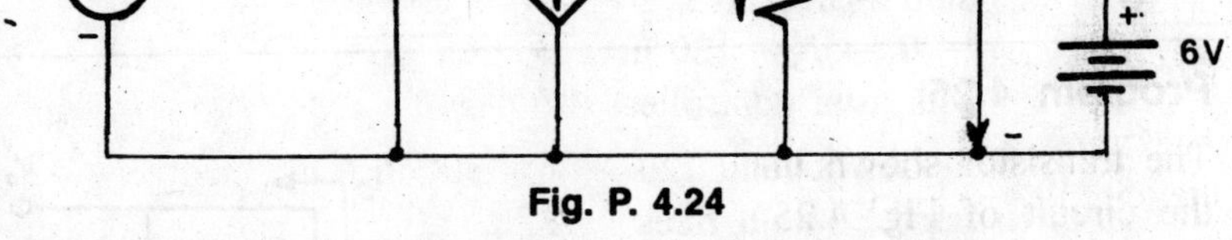

Fig. P. 4.24

Sketch the transfer characteristic of the circuit. Which segment of the characteristic can be used as a switch and which as an amplifier.

Solution:

$$i_1 = \frac{v_s - 0.6}{200 + 50} = (4v_s - 2.4) \text{ mA; when } v_s > 0.6 \text{ V}$$

$$\therefore \quad i_1 = 0; \text{ when } V_s < 0.6 \ V$$

Nodal equation at output terminals is given by

$$60i_1 + \frac{v_2}{20} = \frac{6 - v_2}{4} = 1.5 - 0.25\ v_2$$

$$\therefore \quad v_2 = -\frac{60}{0.3}\ i_1 + 5 = -200\ i_1 + 5$$

(a) $v_s > 0.6$ V

$$v_2 = -200\ (4\ v_s - 2.4) + 5$$

$$\therefore \quad = -(800\ v_s - 485)\ \text{V} \quad \text{...(i)}$$

(b) $v_s < 0.6$ V

$$\therefore \quad v_2 = 5\ \text{V} \quad \text{...(ii)}$$

The transfer characteristic (v_2 *vs* v_s) is drawn in Fig. 4. 24 (a) Note that the scale of v_2 positive is different from v_2 negative. It is observed from the characteristic that.

(i) Segment $v_s < 0.6$ V can be used as a switch; constant output v_2 = 5V.

(ii) Segment $v_s > 0.6$ V can be used as an amplifier. Any small change in v_s in this segment is amplified 800 times (phase reversed).

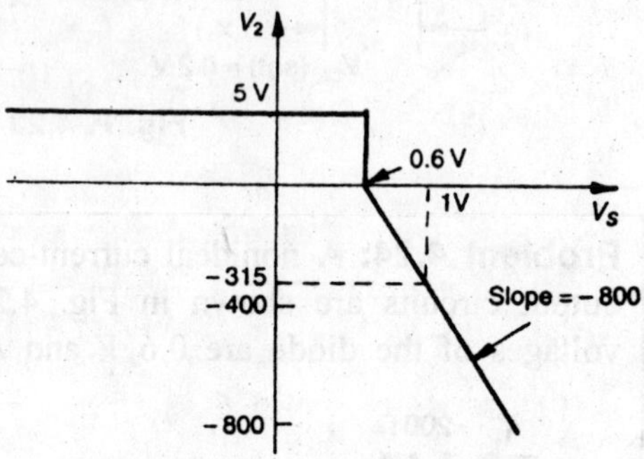

Fig. P. 4.24 (a)

Problem 4.25:

The transistor shown in the circuit of Fig. 4.25 has β = 160 and negligible reverse saturation current.

(a) Determine I_C and V_{CE}.

(b) Repeat part (a) for β = 80.

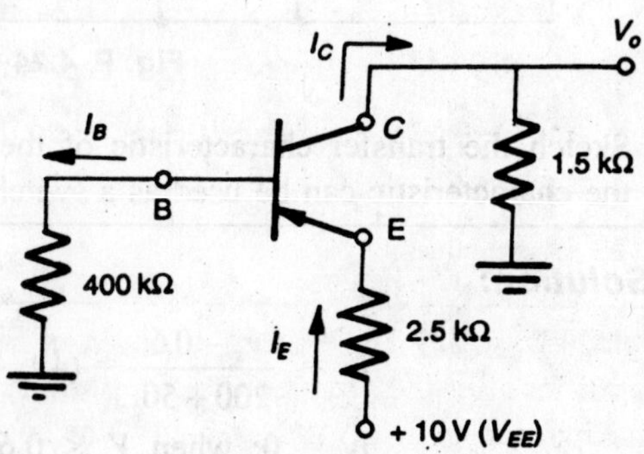

Fig. P. 4.25

Solution:

(a) We have from Fig. 4.25 that the transistor is *p-n-p*.

Take $V_{EB} \approx 0.8$ V.

$$\beta = 160$$

$$I_C = 160\, I_B;\; I_E = I_C + I_B = 161\, I_B$$

For base-emitter loop, we have,

$$10 = 2.5\, I_E + V_{EB} + 400\, I_B \qquad \text{...(i)}$$

Substituting the numerical values we get

$$I_B = \frac{10-0.8}{2.5\times 161+400} \times 10^3 = 11.5\ \mu\text{A}$$

$$I_C = 160 \times 11.5 \times 10^{-3} = 1.84\ \text{mA}$$

$$I_E \approx 1.85\ \text{mA}$$

For collector-emitter loop , we have

$$10 = 2.5\, I_E + V_{EC} + 1.5\, I_C \qquad \text{...(ii)}$$

$\therefore$ $$10 = 2.5 \times 1.85 + V_{EC} + 1.5 \times 1.84$$

$\therefore$ $$V_{EC} = 2.6\ \text{V}$$

or $$V_{CE} = -2.6\ \text{V; It is a } p\text{-}n\text{-}p \text{ transistor}$$

(*b*) $$\beta = 80$$

$$I_C = 80\, I_B, \quad I_E = 81\, I_B$$

Substituting values in eqn. (i)

$$I_B = \frac{10-0.8}{2.5\times 81+400} \times 10^3 = 15.3\ \mu\text{A}$$

$$I_C = 80 \times 15.3 = 1.23\ \text{mA;}$$

$$I_E = 81 \times 15.3 = 1.24\ \text{mA}$$

We have for collector-emitter loop, substituting values in Eq. **(ii)**

$$10 = 2.5 \times 1.24 + V_{EC} + 1.5 \times 1.23$$

or, $$V_{EC} = 5.08\ \text{V}$$

or, $$V_{CE} = -5.08\ \text{V}$$

Problem 4.26: The emitter-follower of Fig. 4.26 is used as a buffer amplifier after the premplifier stage, whose output impedance is 2 kΩ and no-load gain of 100. What is the overall voltage gain V_L/V_S? If the preamplifier input impedance is 1 kΩ, what is the current gain I_L/I_s?

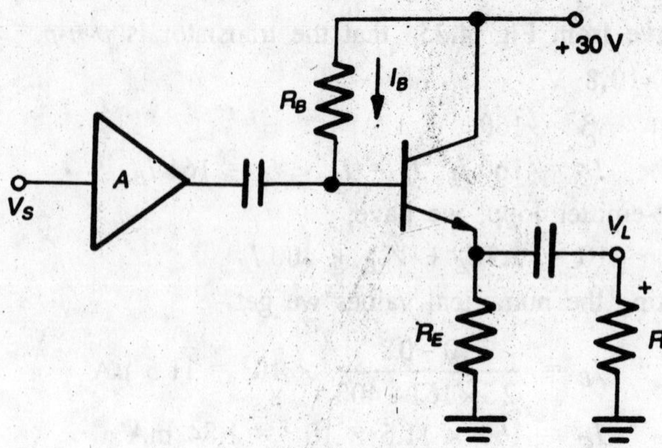

Fig. P. 4.26

Solution: *Given* :

$$R_B = 350\ \text{k}\Omega, \qquad R_E = 2.2\ \text{k}\Omega,$$
$$R_L = 20\ \Omega, \qquad \beta = 150$$

From the I_B loop

$$R_B I_B + V_{BE} + I_E R_E = 30; \qquad I_E = I_C + I_B = (\beta + 1)\, I_B$$
$$350\, I_B + 0.7 + (150 + 1) \times 2.2\, I_B = 30$$
$$I_B = \frac{30 - 0.7}{682.2} = 42.9\ \mu\text{A}$$
$$I_C = 150 \times 42.9 \times 10^{-3} = 6.44\ \text{mA}$$
$$r_\pi = \beta \times \frac{25}{I_C(\text{mA})} = 150 \times \frac{25}{6.44} = 582\ \Omega$$

The circuit model is drawn in Fig. 4.26 (a)

The effective load resistance = 2200 ‖ 20 = 19.8 Ω

Voltage gain. Round the base-emitter loop (as $R_B \gg 2$ kΩ, it is ignored)

$$2000\, I_b + 582\, I_b + 151 \times 19.8\, I_b = 100\, V_s$$
$$\therefore \quad I_b = 17.95\, V_s\ \text{mA}$$
$$\therefore \quad V_L = 151 \times 17.95 \times 19.8 \times 10^{-3}\, V_s$$

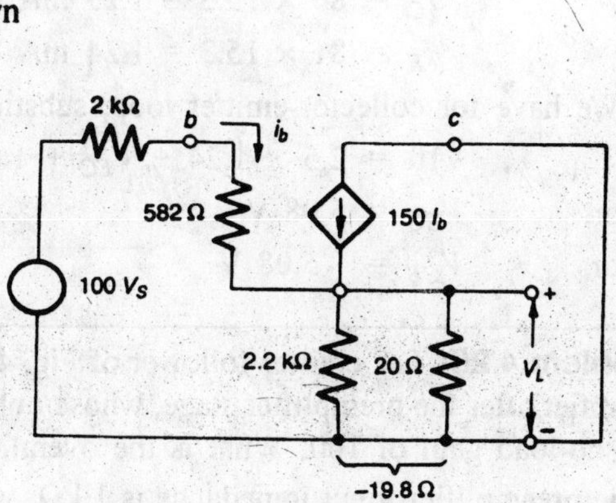

Fig. P. 4.26 (a)

$$A_{VL} = \frac{V_L}{V_s} = 53.7$$

Current gain $I_s = \frac{V_S}{1} = V_s$ mA

$$V_L = 53.7\ V_s$$

$$I_L = \frac{53.7}{20}\ V_s = 2.68\ V_s\ \text{A}$$

$$A_{IL} = \frac{I_L}{I_s} = \frac{2.68}{10^{-3}} = 2680$$

Problem 4.27: For the Darlington pair circuit of Fig. 4.27 $\beta_1 = \beta_2 = 30$. Find the value of R_1 if $R_1 = R_2$.

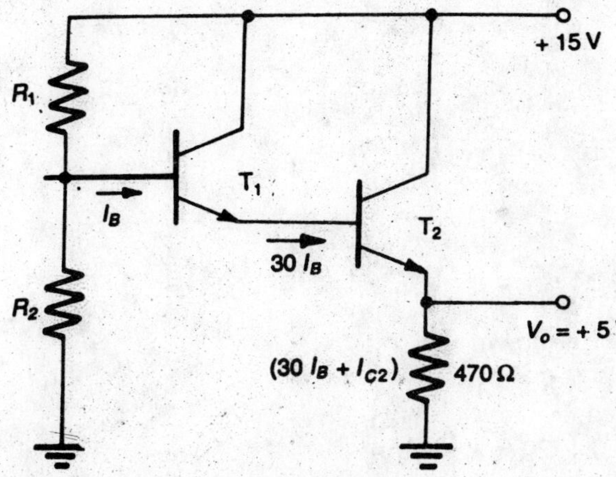

Fig. P. 4.27

Solution:

From Fig. 4.27 for the base-emitter loops of the two transistors

$R_1 = R_2$;

$V_{BB} = 15/2 = 7.5$ V

$R_B = R_1 \parallel R_2 = R_1/2$

$7.5 = (R_1/2)\, I_B + 0.7 \times 2 + (30\, I_B + I_{C2}) \times 0.47 = 0$

$$I_{C2} = 30 \times (30)\, I_B = 900\, I_B$$

$$7.5 = (R_1/2)\, I_B + 1.4 + (30 + 900) \times 0.471\, I_B = 0$$

$$(R_1/2)\, I_B + 437\, I_B = 6.1 \qquad \text{...(i)}$$

The output voltage is given as

$$0.47 \times 930\, I_B = 5 \qquad \text{...(ii)}$$

$$\therefore \qquad I_B = \frac{5}{0.47 \times 930} = 0.0114 \text{ mA}$$

Substituting the value of I_B in Eq. (i)

$$(R_1/2) \times 0.0114 + 437 \times 0.0114 = 6.1$$

$$\therefore \qquad R_1 = \frac{6.1 - 4.98}{0.0114} \times 2 = 196 \text{ k}\Omega$$

Chapter 5

Transistor at Low Frequencies

Note : *Unless otherwise specified, all transistors in these problems are identical, and the numerical values of their h parameters are given in Table 5.1. Also assume that all capacitances are arbitrarily large.*

Problem 5.1: Draw the hybrid model for a CB transistor and write the input and output equations.

Table 5.1

Typical h-parameter values for a transistor (at I_E = 1.3 mA)

Parameter	*CE*	*CC′*	*CB*
$h_{11} = h_i$	1,100 Ω	1100 Ω	21.6 Ω
$h_{12} = h_r$	2.5×10^{-4}	~1	2.9×10^{-4}
$h_{21} = h_f$	50	– 51	– 0.98
$h_{22} = h_0$	24 μA/V	25 μA/V	0.49 μA/V
$1/h_o$	40 K	40 K	2.04 M

Solution:

$$\therefore \quad V_{eb} = h_{ib}i_e + h_{re}V_{cb}$$

$$i_c = h_{fb}i_e + h_{ob}V_{cb}$$

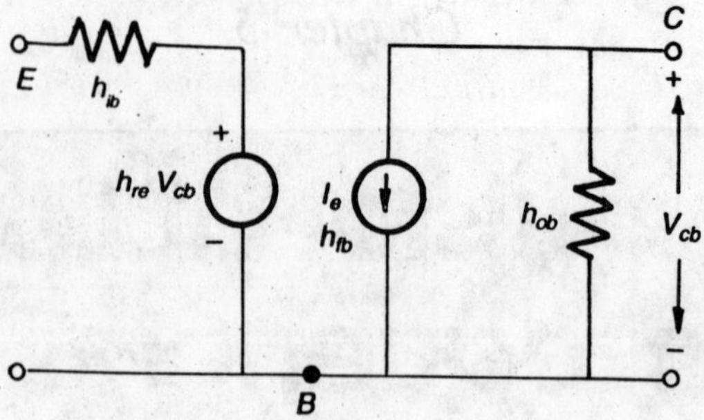

Fig. P. 5.1

Problem 5.2: (a) Describe how to obtain h_{ie} from the CE input characteristics.
(b) Repeat part *a* for h_{re}. Explain why this procedure, although correct in principle, is inaccurate in practice.

Solution: (a) $h_{ie} = \left(\frac{\Delta V_B}{\Delta I_B}\right) = \left.\frac{\partial V_B}{\partial I_B}\right|_{V_C = \text{Constant}}$

Hence, $\left.\frac{V_{B2} - V_{B1}}{I_{B2} - I_{B1}}\right|_{V_C = \text{Constant}}$, is the type of the characteristic corresponding to V_C at the particular operating point.

(b) Also, $h_{re} = \left.\frac{\Delta V_B}{\Delta V_C}\right|_{I_B = \text{Constant}} = \left.\frac{\partial V_B}{\partial V_C}\right|_{I_B = \text{Constant}}$

$\therefore \quad h_{re} = \left.\frac{\Delta V_B}{\Delta V_c}\right|_{I_B = \text{Constant}}$

Generally we note $h_{re} \cong 10^{-4}$ hence $\Delta V_B \ll \Delta V_c$ and consequently this is inaccurate in practice.

Problem 5.3: (a) Find the CC *h* parameters in terms of the CE *h* parameters.
(b) Find the CE *h* parameters in terms of the CC parameters.

Solution : (a)The equation in terms of CE-h parameters :

$$V_{be} = h_{ie}i_b + h_{re}V_{ce}$$
$$i_c = h_{fe}i_b + h_{oe}V_{ce} \quad \text{...(1)}$$

Now, $V_{be} = V_{bc} + V_{ce} = V_{bc} - V_{cc}; \; (\therefore \; V_{ce} = -V_{cc})$

Also $i_c = -(i_b + i_e)$

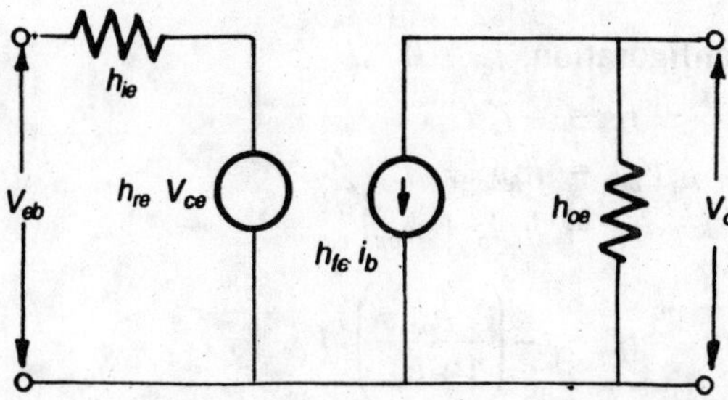

Fig. P. 5.3

Then from (1),

$$\therefore \left.\begin{aligned} V_{bc} - V_{cc} &= h_{fe}i_b - h_{re}V_{cc} \\ -i_e - i_b &= h_{fe}i_b - h_{oe}V_{ce} \end{aligned}\right\} \quad \text{...(2)}$$

$$\left.\begin{aligned} V_{bc} &= h_{ie}i_b + (1 - h_{re})\,V_{ec} \\ i_e &= -(1 + h_{fe})\,i_b + h_{oe}V_{ec} \end{aligned}\right\} \quad \text{...(3)}$$

Also

$$\left.\begin{aligned} V_{bc} &= h_{ic}i_b + h_{re}V_{ce} \\ i_e &= h_{fe}i_b + h_{re}V_{ec} \end{aligned}\right\} \quad \text{...(4)}$$

From (3) and (4),

$$h_{ic} = h_{ie}$$
$$h_{rc} = 1 - h_{re} \cong 1$$
$$h_{fe} = -(1 + h_{fe}),$$
$$h_{oc} = h_{oe} \textbf{ Ans}$$

(b) Starting with set (4)

$$h_{ie} = h_{ic},$$
$$h_{re} = 1 - h_{rc}$$
$$h_{fe} = -(1 + h_{fe});$$
$$h_{oe} = h_{oc} \textbf{ Proved}$$

Problem 5.4: (a) Find the h_{rb} in terms of the CE h parameters.
(b) Find h_{ie} in terms of the CB h parameters.

Solution:

(a) As we know,

$$h_{rb} = \left.\frac{V_{eb}}{V_{cb}}\right|_{i_e=0} = -\frac{V_{be}}{V_{ce} - V_{be}} \quad \ldots(1)$$

For CE configuration, $i_e = 0$

$$\therefore \quad i_b = -i_c$$

$$\therefore \quad V_{be} = h_{ie}i_e + h_{re}V_{ce}$$

$$i_c = h_{fe}i_b + h_{oe}V_{ce}$$

or,
$$i_b = -\left(\frac{h_{oe}}{1+h_{fe}}\right) V_{ce}$$

and
$$V_{be} = \left(-\left(\frac{h_{fe}h_{oc}}{1+h_{fe}}\right) + h_{re}\right) V_{ce}$$

hence from eq (1),

$$h_{rb} = \frac{h_{ie}h_{oe} - h_{re}(1+h_{fe})}{(1-h_{re})(1+h_{fe}) + h_{ie}h_{oe}} \cong -h_{re} + \left(\frac{h_{ie}h_{oe}}{1+h_{fe}}\right)$$

(b)
$$h_{ie} = \left.\frac{V_{be}}{i_b}\right|_{V_{ce}=0} = -\frac{V_{eb}}{i_b} \quad \ldots(ii)$$

But in Case of CB configuration,

$$V_{eb} = h_{ie}i_e + h_{rb}V_{cb}$$

$$i_c = h_{fb}i_e + h_{ob}V_{cb}$$

$$\because \quad V_{ce} = -V_{ce} = 0 \text{ or } V_{ce} = V_{cb} - V_{eb} = 0$$

or,
$$i_e = \left(\frac{1-h_{rb}}{h_{ib}}\right) V_{eb}$$

and
$$-i_b = i_e + i_c$$

$$= (1+h_{f_b}) \left(\frac{1-h_{rb}}{h_{ib}}\right) \times V_{eb} + h_{oe}V_{eb} \quad \ldots(iii)$$

Now, from (ii) and (iii)

$$h_{ie} = \left(\frac{h_{ib}}{h_{ob} \times h_{ib} + (1 - h_{rb})(1 + h_{fb})} \right)$$

$\because \; 1 + h_{fb} \gg h_{ob} \times h_{ib}$ and $1 \gg h_{rb}$

$$\therefore \qquad h_{ie} = \left(\frac{h_{ib}}{1 + h_{fb}} \right) \quad \textbf{Ans}$$

Problem 5.5: For the circuit shown, verify that the modified h parameters (indicated by primes) are

(a) $$h'_{ie} \approx h_{ie} + \frac{(1 + h_{fe})R_e}{1 + h_{oe}R_e}$$

(b) $$h'_{re} = \frac{h_{re} + h_{oe}R_e}{1 + h_{oe}R_e}$$

(c) $$h'_{fe} = \frac{h_{fe} - h_{oe}R_e}{1 + h_{oe}R_e}$$

(d) $$h'_{oe} = \frac{h_{oe}}{1 + h_{oe}R_e}$$

Solution:

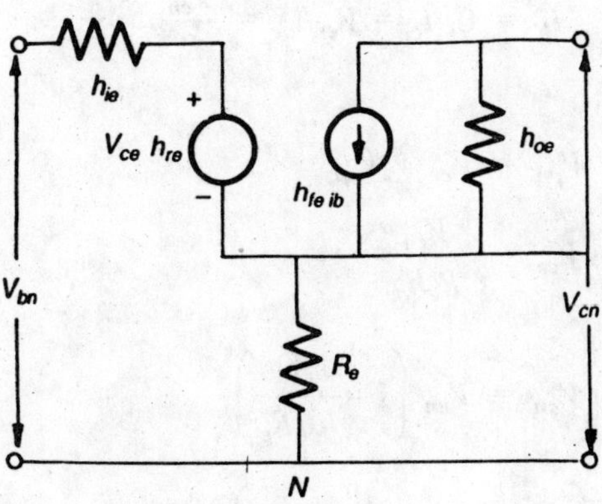

Fig. P. 5.5

(a) $$h'_{ie} = \left.\frac{V_{bn}}{i_b}\right|_{V_{cn}=0}$$

with $$V_{cn} = 0,$$

$$V_{ce} = -(i_b + i_c)\, R_e = -i_bR_e - h_{fe}i_bR_e - V_{ce}h_{oe}R_e$$

Then, $$V_{ce} = -\frac{i_b R_e(1+h_{fe})}{1+h_{oe}R_e} = -V_{en}$$

and $$V_{bn} = i_bh_{ie} + h_{re}V_{ce} + V_{en}$$

$$= i_bh_{ie} + -\frac{i_b R_e(1+h_{fe})}{1+h_{oe}R_e}\,(1 - h_{re})$$

Hence $$h'_{ie} = h_{ie} + \frac{R_e(1+h_{fe})}{1+h_{oe}R_e}\,(1 - h_{re})$$

$$\cong h_{ie} + \frac{R_e(1+h_{ie})}{1+h_{oe}R_e} \quad [\because\ h_{re} \ll 1]$$

(b) $$h'_{re} = \left.\frac{V_{bn}}{V_{cn}}\right|_{i_b=0}$$

Where $$V_{bn} = h_{ie}i_b + h_{re}V_{cn} + V_{en}$$

$\because$ $$i_b = 0,\ i_c = V_{cn}h_{oe} = \frac{V_{cn}}{R_e}$$

Also,

$$V_{cn} = V_{ce} + V_{en}$$

$$V_{ce} = \frac{V_{en}}{R_e h_{oe}};$$

$$V_{cn} = V_{en}\left(1+\frac{1}{h_{oe}R_e}\right)$$

or, $$V_{en} = V_{cn}\left(\frac{h_{oe}\times R_e}{1+h_{oe}R_e}\right)$$

Thus $$V_{bn} = \frac{h_{re}V_{cn}}{1+h_{oe}R_e} + \frac{h_{oe} \times R_e V_{cn}}{1+h_{oe}R_e}$$

$$\therefore \quad h'_{re} = \frac{V_{bn}}{V_{cn}} = \left(\frac{h_{re}+h_{oe}R_e}{1+h_{oe}R_e}\right)$$

(c). $$h'_{fe} = \left.\frac{i_c}{i_b}\right|_{V_{cn}=0}$$

where $$i_c = h_{ie}i_b + h_{oe}V_{ce}$$

$$\therefore \quad V_{ce} = -\frac{i_b R_e(1+h_{fe})}{1+h_{oe}R_e} \text{ [as in part (a)]}$$

$$\therefore \quad i_c = i_b\left(h_{fe} - \frac{h_{oe}R_e(1+h_{fe})}{1+h_{oe}R_e}\right)$$

$$\therefore \quad h'_{fe} = \left(\frac{h_{fe}-h_{oe}R_e}{1+h_{oe}R_e}\right)$$

(d) $$h'_{oe} = \left.\frac{i_c}{V_{cn}}\right|_{i_b=0}$$

Where $$i_e = \frac{V_{en}}{R_e} = \frac{V_{cn}}{R_e} \times \frac{h_{oe}R_e}{1+h_{oe}R_e} = \frac{h_{oe}}{1+h_{oe}R_e} V_{cn}$$

$$\therefore \quad h'_{oe} = \left(\frac{h_{oe}}{1+h_{oe}R_e}\right)$$

Problem 5.6: Given a single-stage transistor amplifier with the h parameters specified in Table, 5.1 calculate A_I, A_v, A_{V_s}, R_i, and R_o for the CC transistor configuration, with $R_s = R_L = 10$ K.

Solution : For the CC configuration,

$$h_{ic} = 1100\ \Omega,\ h_{rc} = 1$$
$$h_{fc} = -51 \text{ and } h_{oc} = 25\ \mu\text{A/V}$$

$$A_1 = \frac{-h_{fc}}{1+h_{oc}R_L} = \frac{51}{1+25\times10^{-6}\times10^4} = 40.8$$

$$R_i = h_{ic} + A_I h_{rc} R_L = 1.1 + 1 \times 40.8 \times 10^4 = 409.1 \text{ K}$$

$$A_r = A_I \frac{R_L}{R_i} = \frac{40.8\times10}{409.1} \cong 0.998$$

Table : 5.1

Parameter	*CE*	*CC*	*CB*
h_i	1100	1100	21.6
h_r	2.5×10^{-4}	1	2.9×10^{-4}
h_f	50	–51	–0.98
h_o	24 μ A/V	25μ A/V	0.49 μ A/V
$\frac{1}{h_o}$	40 K	40 K	2.04 M

$$\gamma_o = h_{oc} - \frac{h_{ic} \times h_{rc}}{h_{ic} + R_s} = 4.625 \times 10^{-3}$$

$$\therefore \quad R_o = 0.0217\times10^3 = 217\ \Omega$$

$$A_{V_s} = \frac{A_V R_i}{R_i + R_s} = \frac{0.998\times409.1}{419.1} = 0.974$$

Problem 5.7: (a) Draw the equivalent circuit for the CE and CC configurations subject to the restriction that $R_L = 0$. Show that the input impedances of the two circuits are identical.

(b) Draw the circuits for the CE and CC configurations subject to the restriction that the input is open-circuited. Show that the output impedances of the two circuits are identical.

Solution:

(a) CE-Connection

$$R_i = \left.\frac{V_{be}}{i_b}\right|_{V_{ce}=0}$$

Similarly for CC-connection,

$$R_i' = \left.\frac{V_{bc}}{i_b}\right|_{V_{ec}=0} = h_{ic}$$

but from definition

$$h_{ie} = \left.\frac{V_{be}}{i_b}\right|_{V_{ce}=0}$$

$$\therefore \quad R_i = h_{ie}$$

As we know, in small hybrid model of transistors,

$$h_{ie} = h_{ic}$$

$$\therefore \quad R_i = R_i' \text{ Proved}$$

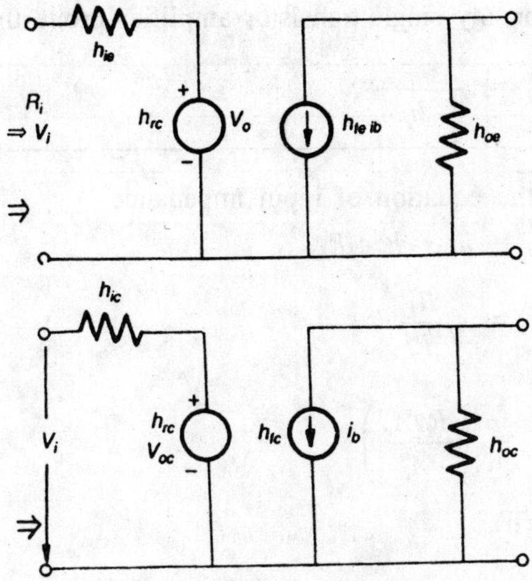

Fig. P. 5.7 (a)

(b) We have that $h_{oe} = h_{oc}$

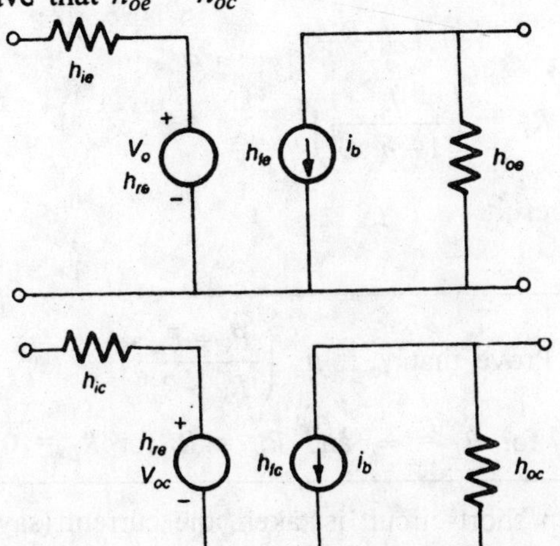

Fig. P. 5.7 (b)

The output impedance of either circuit with the input open circuited i.e., $i_b = 0$ is given by

$$h_{oe} = h_{oc}$$

Hence, the required result. Proved.

Problem 5.8: For any single-transistor amplifier prove that

$$R_i = \frac{h_i}{1 - h_r A_v}$$

Solution : From the equation of input impedance

$$R_i = h_i + h_r A_I R_L \qquad \text{...(i)}$$

Also,
$$A_V = A_I \frac{R_L}{R_i}$$

then,
$$R_L = \left(\frac{R_i\ A_v}{A_I}\right) \qquad \text{...(ii)}$$

From (i) and (ii),

$$R_i = h_i + h_r \times A_I \times \left(\frac{R_i A_v}{A_I}\right)$$

$$= h_i + h_r R_i A_V$$

$$\therefore \quad R_i = \left(\frac{h_i}{1 - h_r A_v}\right)$$

Hence, the result.

Problem 5.9: Prove that $\gamma_o = h_o \left(\frac{R_s + R_{i\infty}}{R_s + R_{io}}\right)$

where $R_{i\infty} \equiv R_i$ for $R_L = \infty$, and $R_{io} \equiv R_i$ for $R_L = 0$

Solution : When short circuit is taken, the current (say) is I_L (when $R_L = 0$)

Thus
$$I_{sc} = h_f I_1;$$

where $I_i = \left(\dfrac{V_s}{h_i + R_s}\right)$

$\because$ $V_2 = 0,$

Therefore $I_{sc} = \left(\dfrac{h_f V_s}{h_i + R_s}\right)$

In case of open circuit, $V_{oc} = V_i$
i.e., when $R_L \to \infty,$

$$V_{oc} = -\frac{h_f I_I}{h_o}$$

When $I_1 = \dfrac{V_s - h_r V_{oc}}{h_i + R_s}$

Then, $V_{oc} = -\dfrac{h_f}{h_o}\left(\dfrac{V_s - h_r V_{oc}}{h_i + R_s}\right)$

or $V_{oc} = \dfrac{-h_f V_s}{h_o(h_i + R_s) - h_f h_r}$

$$\gamma_o = \frac{I_{sc}}{V_{oc}} = h_o - \left(\frac{h_i h_r}{h_i + R_s}\right)$$

Problem 5.10: (a) For a CE configuration, what is the maximum value of R_L for which R_1 differs by no more than 10 percent of its value at $R_L = 0$? Use the transistor parameters given in Table 5.1.

(b) What is the maximum value of R_s for which R_o differs by no more than 10 percent of its value for $R_s = 0$?

(c) For the CB configuration, what is the maximum value of R_L for which R_i does not exceed 50 Ω?

Solution : (a) $R_i = h_{ie} - \dfrac{h_{fe} h_{re}}{h_{oe} + 1/R_L}$

If $R_L \to 0,\ R_i = h_{ie}$

The value of R_L for which

$$R_i = 0.9\ h_{ie},$$

$$0.9\ h_{ie} = h_{ie} - \frac{h_{fe}h_{rc}}{h_{oe} + \frac{1}{R_L}};$$

Then, $$R_L = \frac{0.1\,h_{ie}}{h_{fe} \times h_{re} - 0.1\,h_{ie} \times h_{oe}}$$

Here $$h_{ie} = 1100,$$
$$h_{fe} = 50,$$
$$h_{re} = 2.5 \times 10^4$$
$$h_{oe} = \frac{1}{40 \times 10^3}$$

$$\therefore \quad R_L = \frac{0.1 \times 1100}{50 \times 2.5 \times 10^4 - \frac{1100 \times 0.1}{4.0 \times 10^{-3}}}$$

$$= 11.3\ \text{K}.$$

(b) For the given case of CE-configuration, if R_s increases, there R_o decreases slightly.

If R_o decreases by 10%,

$$\gamma_o = \frac{1}{R_o} \text{ increases by } 11\%$$

$$\gamma_o\Big|_{R_s=0} = h_{oe} - \frac{h_{fe} \times h_{re}}{h_{ie}}$$

$$= 25 \times 10^{-6} - \frac{50 \times 2.5 \times 10^{-4}}{1100} = 13.63 \times 10^{-6}$$

Thus, $$(1.11) \times (13.63 \times 10^{-6}) = 25 \times 10^{-6} - \frac{125 \times 10^{-4}}{(1.1 + R_s)10^3}$$

$$\Rightarrow 9.85 \times 10^{-6} = \frac{12.5 \times 10^{-6}}{1.1 + R_s}$$

$$\therefore \quad R_s = 170\ \Omega$$

(c) $$R_i = h_{ib} - \frac{h_{fb}h_{rb}}{h_{ob} + \frac{1}{R_L}} = 50\ \Omega$$

$$50 - 21.6 = \frac{0.98 \times 2 \times 2.9 \times 10^{-4}}{0.49 + \frac{1}{R_L}}$$

or, Solving we set,

$$R_L = 105.8\ \text{K}$$

Problem 5.11: For the emitter follower with $R_s = 0.5$ K and $R_L = 5$ K, calculate A_I, R_i, A_v, A_{V_s}, R_0. Assume $h_{fe} = 50$, $h_{ie} = 1$ K, $h_{oe} = 25\ \mu$A/V.

Solution : For an emitter follower i.e., common collector,

$$A_I = \left(\frac{1 + h_{fe}}{1 + h_{oe}R_L}\right) = \left\{\left(\frac{1 + h_{fe}}{h_{oe}R_e}\right) \text{for large } R_e\right\}$$

$$R_i = h_{ie} + A_I R_L$$

$$A_I = \frac{51}{1 + 0.185} = 45.3$$

$$R_i = 1 + 226 = 227\ \text{K}$$

$$A_V = 1 - \left(\frac{1}{227}\right) = 0.9956 \quad \left\{\because A_v = 1 - \frac{k_{ie}}{R_i}\right\}$$

$$A_{vs} = A_v \left(\frac{R_i}{R_i + R_s}\right)$$

$$= 0.9956 \times \frac{227}{227 + 0.5} \cong 0.9934$$

$$\gamma_o = 25 + \frac{51}{1.5} = 34.025 \times 10^{-3}\ \text{A/V}$$

$$R_o = 29.4\ \Omega$$

Problem 5.12: (a) Design an emitter follower having R_i = 500 K and R_o = 20 Ω. Assume h_{fe} = 50, h_{ie} = 1 K, h_{oe} = 25 μA/V.
(b) Find A_1 and A_v for the emitter follower of part *a*.

Solution: A common-collector or emitter follower configuration is shown below,

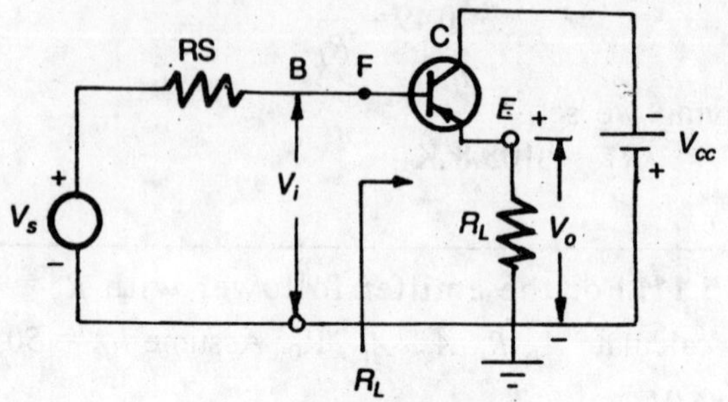

Fig. P. 5.12 (a)

Drawing its equivalent circuit,

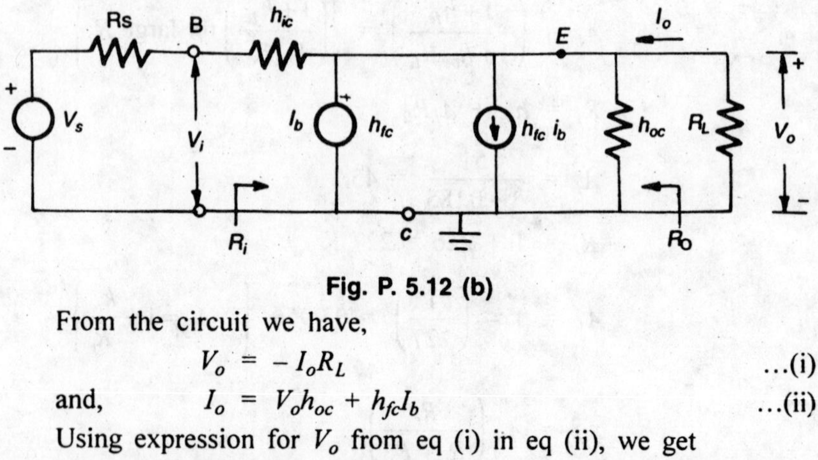

Fig. P. 5.12 (b)

From the circuit we have,

$$V_o = -I_oR_L \quad \ldots(i)$$

and, $$I_o = V_oh_{oc} + h_{fc}I_b \quad \ldots(ii)$$

Using expression for V_o from eq (i) in eq (ii), we get

$$I_o = -I_o\, h_{oc}R_L + h_{fc}I_b$$

or, $$I_o\,[1 + h_{oc}R_L] = h_{fc}I_b$$

or, $$A_I = \frac{I_o}{I_b} = \frac{h_{fc}}{1 + h_{oc}R_L} \quad \ldots(iii)$$

Now using $h_{fc} = -(1 + h_{fe})$ and

$$h_{oc} = h_{oe},$$

We have
$$A_I = \frac{(1+h_{fe})}{1+h_{oe}R_L}$$

$$= \frac{51}{1+25\times10^{-6}\times5\times10^3} = 45.34$$

Again from the circuit,

$$V_i = h_{ic}I_b + h_{rc}V_o$$

Using eq (i),

$$V_i = h_{ic}I_b - h_{rc}I_oR_L$$

From equation (iii),

$$V_i = h_{ic}I_b + h_{rc}R_LA_II_b$$

or, Input resistance,

$$R_i = \frac{V_i}{I_b} = h_{ic} + h_{rc}A_IR_L$$

putting $h_{ic} = h_{ie} = 1$

and $h_{rc} = 1$

then, $R_i = 1 + 1 \times 45.34 \times 5$

$= 227.6$ K

Voltage gain,

$$(A_v) = \frac{V_o}{V_i} = \frac{-I_oR_L}{I_iR_i}$$

$$= A_I \times \frac{R_L}{R_i} = (45.34) \times \frac{5}{227.67} = 0.995$$

$$A_{V_S} = A_v \times \frac{R_i}{R_i + R_s}$$

$$= 0.995 \times \frac{227.6}{227.6+0.5} = 0.992$$

For R_o, shorting the voltage source to zero and setting $R_L = \infty$ and driving the output terminal from a generator V_2. If current drawn from V_2 is I_2, then,

$$\gamma_o = \frac{1}{R_o} = \frac{I_2}{V_2}$$

From equation (ii),

$$I_2 = V_2 h_{oc} + h_{fc} I_b$$

or,

$$\frac{I_2}{V_2} = h_{oc} + h_{fc} \times \frac{I_b}{V_2}$$

Also, from the circuit when $V_s = 0$, We have

$$-R_s I_b - h_{ic} I_b - h_{rc} V_o = 0$$

$$\frac{I_b}{V_o} = \frac{-h_{rc}}{R_s + h_{ic}}$$

Then from equation (iv),

$$\gamma_2 = \frac{I_2}{V_2} = h_{oc} - \frac{h_{fc} h_{rc}}{R_s + h_{ic}} = h_{oc} + \frac{(1 + h_{fc}) \times 1}{R_s + h_{ic}}$$

$$= (25 \times 10^{-6}) + \frac{51}{1.5 \times 10^3}$$

$$= 34.025 \times 10^{-3}$$

$$R_o = \frac{1}{\gamma_D} = 29.4 \text{ K}\Omega$$

Problem 5.13: The transistor amplifier shown uses a transistor whose h parameters are given in Table 5.1. Calculate $A_I = I_o/I_i$, A_v, A_{V_S}, R_o, and R_i.

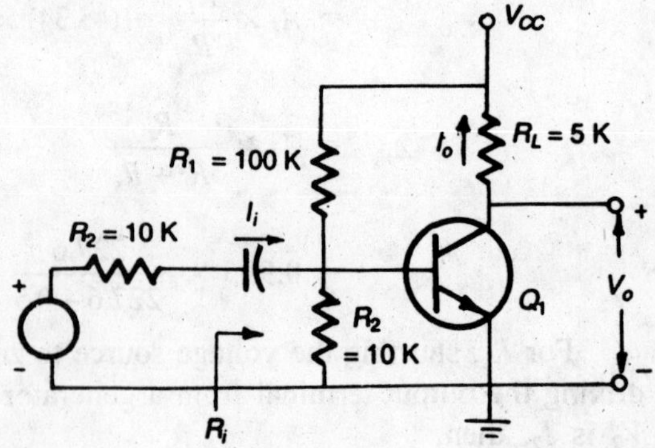

Fig. P. 5.13

Solution : Drawing the equivalent circuit for the given amplifier,

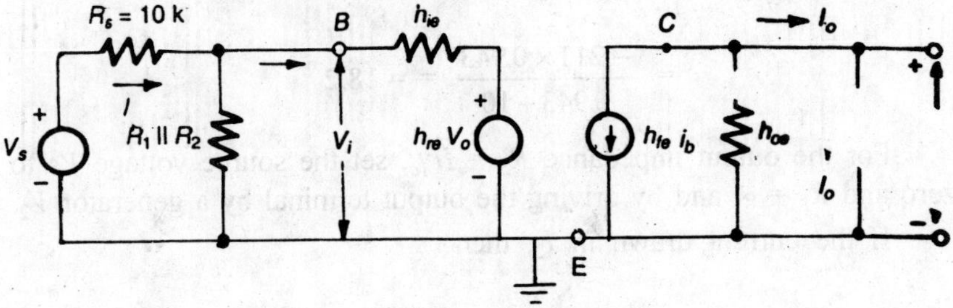

Fig. P. 5.13 (a)

From the above circuit we have, at collector junction,

$$-h_{fc}I_b = V_o\left[\frac{1}{R_L} + h_{oc}\right]$$

or, $$-h_{fc}I_b = I_oR_L\left[h_{oc} + \frac{1}{R_L}\right]$$

Then, $$A_i = \frac{I_o}{I_b} = \frac{-h_{fc}}{1 + h_{oc}R_L} = \frac{-50}{1 + 5/40} = -44.5$$

Now, $$V_i = I_bh_{ie} + h_{re}V_o = h_{ie}I_b + h_{re}I_oR_L$$
$$= h_{ie}I_b + h_{re}A_i'I_bR_L$$

Then, $$R_i' = V_i/I_b = h_{ie} + h_{re}A_i'R_L$$
$$= [1.1 + (2.5 \times 10^{-4} \times (-44.5) \times 5]\ \text{K}$$
$$= 1.05\ \text{k}\Omega$$

Current gain, $$A_I = \frac{I_o}{I_i} = \frac{I_O}{I_b} \times \frac{I_b}{I_i} = A_i' \times \frac{(R_1 \parallel R_2)}{(R_1 \parallel R_2) + R_i'}$$

$$= (-44.5) \times \frac{9.1\ \text{K}}{(9.1 + 1.05)\ \text{K}} = -39.8$$

$$R_i = (R_1 \parallel R_2) \parallel R_i' = (9.1\ \text{K}) \parallel (1.05\ \text{K})$$
$$= 0.943\ \text{K}\Omega$$

Voltage gain, $$A_v = \frac{A_I R_L}{R_i} = \frac{-39.8 \times 5\ \text{K}}{0.943\ \text{K}} = -211$$

$$A_{V_S} = \frac{A_v R_L}{R_i + R_s}$$

$$= \frac{-211 \times 0.943}{0.943 + 10} = -18.2$$

For the output impedance $R_o = I/\gamma_o$, set the source voltage Vs to zero and $R_L = \infty$ and by driving the output terminal by a generator V_2. If the current drawn in I_2, then

$$R_o = I/\gamma_o = \frac{V_2}{I_2}$$

Now, the current will be,

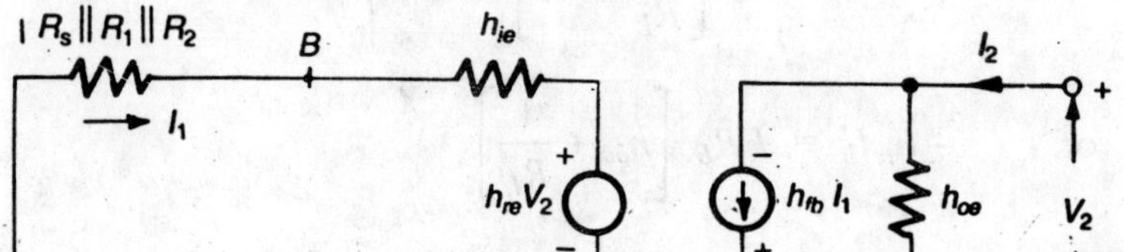

Fig. P. 5.13 (b)

Applying KVL we get

$(R_s \| R_1 \| R_2)\ I_1 + h_{ie}I_1 + h_{re}V_2 = 0$

and $\quad I_2 = V_2 h_{oe} + h_{fc}I_1$

or, $$\frac{I_2}{V_2} = h_{oe} + h_{fc}\frac{I_1}{V_2}$$

From equations (i) and (ii),

$$\gamma_o = h_{oe} - \frac{h_{fe}h_{re}}{h_{ie} + (R_s \| R_1 \| R_2)}$$

$$= 25 \times 10^{-6} - \frac{50 \times 2.5 \times 10^{-4}}{1.1\text{ K}(10119.1)\text{ K}} = 22.9 \times 10^{-6}\ \Omega$$

$$R_o = \frac{1}{\gamma_o} = 43.7\text{ K}\Omega$$

Problem 5.14:

(a) In the circuit shown, find the input impedance R_i in terms of the CE h parameters, R_L and R_e.

(b) If $R_L = R_e = 1$ K and the h parameters are as given in Table 5.1, what is the value of R_i?

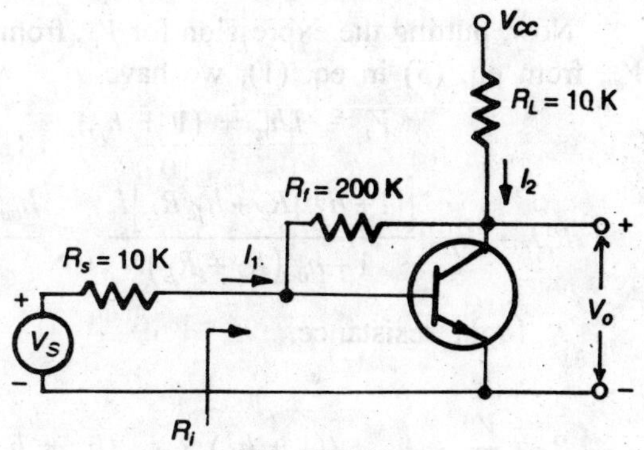

Fig. P. 5.14

Solution: Drawing the equivalent circuit of the given circuit,

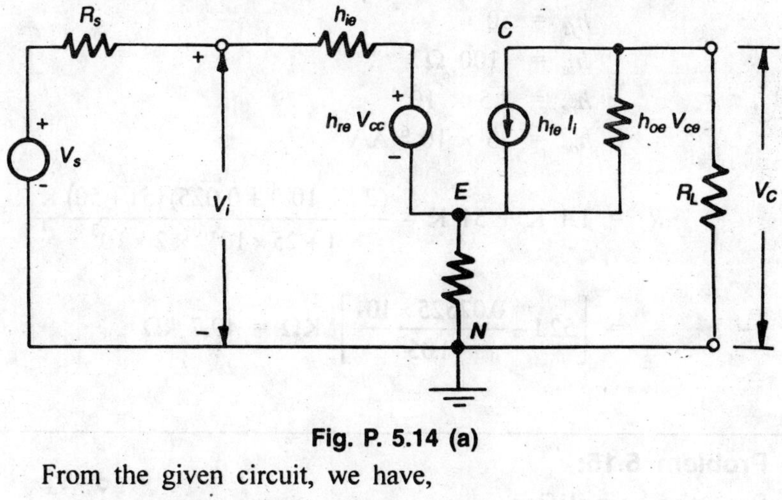

Fig. P. 5.14 (a)

From the given circuit, we have,

$$V_i = I_i h_{ie} + h_{re} V_{ce} + V_{en} \quad \text{...(i)}$$

$$V_{en} = I_i + h_{fe} R_e I_i + V_{ce} h_{oe} R_e \quad \text{...(ii)}$$

and,

$$V_{ce} = -V_{en} - (V_{ce} h_{oc} + h_{fe} I_i) R_L \quad \text{...(iii)}$$

From eq. (ii),

$$V_{ce} = -I_i (1 + h_{fe} R_e) - V_{ce} h_{oc} R_e - V_{ce} h_{oe} R_L - h_{fe} R_L I_i$$

or,

$$V_{ce} [1 + h_{oe} (R_e + R_L)] = -[(1 + h_{fe}) R_e + R_L h_{fe}] I_i \quad \text{...(iv)}$$

Putting the expression of V_{ce} in eq. (ii), we have,

$$V_{en} = I_i \left[\left((1 + h_{fe}) R_e \right) - \frac{R_e h_{oe} \left[(1 + h_{fe}) R_e + h_{fe} R_L \right]}{1 + h_{oc} (R_e + R_L)} \right] \quad \text{...(v)}$$

Now, putting the expression for V_{ce} from eq (4) and expression for V_{en} from eq. (5) in eq. (1), we have

$$V_i = I_i h_{ie} + (1 + h_{fe})$$

$$R_c I_i - h_{re} \frac{\left[(1+h_{fe})R_e + h_{fe}R_L\right]I_i}{1+h_{oe}(R_e+R_L)} - \frac{h_{oe}R_e\left[(1+h_{fe})R_e + h_{fe}R_L\right]I_i}{1+h_{oe}(R_e+R_L)}$$

$\therefore$ Input resistance,

$$R_i = \frac{V_i}{I_i} = h_{ie} + ((1 + h_{fe})\ R_e) - (h_{re} + h_{oe}R_e) \frac{\left[(1+h_{fe})R_e + h_{fe}R_L\right]}{1+h_{oe}(R_e+R_L)}$$

(b) When $R_L = R_e = 1$ K;

$h_{fe} = 50$

$h_{ie} = 1100\ \Omega$

$h_{re} = 2.5 \times 10^{-4}$

$h_{oe} = 25 \times 10^{-6}$ A/V

$$R_i = 1.1 \text{ K} + 51 \text{ K} - \frac{(2.5\times10^{-4} + 0.025)\,(51+50)\text{ K}}{1+25\times10^{-6}\times2\times10^{3}}$$

$$= \left[52.1 - \frac{0.02525\times101}{1.05}\right] \text{K}\Omega = 49.7\ \text{k}\Omega$$

Problem 5.15: For the amplifier shown, using a transistor whose parameters are given in Table 5.1 compute

$A_I = I_o/I_i$

A_v, A_{V_s},

and R_i.

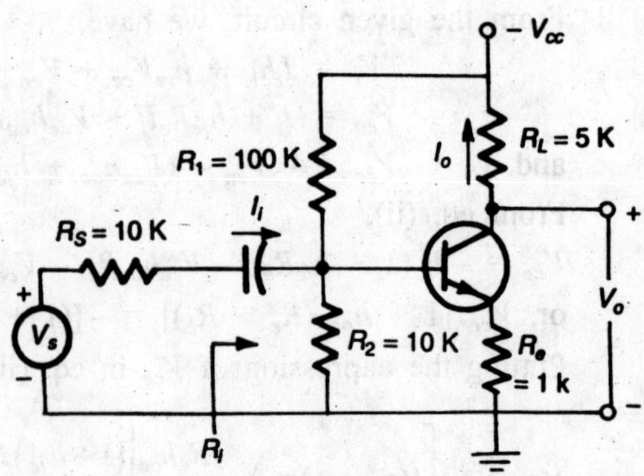

Fig. P. 5.15

Solution: Drawing the equivalent circuit,

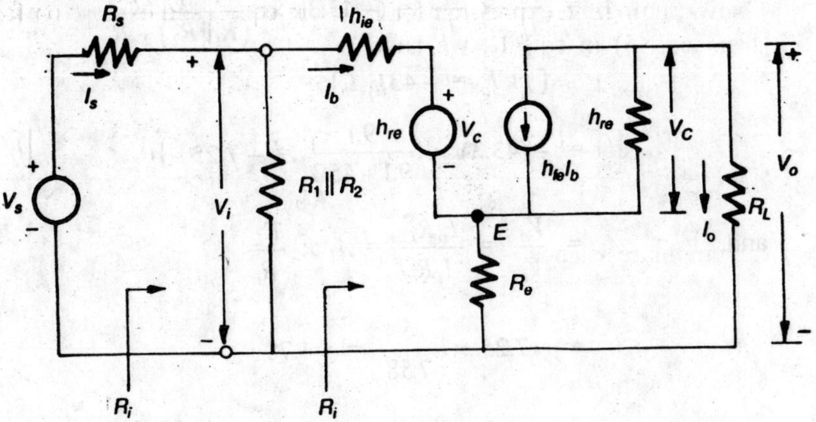

Fig. P. 5.15 (a)

From the circuit, we have,

$$-I_o = h_{fe}I_b + h_{oe}V_c \quad \text{...(i)}$$

$$V_c = V_o - (I_b - I_o) R_e = V_o - (I_b - I_o) \times 10^3 \quad \text{...(ii)}$$

and, $$V_o = + I_oR_L \quad \text{...(iii)}$$

Substituting equations (ii) and (iii) in eq (i), we get

$$-I_o = 50\, I_b + 25 \times 10^{-6}\, (0.5 \times 10^3 \times I_o - 10^3\, I_b + 10^3\, I_o)$$

$$= 49.75\, I_b + 0.15\, I_o$$

or, $$I_o = -43.3\, I_b \quad \text{...(iv)}$$

Also, $$V_i = h_{ie}I_b + h_{re}V_c + (I_b - I_o) R_e \quad \text{...(v)}$$

From eq. (ii)

$$V_c = I_oR_L - (I_b - I_o) \times 10^3$$

$$= -43.3 \times 5 \times 10^3\, I_b + 44.3 \times 10^3\, I_b$$

then from eq. (v),

$$V_i = 1.1 \times 10^3\, I_b + 2.5 \times 10^{-4}\, (-26.1 \times 10^4\, I_b) + 44.3 \times 10^3\, I_b$$

$$= 45.3 \times 10^3\, I_b$$

Or, $$R_i' = \frac{V_i}{I_b} = 45.3 \text{ K}$$

then, $$R_i = R_i' \,\|\, (9.1 \text{ K}) = \frac{45.3 \text{ K} \times 9.1 \text{ K}}{54.4 \text{ K}} = 7.58 \text{ K}$$

$$A_I = \frac{I_o}{I_b} \times \frac{I_b}{I_i} = (-43.3) \times \frac{(R_1 \| R_2)}{(R_1 \| R_2) + R_i'}$$

$[\because\ I_o = -43.3\ I_b]$

$$= -(43.3) \times \frac{9.1}{9.1 + 45.3} = -7.25$$

and, $$A_v = \frac{V_o}{V_i} = \frac{I_o R_L}{I_i R_i} = A_I \times \frac{R_L}{R_i}$$

$$= -7.25 \times \frac{5}{7.58} = -4.78$$

$$A_{V_S} = A_v \times \frac{R_i}{R_i + R_s} = (-4.78) \times \frac{7.58\ \text{K}}{17.58\ \text{K}} = -2.05$$

Problem 5.16: (a) Calculate R_i, A_v, and $A_I = -I_o/I_i$ for the circuit shown. Use the h-parameter values given in Table 5.1.

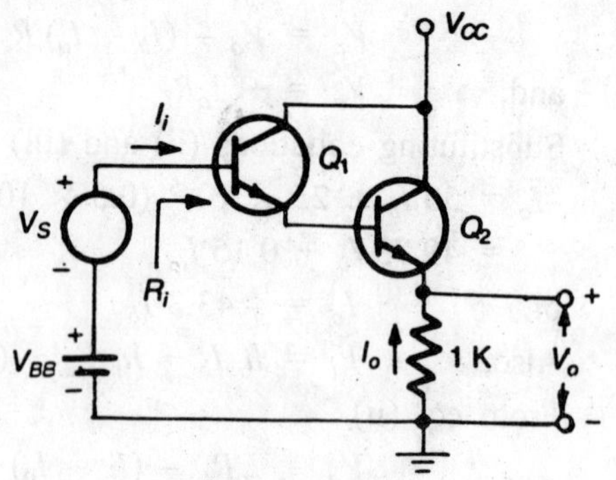

Fig. P. 5.16

Solution: From the equivalent circuit, We have

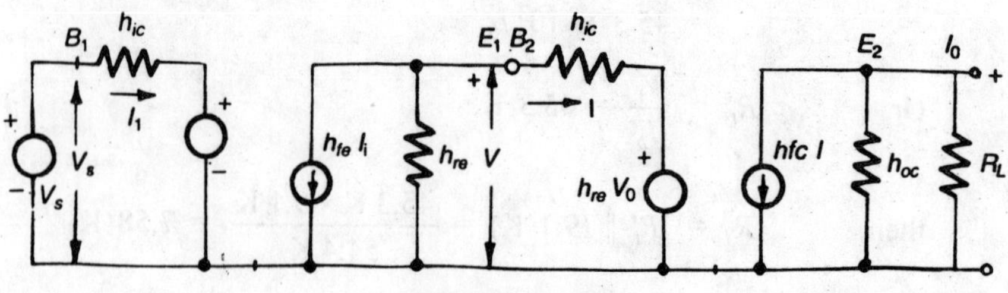

Fig. P. 5.16 (a)

$$I_o = h_{fc}I + h_{oc}V_o$$

and $\quad V_o = -I_oR_L \quad$ ...(ii)

Then, $\quad I_o = h_{fc}I - h_{oc}I_oR_L$

$$\Rightarrow \quad -\frac{I_o}{I} = AI_1 = \frac{h_{fc}}{1+h_{oc}R_L} \quad \text{...(iii)}$$

[–ve indicates both I_0 and I are in opposite direction]

Also, $\quad I = -[h_{fc}I_i + h_{oc}V] \quad$ ...(iv)

and, $\quad V = h_{ic}I + h_{rc}V_o \quad$ ...(v)

Substituting eq. (ii) in eq. (v), we get

$$V = h_{ic}I - h_{rc}I_oR_L$$

Using eq. (iii), we have

$$V = h_{ic}I - h_{rc}AI_IR_LI$$

$$\frac{V}{I} = [h_{ic} - h_{rc}AI_I R_L I] \quad \text{...(vi)}$$

Substituting expression for V in eq. (iv), we get

$$I = -[h_{fc}I_i + h_{oc}(h_{ic} - h_{rc}AI_IR_L)I]$$

$$\text{or} \quad \frac{I}{I_i} = A_{I_2} = \frac{-h_{fc}}{h_{oc}\left[h_{ic} - h_{rc}A_I R_L\right]+1} \quad \text{...(vii)}$$

Now, overall current gain, $A_I = A_{I_1} \times A_{I_2}$

$$\text{or,} \quad \frac{-I_e}{I_i} = \frac{-I_e}{I} \times \frac{I}{I_i}$$

From eq. (iii) and (vii), we have

$$A_I = \frac{-h_{fe}}{1+R_Lh_{oc}} \times \frac{-h_{fc}}{h_{oc}\left[h_{ic} - h_{rc}A_{I_1}R_L\right]+1}$$

$$= \frac{51}{1+10^3 \times 25 \times 10^{-6}} \times \frac{51}{25 \times 10^{-6}\left[1.1 \times 10^3 + \dfrac{h_{rc}R_Lh_{fc}}{1+R_Lh_{oc}}\right]+1}$$

$$= 49.8 \times 22.5 = 1120$$

$$R_i = \frac{V_s}{I_i} = \frac{I_ih_{ic} + h_{rc}V}{I_i}$$

Using eq. (vi), we have

$$R_i = \frac{h_{ic}I_i + h_{rc}\left(h_{ic} - h_{rc}A_{I_1}R_L\right)I}{I_i}$$

From equation (viii),

$$I = A_{I_2} \times I_i$$

Then, $R_i = h_{ic} + h_{rc}\,[h_{ic} - h_{rc}AI_1R_L]\,A_{I_2}$

$$= [1.1 + 50.9 \times 22.5] \times 10^3 = 1.146\ \text{M}\Omega$$

$$\therefore \text{ Voltage gain} = \frac{V_o}{V_s} = \frac{R_e I_o}{R_i I_i} = A_I \times \frac{R_c}{R_i} = \frac{1120\times10^3}{1.146\times10^3} = 0.98$$

Problem 5.17:
For the circuit shown, with the transistor parameters specified in Table 5.1, calculate

$A_I = \dfrac{I_o}{I_i}$, A_V, A_{V_s}, and R_i.

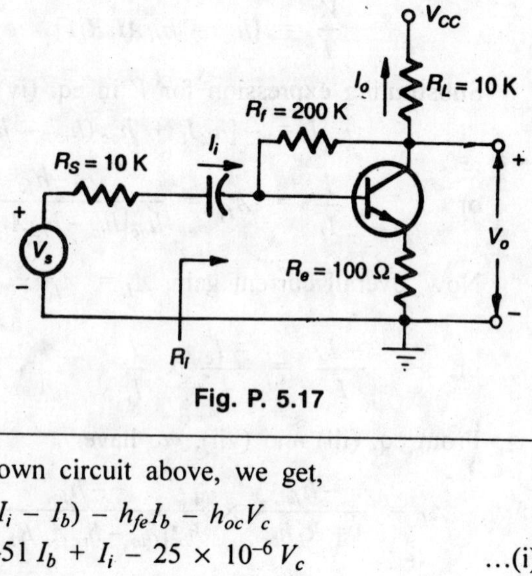

Fig. P. 5.17

Solution: From the shown circuit above, we get,

$$I_o = (I_i - I_b) - h_{fe}I_b - h_{oc}V_c$$

$$= -51\,I_b + I_i - 25 \times 10^{-6}\,V_c \qquad \ldots\text{(i)}$$

$$V_c = V_o + (I_o - I_i)\,R_e$$

$$= V_o + 100\,I_o - 100\,I_1 \qquad \ldots\text{(ii)}$$

and $\qquad V_o = I_oR_L = 10^4\,I_o \qquad \ldots\text{(iii)}$

and eq. (ii) and (iii),

$$I_o = -51\,I_b + 1.0025\,I_c - 0.2525\,I_o$$

$$(1.2525)\,I_o = -51\,I_b + 1.0025\,I_i$$

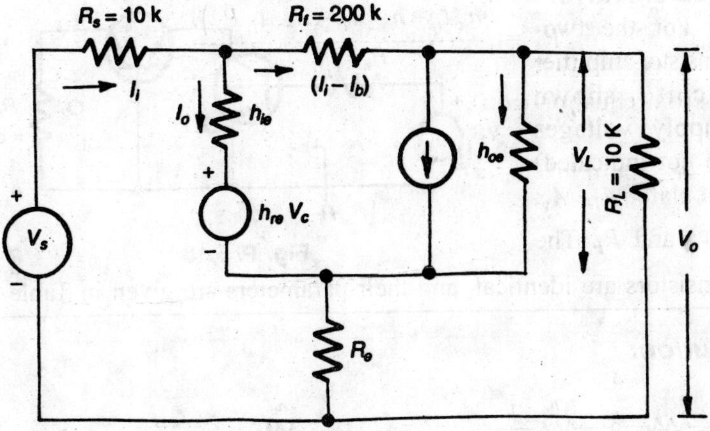

Fig. P. 5.17 (a)

Also from the circuit,

$$V_i = h_{ie}I_b + h_{re}V_c + (I_I - I_o)\, R_e \quad \text{...(iv)}$$

and $$V_i = R_f(I_i - I_b) + V_o \quad \text{...(v)}$$

From eq. (iv) & (v), solving we get,

$$I_o = -\,13.0\, I_i \quad \text{...(vi)}$$

and $$I_b = 0.339\, I_i \quad \text{...(vii)}$$

$$\therefore \quad A_I = \frac{I_o}{I_i} = -\,13.0$$

from eq. (v) and (vii),

$$V_i = 200 \times 10^3\ (1 - 0.339)\, I_i + 10^4\ (-\,13\, I_i)$$
$$= 2.2 \times 10^3\, I_i$$

$$\therefore \quad R_i = \frac{V_i}{I_i} = 2.2\text{ K}$$

and $$A_V = \frac{A_I R_2}{R_i} = \frac{-13 \times 10}{2.2} \cong -\,59$$

$$A_{V_s} = \frac{-\,59 \times 2.2}{2.2 + 10} = -\,10.65$$

Problem 5.18:
(a) For the two-transistor amplifier circuit shown (supply voltages are not indicated) calculate A_I, A_v, A_{V_s}, and R_i. The transistors are identical, and their parameters are given in Table 5.1.

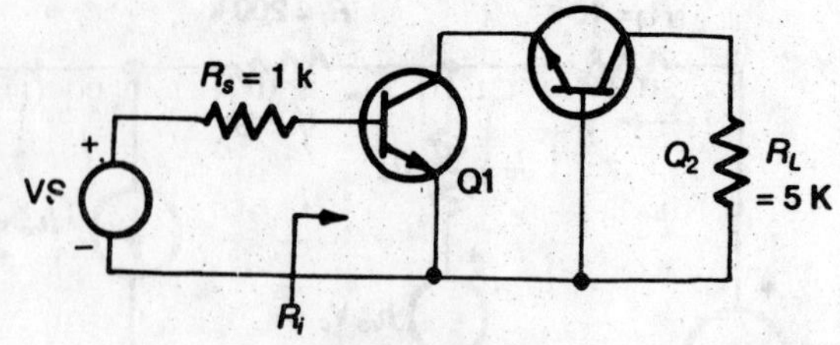

Fig. P. 5.18

Solution:

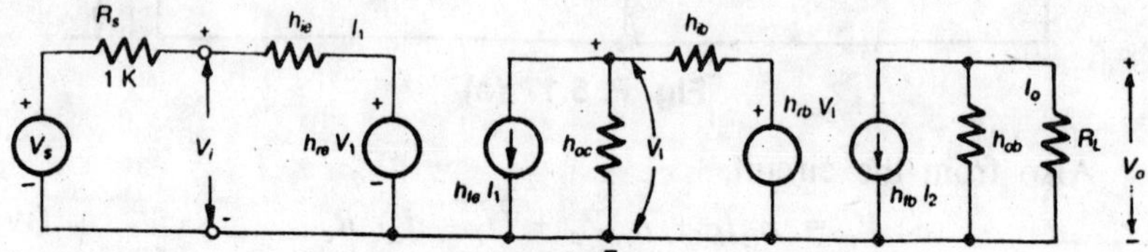

Fig. P. 5.18 (a)

From the circuit,

$$V_i = h_{ie}I_1 + h_{re}V_1 \quad \text{...(i)}$$

$$V_1 = h_{ib}I_2 + h_{rb}V_o \quad \text{...(ii)}$$

$$V_o = -I_oR_L = -h_{fb}I_2R_L \ [\because \frac{1}{h_{ob}} \gg R_L] \quad \text{...(iii)}$$

and, $$I_2 = -h_{oe}V_1 - h_{fe}I_1 \quad \text{...(iv)}$$

Substituting equation (ii) and (iii) in eq. (iv), we have

$$I_2 = -h_{fe}I_1 - h_{oe}\,(h_{ib}I_2 + h_{rb}V_o)$$

$$I_2 = -h_{fc}I_1 - h_{oe}\,(h_{ib}I_2 - I_oR_Lh_{rb})$$

$$= -50\,I_1 - h_{oe}\,[h_{ib}I_2 - h_{fb}I_2R_Lh_{rb}]$$

Putting,

$$h_{oe} = 24 \times 10^{-6}\ \mu A/V$$

$$h_{ib} = 21.6\ \Omega$$

$$h_{fb} = -0.98$$

$$h_{rb} = 2.9 \times 10^{-4}$$

$$R_L = 5K$$

$$h_{ie} = 1.1\ K$$

$I_2 = -50I_1, -24 \times 10^{-6}\,[21.6I_2 + 0.98 \times 2.9\ 10^{-4} \times 5 \times 10^3 \times I_2]$
$= -49.97\ I_1$

Now, substituting eq. (ii) and (iii) in eq (i), we have,

$$V_i = h_{ie}I_1 + h_{re}\,(h_{ib}I_2 + h_{rb}V_o)$$
$$= h_{ie}I_1 + h_{re}\,(h_{ib}I_2 + h_{rb}\,(-h_{fe}I_2R_L))$$

Using eq. (v),

$V_i = h_{ie}I_1 + h_{re}\,[h_{ib} - h_{rb}h_{fe}\,R_L] \times (-49.97)\,I_1$

or, $$R_i = \frac{V_i}{I_i} = h_{re} + h_{re}\,[h_{ib} - h_{rb}h_{fe}R_L] \times (-49.97)$$

$= 1.1 \times 10^3 + 2.5 \times 10^{-4}\,[21.6 - 2.9 \times 10^{-4} \times 50 \times 10^3] \times (-49.97)$

$R_i = 1.1 \times 10^3$

Now, $$A_I = -\frac{I_o}{I_1} = -\frac{I_o}{I_2} \times \frac{I_2}{I_1} = -h_{fb} \times (-49.97) \approx 49$$

$\therefore$ Voltage gain,

$$A_V = \frac{A_I R_L}{R_i} = \frac{49 \times 5}{1.1} \approx 222$$

and, $$A_{V_S} = A_v \times \frac{R_i}{R_i \times R_s} = 222 \times \frac{1.1}{1.1 \times 1} \approx 116$$

> **Problem 5.19:** (a) Consider a CB connection with $R_s = 2$ K and $R_L = 4$ K. Find the exact and approximate values of A_I, A_v, A_{V_S}, R_i and R_o'.
> (b) Repeat part *a* for the CE connection.
> (c) Repeat part *a* for the CC connection.

Solution: Exact analysis :

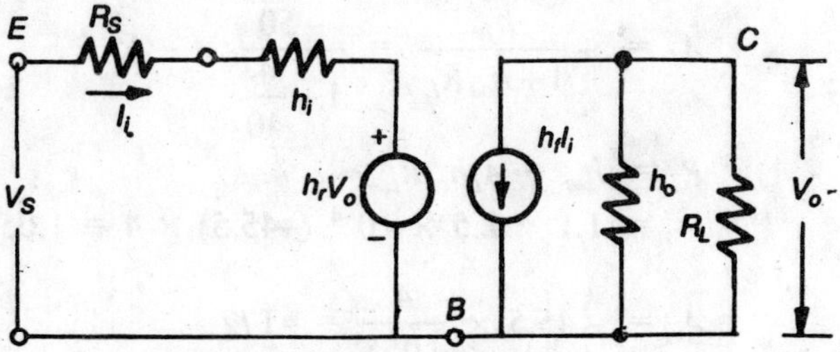

Fig. P. 5.19

$$A_I = -\frac{h_{fb}}{1+h_{ob}R_L} = -\frac{(0.98)}{1+0.49\times10^{-6}\times4\times10^{3}} \cong 0.98$$

$$R_i = h_{ib} + h_{rb}A_I R_L$$
$$= 21.6 + 2.9 \times 10^{-4} \times 0.98 \times 4 \times 10^3 = 22.7\ \Omega$$

$$A_v = A_i\frac{R_L}{R_i} = 0.98 \times \frac{4000}{22.7} = 173$$

$$\gamma_o = h_{ob} - \frac{h_{fb}\times h_{rb}}{h_{fb}+R_s} = 0.627 \times 10^{-6};$$

$$R_o = \frac{1}{\gamma_o} = 1.59\ \text{M}$$

$$\therefore \quad R_o' = 1.59 \parallel 4\ \text{K} \cong 4\ \text{K}$$

Now, $$A_{V_S} = A_v\frac{R_i}{R_i+R_s} = 173 \times \frac{22.7}{2022.7} = 1.94$$

Approx. solution

$$A_I = -h_{fb} = 0.98$$
$$R_i = h_{ib} = 21.6\ \Omega,$$

$$A_v = -h_{fb}\frac{R_L}{h_{ib}} = 0.98 \times \frac{4000}{21.6} = 182$$

$$R_o \to \infty \text{ and } R_i' = 4\text{K};$$

$$A_{V_s} = \frac{182\times21.6}{2021.6} = 1.94$$

(b) For CE-connection :

Exact analysis :

$$A_I = -\frac{h_{fe}}{1+h_{oe}R_L} = \frac{-50}{1+\frac{4}{40}} = -45.5$$

$$R_i = h_{ie} + A_I h_{re} R_L$$
$$= 1.1 + 2.5 \times 10^{-4}\ (-45.5) \times 4 = 1.055\ \text{K}$$

$$A_v = -45.5 \times \frac{4}{1.055} = -172$$

$$\gamma_o = 2.5 \times 10^{-4} - \frac{50 \times 2.5 \times 10^{-4}}{1100 + 2000}$$

$$= 2.1 \times 10^{-4}$$

$$R_o = 47.6\ \text{K};$$

$$R'_o = 47.6 \parallel 4\ \text{K} = 3.7\ \text{K}$$

$$A_{V_S'} = \frac{-172 \times 1055}{3055} = -59.5$$

Approximate solution :

$$A_I = -h_{fe} = -50,$$

$$R_i = h_{ie} = 1.1\ \text{K}$$

$$A_v = -50 \times \frac{4}{1.1} = -182$$

$$A_{V_S} = -182 \times \frac{1.1}{3.1} = -64.5;$$

$$R_o = \infty,$$

$$R'_o = 4\text{K}$$

(c) CC-connection :

Exact analysis :

$$A_I = -\frac{h_{fc}}{1 + h_{oc} R_L} = 46.3$$

$$R_i = 1.1 + 1 \times 46.3 \times 4 = 186.3\ \text{K}$$

$$A_v = 1 - \frac{h_{ie}}{R_i} = 1 - 0.006 = 0.994$$

$$\gamma_o = 2.5 \times 10^{-6} + \frac{51 \times 1}{3100}$$

$$= 1.67 \times 10^{-2};$$

$$R_o = \frac{1}{1.67 \times 10^{-2}} = 60\ \Omega$$

$$R'_o = 60 \parallel 4000\ \Omega \cong 60\ \Omega$$

$$A_{V_S} = 0.994 \times \frac{186.3}{188.3} \cong 0.984$$

Approx analysis :

$$R_i = 1.1 \text{ K} + 51 \times 4 = 205 \text{ K}$$

$$A_I = (1 + h_{fe}) = 51$$

$$A_v = 1 - \frac{1.1}{205} = 0.995,$$

$$R_o = \frac{R_s + h_{ie}}{1 + h_{fe}} = 60.8 \ \Omega$$

$$A_{V_S} = 0.995 \times \frac{205}{207} = 0.985$$

$$R_o' = 60.8 \parallel 4000 \ \Omega \cong 60 \ \Omega$$

Problem 5.20: For the circuit shown, compute A_1, A_v, A_{V_S}, R_i, and R_o',

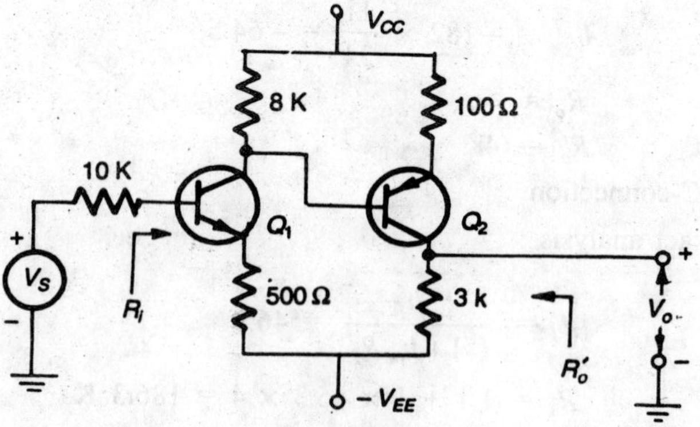

Fig. P. 5.20

Solution: Here it is advantageous to start calculation from the stage II side.

Here $\quad (R_{e_2} + R_{e_2})\, h_{oe} = 3.1 \times 10^3 \dfrac{1}{40 \times 10^3}$

$$\cong 0.0775 \ll 0.1$$

Hence, the approximate analysis can be used

Stage II :

$$A_{I_2} = -h_{fe} = -50$$

$$R_{i_2} = h_{ie} + (1 + h_{fe}) R_e$$

$$= 1.1 + 51 \times 0.1 = 6.2 \text{ K}$$

$$A_{V_2} = -\frac{h_{fe} R_{L_2}}{Ri_2}$$

$$= -\frac{50 \times 3}{6.2} = 24.2;$$

$$R_{o_2} = \infty$$

Now **Stage I :**

$$R_{L_1} = 8 \text{ K} \parallel 6.2 \text{ K} = 3.49 \text{ K};$$

$$(R_{L_1} + R_{e_1}) h_{oe} = 0.0999 \ll 0.1$$

Hence, approximate solution can be used.

$$A_{I_1} = -50$$

$$R_{i_1} = 1.1 \text{ K} + 51 \times 0.5 \text{ K} = 26.6 \text{ K}$$

$$A_{V_2} = -\frac{50 \times 3.40 \text{ K}}{26.6 \text{ K}} = -6.56,$$

$$R_{o_1} = \infty$$

Now calculating the parameters of whole stage

Entire amplifier :

$$A_v = A_{v_1} \times A_{v_2} = (-6.56)(-24.2) \cong 159$$

$$R_i = R_{i_1} = 26.6 \text{ K},$$

$$R_o' = 3\text{K}$$

$$A_I = \frac{A_v \times R_i}{R_{L_2}} = \frac{159 \times 26.6 \text{ K}}{3 \text{ K}}$$

$$= 1410$$

$$A_{V_S} = A_v \frac{R_i}{R_i + R_s} = \frac{159 \times 26.6}{36.6}$$

$$= 115$$

Problem 5.21: For the two-stage amplifier shown calculate A_V, A_{V_S}, R_i, and R_o'. Neglect the effect of all capacitances.

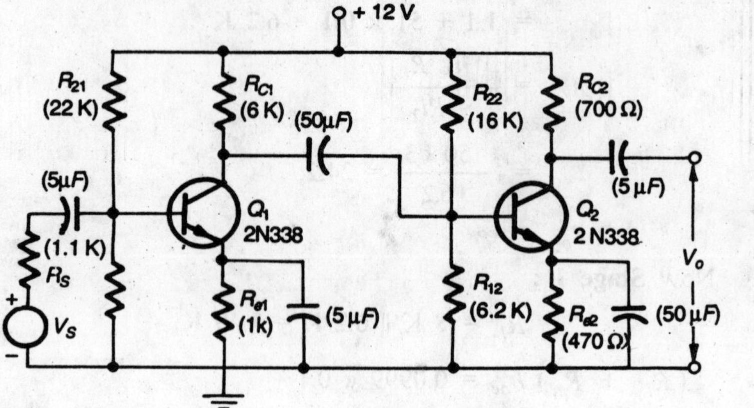

Fig. P. 5.21

Solution:

Stage II

Since $h_{oc}R_{L_2'} = \dfrac{0.7}{40} \leq 0.1$

So, we can use approximate analysis.

Now, drawing the equivalent circuit.

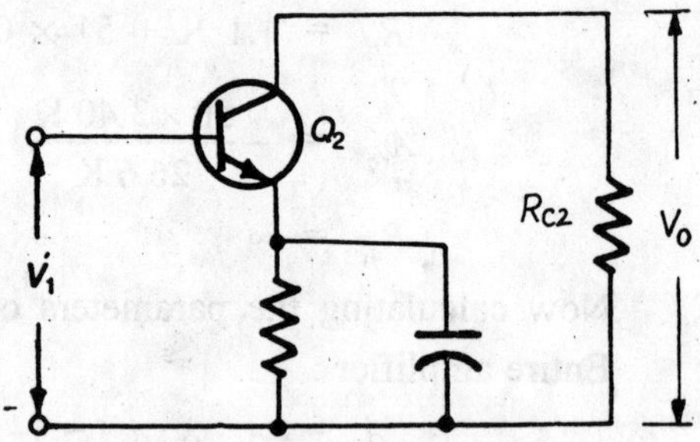

Fig. P. 5.21 (a)

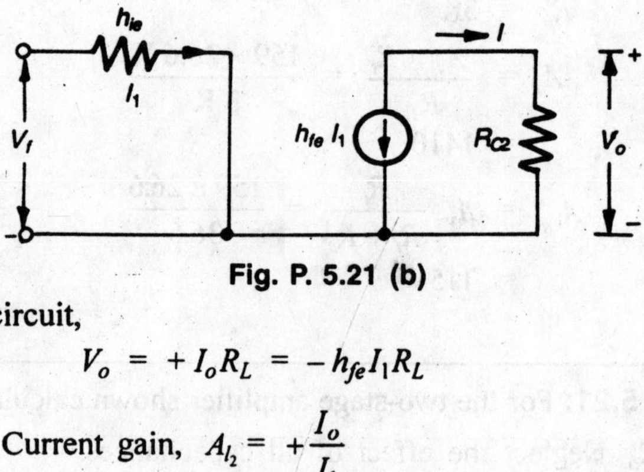

Fig. P. 5.21 (b)

From the circuit,

$$V_o = +I_oR_L = -h_{fe}I_1R_L$$

Then, Current gain, $A_{I_2} = +\dfrac{I_o}{I_1}$

$$= -h_{fe} = -50$$

and, $\quad V_1 = h_{ie} I_1$

or, $\quad R_{i_2} = \dfrac{V_1}{I_1} = h_{ie}$

and, voltage gain for stage II

$$A_{V_2} = \frac{A_{I_2} \times R_L}{R_{i_2}} = \frac{-50 \times 0.7}{1.1} = -31.9$$

Now, $\quad \gamma_o = h_{oe} - \dfrac{h_{fe} h_{re}}{h_{ice} + R_s} = 0$

then, $\quad R_{o_2} = \dfrac{1}{\gamma_o} = \infty$

Stage I: In this stage, circuit drawn is shown below.

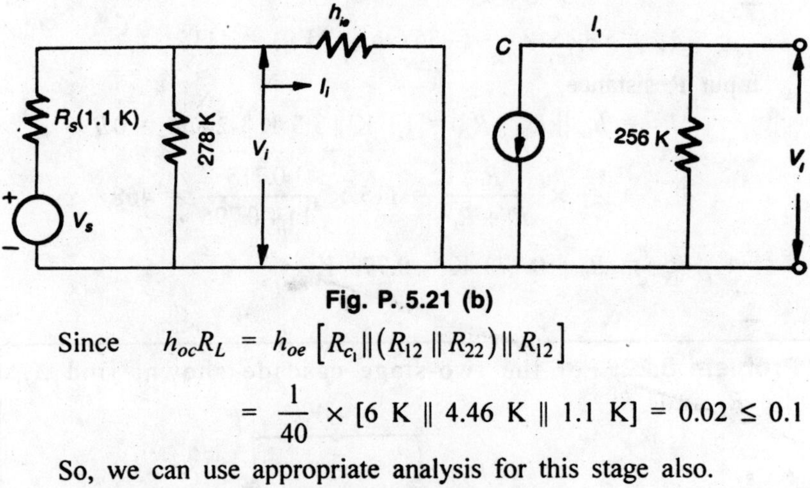

Fig. P. 5.21 (b)

Since $\quad h_{oc} R_L = h_{oe} \left[R_{c_1} \| (R_{12} \| R_{22}) \| R_{12} \right]$

$$= \frac{1}{40} \times [6 \text{ K} \| 4.46 \text{ K} \| 1.1 \text{ K}] = 0.02 \leq 0.1$$

So, we can use appropriate analysis for this stage also.

Now drawing the equivalent circuit

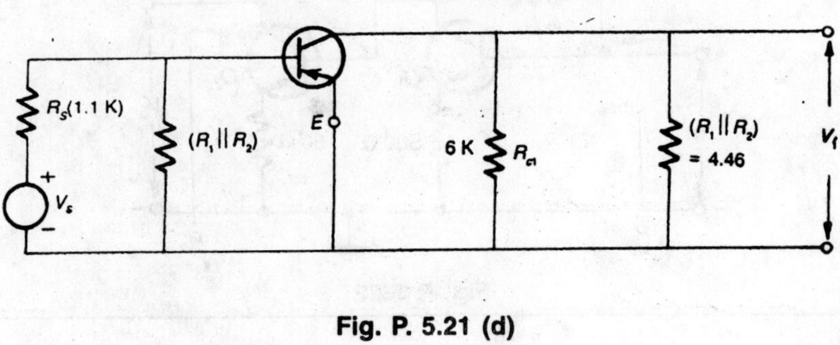

Fig. P. 5.21 (d)

From the circuit,

$$A_{I_1} = \frac{I_1}{I_i} = -h_{fe} = -50$$

$$R_{i_1} = h_{ie} = 1.1 \text{ K}$$

and $$R_o = \frac{1}{\gamma_o} = \frac{1}{o} = \infty$$

$$A_{V_1} = \frac{A_{I_1} \times R_{L_1}}{R_{i_1}} = \frac{-50 \times [(16 \text{ K}) \| (4.46 \text{ K}) \| (1.1 \text{ K})]}{1.1 \text{ K}}$$

$$= \frac{-50 \times 0.773 \text{ K}}{1.1} = -35$$

∴ Overall voltage gain,

$$A_V = A_{V_1} \times A_{V_2} = (-35) \times (-31.9) = 1115$$

Input Resistance,

$$R_i = h_{ie} \| R_{11} \| R_{21} = 1.1 \text{ K} \| 3.3 \text{ K} \| 22 \text{ K} = 0.795 \text{ K}$$

$$A_{V_S} = A_V \times \frac{R_i}{R_i + R_s} = 115 \times \frac{0.715}{1.1 + 0.795} = 468$$

$$R'_o = R_o \| 0.773 \text{ K} = 0.773 \text{ K}$$

Problem 5.22: For the two-stage cascade shown, find A_I, A_V, R_i and R'_o.

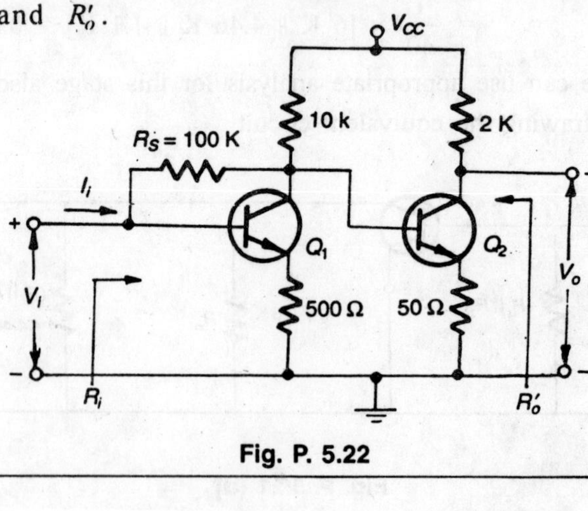

Fig. P. 5.22

Solution:

Stage II:

Here, $h_{oe}\ (R_{L_2} + R_{e_2}) = 2.5 \times 10^{-6}\ (2000 + 50) = 0.051 < 0.1$

So, for stage II, use approximate analysis.
Drawing equivalent circuit for stage II
From the circuit,

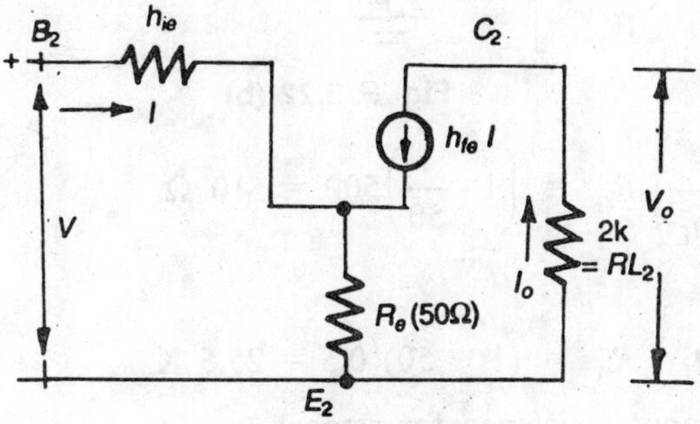

Fig. P. 5.22 (a)

$$A_{I_2} = \frac{-I_o}{I} = \frac{-h_{fe} I}{I} = -h_{fe} = -50 \quad \text{...(i)}$$

$$R_{i_2} = \frac{V}{I} = h_{ie} + (1 + h_{fe})\ R_e \quad \text{...(ii)}$$

$$= 1.1\ \text{K} + 51 \times 0.05\ \text{K} = 3.65\ \text{K}$$

$$A_{v_2} = \frac{V_o}{V} = \frac{-I_o \times R_{L_2}}{1\left[h_{ie} + (1 + h_{fe}) R_e\right]}$$

$$= A_{I_2} \times \frac{2}{3.65} = -50 \times \frac{2}{3.65} = -27.4$$

For stage I : Using Miller's theorem drawing at first stage,
Since,

$h_{oe}\ (R_{12} \parallel 10\ \text{K}) = 0.067 < 0.1$,
We can use approximate analysis.

$$A_{i_1} = -h_{fe} = -50$$

gives,

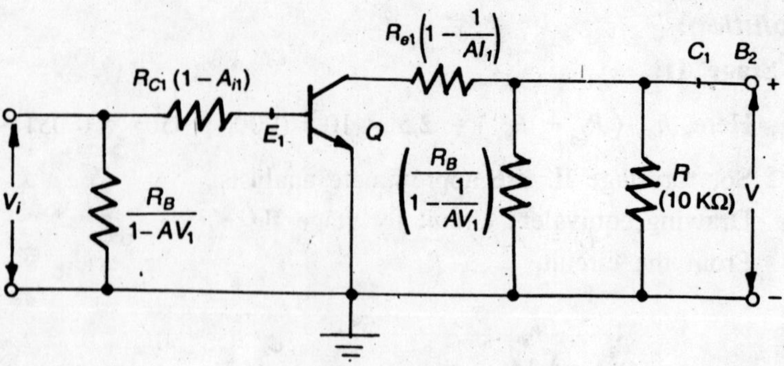

Fig. P. 5.22 (b)

$$\left(1-\frac{1}{A_{I_1}}\right) R_{e_1} = \left(1+\frac{1}{50}\right) 500 = 510\ \Omega$$

and,

$$\left(1-A_{I_1}\right) R_{e_1} = (1+50)\ 0.5 = 25.5\ \text{K}$$

Then, input resistance for stage I,

$$R_{i_1} = h_{ie} + \left(1-A_{I_1}\right) R_{e_1} = 1.1 + (1+50)\ 0.5 = 26.6\ \text{K}$$

$$A_{V_1} = A_{I_1} \times \frac{R_{L_1}}{R_{i_1}} = -50\ \frac{\left[2.67 \,\|\left(R_B \Big/ \left(1-\frac{1}{A_v}\right)\right)\right]\text{K}}{26.6\ \text{K}}$$

$$= \frac{-50}{26.6} \times \frac{2.67 \times 100 \times \frac{1}{1-\frac{1}{A_{V_1}}}}{2.67 + \frac{100}{1-\frac{1}{A_{V_1}}}}$$

or,

$$71\ A_{v_1} + 2660\ A_{v_1} - 71 = -13{,}350$$

$$A_{V_1} = -\frac{13{,}279}{2731} = -4.86$$

Thus, $\dfrac{R_B}{1-A_{V_1}} = \dfrac{100}{1+4.86} = 17\ \text{K}$

and, $R_i = 17\ \text{K} \,\|\, 26.6\ \text{K} = 10.4\ \text{K}$

$$R'_o = 2\text{ K} \parallel R_{o_2} = 2\text{ K}$$

Voltage gain overall,

$$A_V = A_{V_1} \times A_{V_2} = (-4.86)\ (-27.4) = 133$$

and, overall current gain,

$$A_I = A_V \times \frac{R_i}{R_{L_2}} = 133 \times \frac{10.4}{2} = 690$$

Problem 5.23:
For the circuit shown, find the voltage gain V_o/V_s and input impedance as a function of R_s, b, R_e, and R_L. Assume that $h_{oe}(R_e + R_L) \leq 0.1$

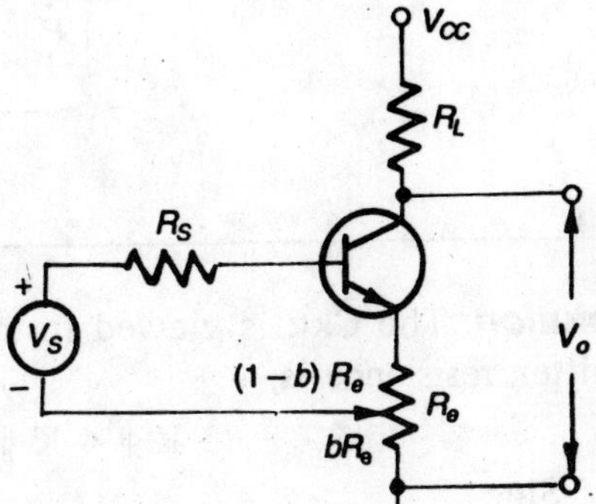

Fig. P. 5.23

Solution: Since, $h_{oe}(R_L + R_e) \leq 0.1$, we use approximate model of the transistor. The resulting small signal equivalent circuit shown below. From the circuit, we have,

$$V_s = [R_s + h_{ie} + (1 - \beta)\ R_e\ (1 + h_{fe})]\ I \quad \ldots\text{(i)}$$

then,

$$A_V = \frac{-h_{fe} I R_L}{I\left[R_s + h_{ie} + (1+h_{fe})(1-\beta)\ R_e\right]}$$

$$= \frac{-h_{fe} R_L}{R_s + h_{ie} + (1+h_{fe})(1-\beta)\ R_e} \quad \ldots\text{(ii)}$$

from eq (i), input resistance,

$$R_i = \frac{V_s}{I}$$

$$= R_s h_{ie}\ (1 + h_{fe})\ (1 - \beta)\ R_e$$

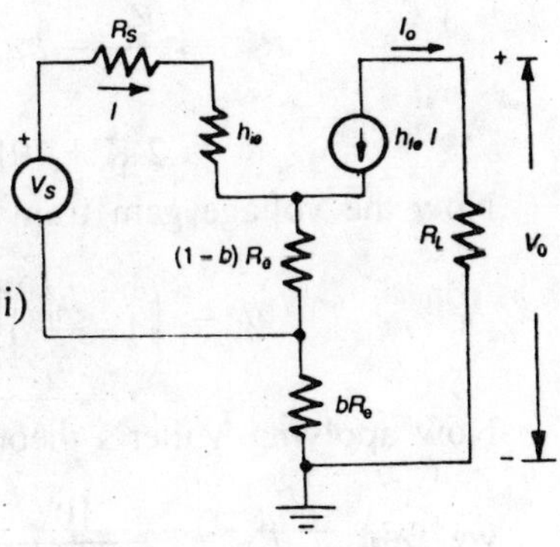

Fig. P. 5.23 (a)

Problem 5.24:
For the bootstrap circuit shown, calculate $A_I \equiv I_o/I_i$, R_i and A_v. The transistor parameters are h_{ie} = 2 K, h_{fe} = 100, $1/h_{oe}$ = 40 K, and h_{re} = 2.5 × 10^{-4}.

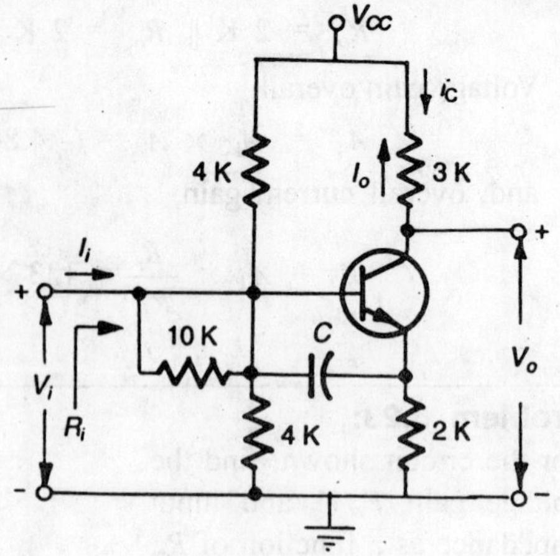

Fig. P. 5.24

Solution: The Ckt. is viewed from emitter side, and the effective emitter resistance is,

$$R'_e = (2\text{ K} \parallel 4\text{ K} \parallel 4\text{ K}) = 1\text{ K}$$

Since,

$$h_{oe}\,(R_c + R'_e) = \frac{4}{40} = 0.1$$

Hence, we can use the approximate analysis for the convenience

Thus, $$A'_I = \frac{I_o}{I_b} = -h_{fe} = -100$$

$$R'_i = \frac{V_i}{i_b} = h_{ie} + (1 + h_{fe})\,R'_e$$

$$= 2\text{ K} + 101 \times 1 = 103\text{ K}$$

Now the voltage gain from base to emitter :

$$A'_v = \left(1 - \frac{h_{ie}}{R_i}\right) = 1 - \frac{2}{103} = 0.9806$$

Now applying Miller's theorem to 10 K

we find, $$R_1 = \frac{10}{1 - 0.9806} = 515\text{ K}$$

and $$R_2 = \frac{10K}{1 - \frac{1}{0.9806}} = -500 \text{ K}$$

Here the value of R_2 calculate should be connected in parallel to 2 K, which is neglected above.

$$\therefore \quad R_i = (103 \text{ K} \parallel 515 \text{ K}) = 85.6 \text{ K}$$

$$\therefore \quad A_I = \frac{I_o}{I_i} = \frac{I_o}{I_b} \times \frac{I_b}{I_i}$$

$$= -100\left(\frac{515}{515+103}\right) = -83.2$$

and $$A_V = -\frac{R_2}{R'_e} = -\frac{3 \text{ K}}{1 \text{ K}} = -3$$

Problem 5.25:
For the BJT circuit shown in Fig. 5.25 draw the small-signal model. Also derive the expression for the resistance seen between base and ground. Calculate its value.

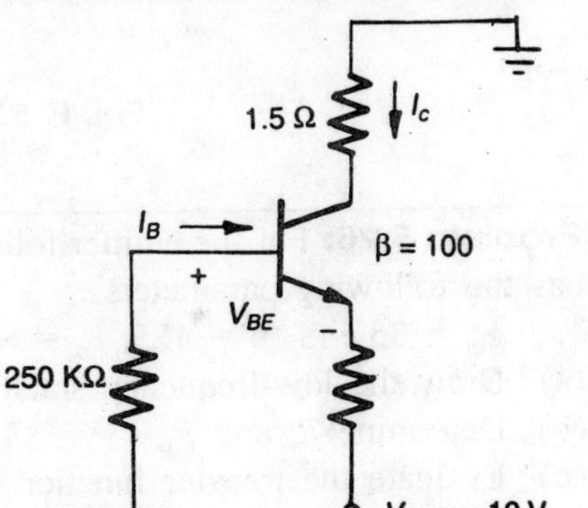

Fig. P. 5.25

Solution: We have from base-emitter loop :

$$250\, I_b + V_{BE} + (I_C + I_B) = 10$$

$$V_{BE} = 0.7\ V,\ I_C + I_B = 101\ I_B$$

$$\therefore \quad I_B = 26.5\ \mu\text{A}$$

$$I_c = 100 \times 26.5 \times 10^{-3} = 2.65 \text{ mA}$$

$$r_\pi = \beta \times \frac{25}{I_{c(mA)}} = 100 \times \frac{25}{2.65} = 943\ \Omega$$

From small-signal model in Fig. 5.25 (a) :

$$V_i = (r_x + r_\pi) I_b + (1 + \beta) R_E I_b$$

$$R_i = \frac{V_i}{I_b} = r_x + r_\pi + (1 + \beta) R_E$$

Substituting values ($r_x = 0$)

$$R_i = 943 + (1 + 100) \times 1 = 101.943 \text{ k}\Omega \approx 102 \text{ k}\Omega$$

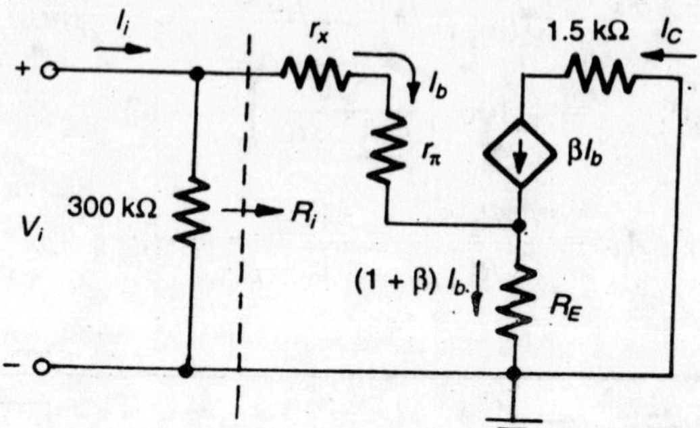

Fig. P. 5.25 (a)

Problem 5.26: For the emitter follower circuit of Fig. 5.26 the BJT has the following parameters :

$$g_m = 35 \text{ mS}, \ \beta = 125, \ r_o = \infty, \ r_x = 0$$

(a) Draw the low-frequency small-signal model of this circuit
(b) Determine R_i and R_o
(c) Evaluate the transfer function V_o/V_i.

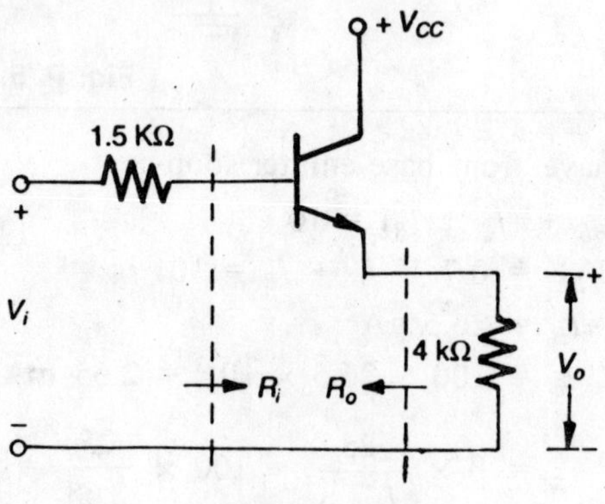

Fig. P. 5.26

Solution:

$$r_\pi = \frac{\beta}{g_m} = \frac{125}{35} = 3.57 \text{ k}\Omega$$

(a) From small-signal model in Fig. 5.26 (a)

$$R_i = r_\pi + (1 + \beta) R_E = 3.57 + 126 \times 4 = 508 \text{ k}\Omega$$

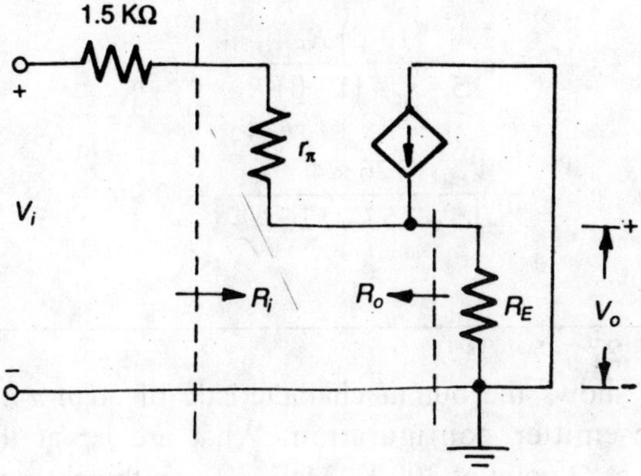

Fig. P. 5.26 (a)

(b)
$$R_o = \frac{V_o}{I_{SC}}$$

$$V_o = (1 + \beta) R_E I_b$$

$$I_b = \frac{V_i}{1.5 + r_\pi + (1+\beta)R_E}$$

$$V_o = (1 + \beta) R_E \times \frac{V_i}{1.5 + r_\pi + (1+\beta)R_E}$$

∴ Short circuiting output

$$I_b = \frac{V_i}{(1.5 + r_\pi)}$$

$$I_{SC} = \frac{(1+\beta) V_i}{(1.5 + r_\pi)}$$

$$R_o = \frac{V_o}{I_{SC}} = \frac{(1+\beta) R_E}{1.5 + r_\pi + (1+\beta) R_E} \times \frac{(1.5 + r_\pi)}{(1+\beta)}$$

$$\therefore \quad R_o = \frac{(1.5 + r_\pi)\, R_E}{(1.5 + r_\pi) + (1+\beta)\, R_E}.$$

Substituting values

$$R_o = \frac{(1.5 + 3.57) \times 4}{(1.5 + 3.57) + (1 + 125) \times 4} = 0.0398 \text{ k}\Omega = 39.8\ \Omega$$

(c)

$$V_o/V_i = \frac{(1+\beta)\, R_E}{1.5 + r_\pi + (1+\beta)\, R_E}$$

$$= \frac{126 \times 4}{1.5 + 3.57 + 126 \times 4} = 0.99$$

Problem 5.27:

Figure 5.27 shows the output characteristic of an *n-p-n* transistor in common-emitter configuration. What are h_{fe} at the Q-point. Estimate h_{ie} at Q-point at 300 K. Mark V_{CC} on the characteristic and identify AC and DC load lines. What are the values of R_C and R_E if the load lines are for the amplifier (CE) of Fig. 5.28 with $R_L = \infty$. What is the maximum unloaded, undistorted voltage swing approximately? What is the optimal Q-point for a load of 2 kΩ which is capacitively coupled to the output?

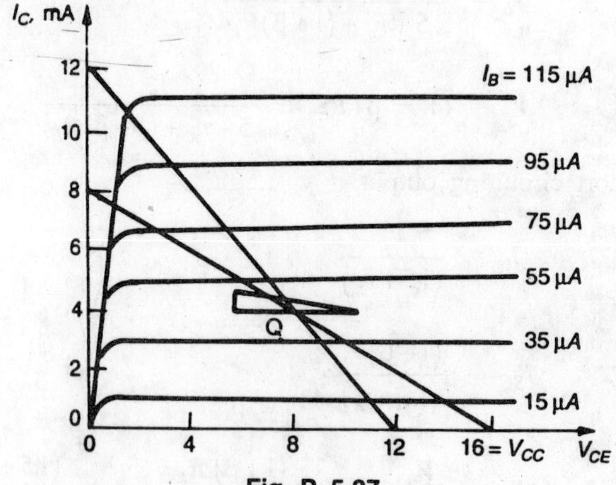

Fig. P. 5.27

Solution: We have :

$$h_{ie} = \left.\frac{\partial V_{BE}}{\partial I_B}\right|_{V_{CE}=const} = r_\pi, \quad h_{fe} = \left.\frac{\partial I_C}{\partial I_B}\right|_{V_{CE}=const.} = \beta$$

$$h_{oe} = \left.\frac{\partial I_C}{\partial V_{CE}}\right|_{I_B=const.} = r_o$$

We have for small increments about the Q-point,

$$h_{fe} \approx \frac{\Delta I_C}{\Delta I_B} = \frac{2\text{ mA}}{20\ \mu\text{A}} = \frac{2\times 10^3}{20} = 100$$

$$h_{oe} \approx \frac{\Delta I_C}{\Delta V_{CE}} = \frac{2\text{ mA}}{2\text{ V}} = \frac{2000\ \mu\text{A}}{2\text{ V}} = 1000\ \mu\text{S}$$

Note : I_C cannot be read accurately from the given characteristics.

Smaller slope indicates DC load line from the figure

$$V_{CC} = 16\text{ V}$$

$$R_{dc} = R_C + R_E =$$

$$\frac{16\text{ V}}{8\text{ mA}} = 2\text{ k}\Omega$$

As $\quad R_L = \infty$

$$R_{ac} = R_C = \frac{12\text{ V}}{12\text{ mA}} = 1\text{ k}\Omega$$

$$R_E = 2 - 1 = 1\text{ k}\Omega$$

For a load of 2 kΩ

$$R_{ac} = R_C \parallel R_L = \frac{1\times 2}{1+2} = \frac{2}{3} = 0.667\text{ k}\Omega$$

For optimal Q,

$$I_{CQ} = \frac{V_{CC}}{R_{ac}+R_{dc}} = \frac{16}{(2/3)+2} = \frac{16}{\frac{8}{3}} = 6\text{ mA}$$

$$V_{CEQ} = V_{CC} - I_{CQ}R_{dc}$$

$$= 16 - 6\times 2 = 4\text{ V}$$

Problem 5.28: For the CE amplifier of Fig. 5.28 place the Q-point optimally so that the load voltage has maximum undistorted swing. Deduce that the optimal Q–point is at

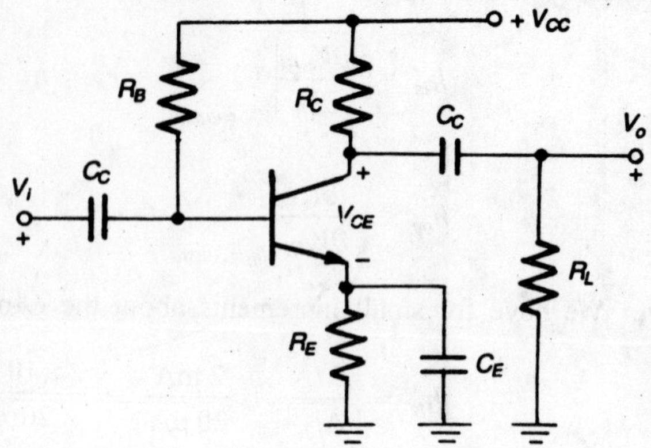

Fig. P. 5.28

$I_{CQ} = V_{CC}/R_{ac} + R_{dc}),$

$V_{CEQ} = V_{CC} - I_{CEQ}\, R_{dc}$

where $\quad R_{ac} = R_C \parallel R_L$

and $\quad R_{dc} = R_C + R_E.$

Note : Under ac conditions the capacitors act as short circuits, so that $R_{ac} = R_C \parallel R_L$ as different from $R_{dc} = R_C + R_E$.

Solution: Under ac conditions the bypass capacitor C_E short circuits R_E and the C_C also offers zero reactance thus connecting load R_L in parallel to R_C.

$\therefore \quad R_{dc} = R_C + R_E$

But $\quad R_{ac} = R_C \parallel R_L$

As $\quad R_{ac} < R_{dc}$

The slope of the ac load line is larger than that of dc load line. But the ac load line must pass through the Q-point on the dc load line.

Optimal Q-point means maximum undistorted ac swing. This would happen when V_{CEQ} is located at the mid-point of the intercept of ac load line on V_{CE}-axis. Of course the ac load line passes through the Q-point located on the dc load line. Such dc and ac load lines are drawn in Fig. 5.28 (a)

From figure 5.28 (a) for optimal Q,

$$OB = BD$$

$$OB = V_{CEQ} = V_{CC} - I_{CQ}R_{dc} \quad \text{...(i)}$$

$$DB = I_{CQ}R_{ac} \quad \text{...(ii)}$$

from the eqns. (i) and (ii) we get

$$V_{CC} - I_{CQ}R_{dc} = I_{CQ}R_{ac}$$

$$\therefore \qquad I_{CQ} = \frac{V_{CC}}{R_{dc} + R_{ac}} \qquad \text{...(iii)}$$

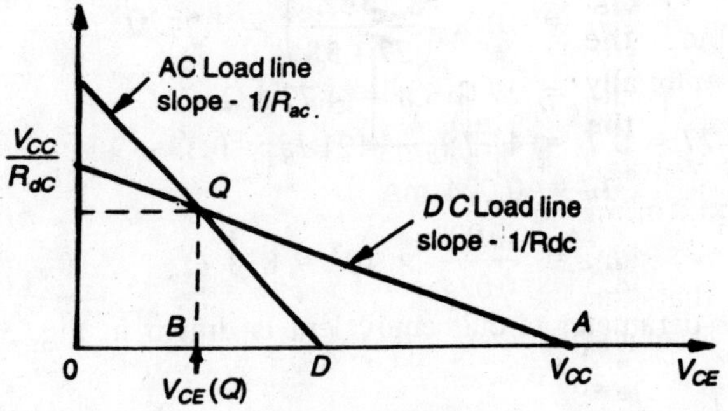

Fig. P. 5.28 (a)

Problem 5.29: The transistor of *CE* amplifier circuit of Fig. 5.29 has the following parameters

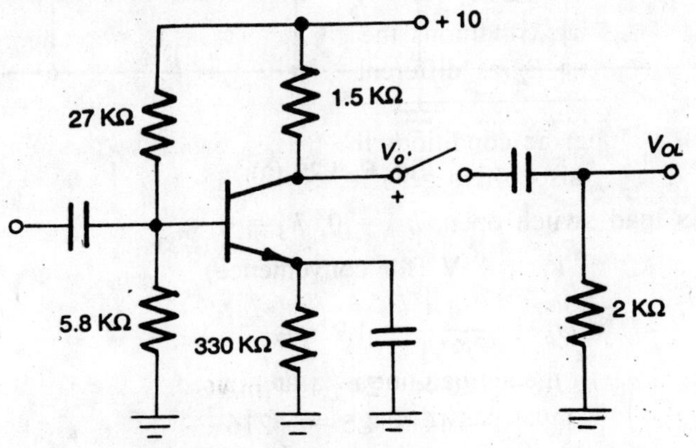

Fig. P. 5.29

$$h_{fe} = 120,$$
$$h_{ie} = 0.02/I_B$$
$$h_{re} = h_{oe} = 0$$

(a) Find h_{ie} and use the *h*-parameter circuit model to find the gain

$$A_{V_o} = \frac{V_o}{V_i};$$

V_o = signal component of the voltage at collector

(b) Find A_{VL} when the output switch is on.

If h_{oe} = 50 μS, and the source has an internal resistance of 2 kΩ, recalculate A_{VL}.

Solution: Using Thevenin equivalent of the potential divide in the circuit of Fig. 5.29

$$V_{BB} = 10 \times \left(\frac{5.8}{27+5.8}\right) = 1.77 \text{ V}$$

$$R_B = 27 \parallel 5.8 = 4.77 \text{ k}\Omega$$

$$1.77 - 0.7 = 4.77\, I_B + 121\, I_B \times 0.33$$

or
$$I_B = 0.024 \text{ mA}$$

$$h_{ie} = \frac{0.02}{0.024} \times 10^3 = 833 \ \Omega$$

The h-parameter circuit equivalent is drawn in Fig. 5.29 (a)

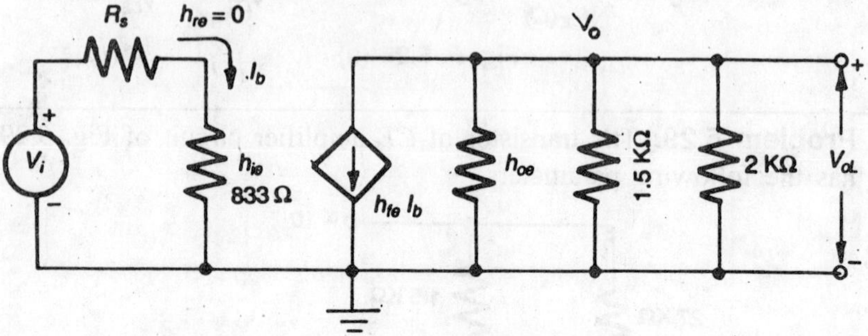

Fig. P. 5.29 (a)

(a) As load switch open, $h_{oe} = 0$, $R_S = 0$

Let
$$V_i = 1 \text{ V (for convenience)}$$

$$I_b = \frac{1}{0.833} = 1.2 \text{ mA}$$

$$h_{fe}I_b = 120 \times 1.2 = 144 \text{ mA}$$

$$A_{vo} = -144 \times 1.5 = -216$$

(b) As load switch on, $h_{oe} = 50\ \mu\text{S}$, $R_S = 2 \text{ k}\Omega$

$$\therefore \quad 1/h_{oe} = \frac{1}{50} \times 1000 = 20 \text{ k}\Omega$$

Effective resistance at load terminal is given by

$20 \parallel 1.50 \parallel 2 = 0.822 \text{ k}\Omega$

$$I_b = \frac{1}{2+0.833} = 0.353 \text{ mA}$$

$$h_{fe}I_b = 120 \times 0.353 = 42.36$$

$$A_{VL} = V_L = -42.36 \times 0.822 = -34.8$$

Problem 5.30: The amplifier circuit of Fig. 5.30 uses a silicon BJT with $\beta = 100$ with the values of other circuit elements given as under:

$R_1 = 60\ k\Omega$,
$R_2 = 140\ k\Omega$,
$R_s = 4\ k\Omega$,
$R_E = 3\ k\Omega$,
$R_C = 4\ k\Omega$;

the capacitors are assumed to be large; $V_{CC} = 10$ V

(a) Estimate the dc bias current I_C and V_{CE}.

(b) The amplifier is required to give a small-signal gain of 30. Specify the load resistance R_L (to be connected across V_o).

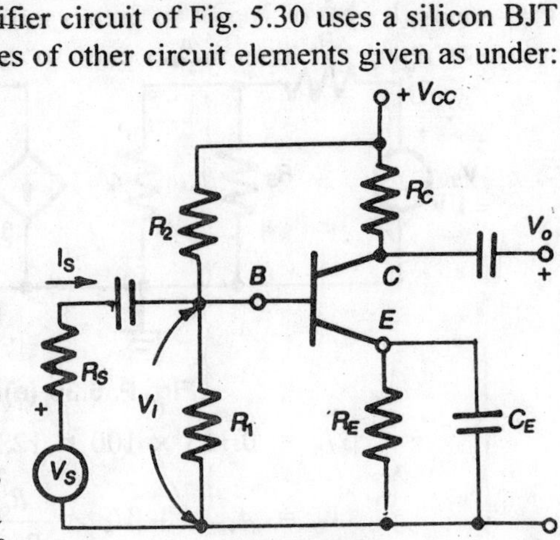

Fig. P. 5.30

Solution: (a)Applying DC analysis

From Fig. 5.30 all capacitors are taking as open circuit.

$$R_B = R_1 \parallel R_2 = 60 \parallel 140 = 42\ k\Omega$$

$$V_{Th} = \frac{10 \times 60}{60 + 140} = 3\ V$$

Round the BE loop

$$3 = 42\ I_B + 0.7 + (100 + 1) \times 3\ I_B$$

$$\therefore \quad I_B = 6.67 \times 10^{-3}\ mA$$

$$I_C = \beta_{IB} = 100 \times 6.67 \times 10^{-3} = 0.667\ mA$$

Round the CE loop

$$10 = 4 \times 0.667 + V_{CE} + 101 \times 3 \times 6.67 \times 10^{-3}$$

$$V_{CE} = +5.31\ V$$

(b) For small-signal analysis

The circuit model is drawn in Fig. 5.30 (a)

$$r_\pi = \frac{25 \times 100}{0.667} \times 10^{-3}$$

$$= 3.75\ k\Omega,\ R_B \parallel r_\pi = 42 \parallel 3.75 = 3.44\ k\Omega$$

$$I_b = 1 \times \frac{(R_B \| r_\pi)}{\left[R_S + (R_B \| r_\pi)\right] r_\pi} = \frac{3.44}{(4 + 3.44) \times 3.75} = 0.123\ mA$$

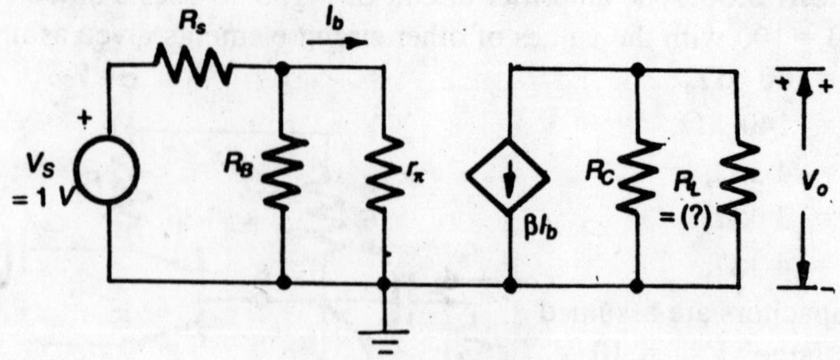

Fig. P. 5.30 (a)

$$\beta I_b = 0.123 \times 100 = 12.3 \text{ mA}$$

$$V_o = A_{vo} = -\beta I_b \times \frac{R_C R_L}{R_C + R_L}$$

$$-30 = -12.3 \times \frac{4R_L}{4 + R_L}$$

$$\therefore \quad R_L = 6.25 \text{ K}\Omega$$

Problem 5.31: In the amplifier circuit of Fig. 5.30 $\beta = 100$, $R_1 = 25$ kΩ, $R_2 = 100$ kΩ, $R_E = 2$ kΩ, $V_{CC} = +10$ V

(a) Estimate I_C (bias current) and therefrom calculate r_π.

(b) Specify R_C for midfrequency gain of $V_o/V_i = 160$.

(c) With R_C calculated as in part (b) and $R_s = 1$ kΩ, calculate gain V_o/V_s. All capacitors act as short circuits.

Solution: (a) From DC analysis :

All capacitors in Fig. 5.30 act as open circuits.

$$R_B = \frac{25 \times 100}{25 + 100} = 20 \text{ k}\Omega$$

$$V_{Th} = \frac{10 \times 25}{125} = 2 \text{ V}$$

Round the BE loop

$$2 = 20\, I_B + 0.7 + 101 \times 2\, I_B$$

or

$$I_B = 5.85 \text{ } \mu\text{A}$$

$$I_C = \beta I_B = 100 \times 5.85 \times 10^{-3} = 0.585 \text{ mA}$$

$$r_\pi = \frac{25 \times 100 \times 10^{-3}}{0.585} = 4.27 \text{ k}\Omega$$

(b) From small-signal analysis :

The small-signal circuit is drawn in Fig. 5.31

$$I_b = 1/4.27 = 0.234 \text{ mA}$$

$$-\beta I_b R_C = V_o = A_{vo}$$

$$-23.4 \times R_C = -160$$

$$R_C = 160/23.4 = 6.84 \text{ K}\Omega$$

(c) $R_C = 6.84$ kΩ,

$$R_s = 1 \text{ k}\Omega$$

Take $V_s = 1$ V

$$20 \parallel 4.27 = 3.52 \text{ k}\Omega$$

$$I_b = 1 \times \frac{3.52}{(1+3.52) \times 4.27} = 0.182 \text{ mA}$$

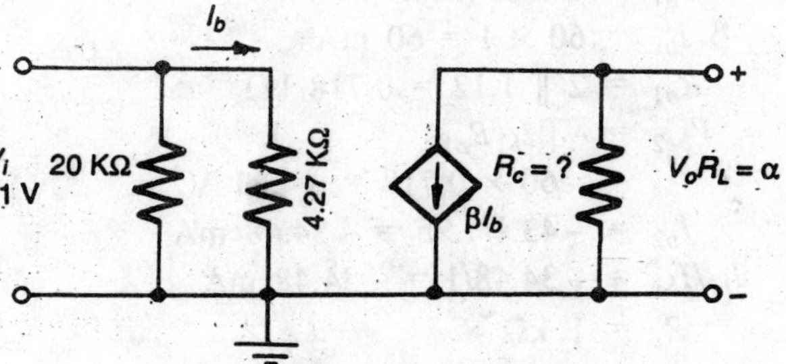

Fig. P. 5.31

$$\beta I_b = 18.2 \text{ mA (as calculated in part (c))}$$

$$A_{Vo} = V_o/V_s = -\beta I_b \times (R_C \parallel R_L);\ R_C \parallel R_L = 4.42$$

$$= -18.2 \times 4.42 = -80.4$$

Problem 5.32: For the amplifier circuit of Fig. 5.30 assume the following values of various parameters :

$\beta = 60$, $R_1 = 12$ kΩ, $R_2 = 100$ kΩ, $R_C = 2$ kΩ, $I_c = 1.2$ mA

(a) Find the mid-frequency current gain I_{b2}/I_{b1}, if the stage that follows employs a similar transistor with same circuit parameters.

(b) Predict the overall voltage gain of the first stage if $R_s = 1$ kΩ.

Solution: (a) From small-signal circuit model is drawn in Fig. 5.32

$$r_\pi = \frac{25 \times 60}{1.2} \times 10^{-3} = 1.25 \text{ k}\Omega$$

$$\therefore \quad R_L = \infty$$

$$R_B = R_1 \parallel R_2 = \frac{12 \times 100}{112} = 10.7 \text{ k}\Omega$$

$$R_B \parallel r_\pi = 10.7 \parallel 1.25 = 1.12 \text{ k}\Omega$$

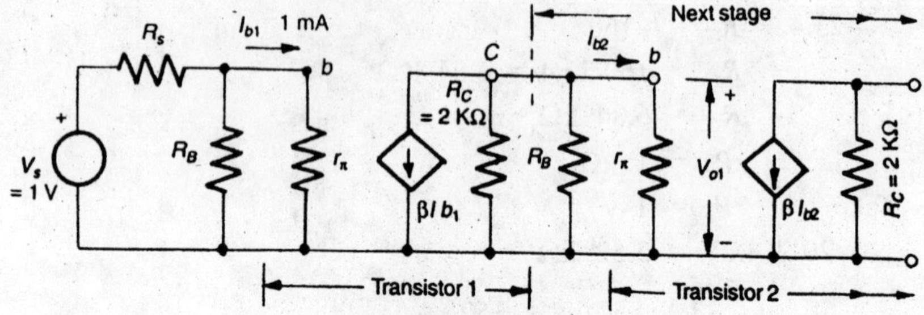

Fig. P. 5.32

$$I_{b1} = 1 \text{ mA (say)}$$

$$\beta I_{b1} = 60 \times 1 = 60 \text{ mA}$$

$$R_{o1} = 2 \parallel 1.12 = 0.718 \text{ k}\Omega$$

$$V_{be2} = -\beta I_{b1} R_{o1}$$

$$= -60 \times 0.718 = -43.1 \text{ V}$$

$$I_{b2} = -43.1/1.25 = -34.48 \text{ mA}$$

$$I_{b2}/I_{b1} = -34.48/1 = -34.48 \text{ mA}$$

(b)

$$R_s = 1 \text{ k}\Omega$$

$$V_s = 1 \text{ V (say)}$$

$$I_{b1} = \frac{1 \times 1.12}{(1 + 1.12)} \times \frac{1}{1.25} = 0.423 \text{ mA}$$

$$I_{b2} = -34.28 \times 0.423 = -14.58 \text{ mA}$$

$$V_{o1} = I_{b2} \times r_\pi = -14.58 \times 1.25 = -18.23 \text{ V}$$

$$A_{Vo1} = -18.23$$

Problem 5.33: For the transistor amplifier of Fig. 5.30
$\beta = 50$, $R_1 = 25$ kΩ, $R_2 = 75$ kΩ, $I_c = 1.25$ mA, $V_{CC} = 10$ V
The input signal strength is 10 mV with an output impedance of 0.5

kΩ (R_s). The load at output terminal on account of the following stage is 12 kΩ.
It is desired to have $V_V = 1$ V in mid-frequency region; Specify R_C, R_E.

Solution: From small-signal model is drawn in Fig. 5.33

$$r_\pi = \frac{25 \times 50}{1.25} \times 10^{-3} = 1 \text{ k}\Omega$$

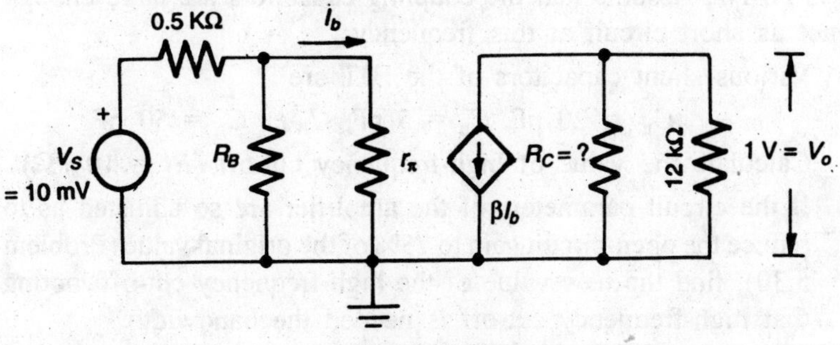

Fig. P. 5.33

$$R_B = 25 \parallel 75 = 18.75 \text{ k}\Omega$$
$$R_B \parallel r_\pi = 18.75 \parallel 1 = 0.95 \text{ k}\Omega$$
$$A_{Vo} = 1 \times 1000/10 = 100$$

Let $V_s = 1$ V (for convenience)

$$I_b = \frac{1 \times 0.95}{(0.5 + 0.95) \times 1} = 0.655 \text{ mA}$$

$$\beta I_b = 50 \times 0.655 = 32.75 \text{ mA}$$

$$A_{Vo} = V_o = -32.75 \times \frac{R_C \times 12}{R_C + 12} = -100$$

$\therefore$ $R_C = 5.1$ kΩ

From D.C. Analysis

$$V_{Th} = 10 \times \frac{25}{25 + 75} = 2.5 \text{ V}$$

Round the BE loop

$$2.5 = 18.75\, I_B + 0.7 + (50 + 1)\, R_E I_B$$
$$I_B = I_C/\beta$$

$$= 1.25/50 = 0.025 \text{ mA}$$

Substituting values

$$2.25 = 18.75 \times 0.025 + 0.7 + 51 \times 0.025\, R_E$$

or $$R_E = 0.848 \text{ k}\Omega$$

Problem 5.34: (a) For the BJT amplifier of Fig. 5.30 calculate the value of the bypass capacitor C_E such that the low-frequency cut-off is 20 Hz. Assume that the coupling capacitors are large enough to act as short circuit at this frequency.

(b) Various shunt capacitors of the BJT are

$$C_\pi = 20 \text{ pF},\ C_\mu = 5 \text{ pF},\ C_W + C_L = 50 \text{ pF}$$

Calculate the value of high-frequency cut-off ($R_L = 10$ kΩ).

(c) If the circuit parameters of the amplifier are so adjusted as to reduce the open-circuit gain to 75% of the original value (Problem 5.30), find the new value of the high-frequency cut-off; noting that high-frequency cut-off is needed the bandwidth.

Solution:

(a) Low-frequency **small-signal circuit** on the input side is drawn in Fig. 5.34 (a) assuming **input/output** coupling capacitors to act as short circuits.

To find the cut-off, short circuiting V_s in the circuit of Fig. 5.34 (a) and combining resistances we get,

$$R_{\text{eqv}} = [(4 \parallel 42) + 3.75] \parallel 303 = 7.22 \text{ k}\Omega$$

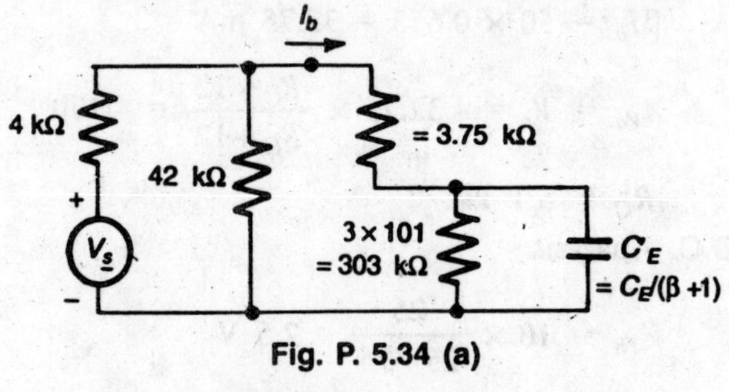

Fig. P. 5.34 (a)

Now, $$\omega_L = 1/(C'_E\, R_{\text{eq}})$$

or $$2\pi \times 20 = (101 \times 10^{-3})/(C_E \times 7.22)$$

or $$C_E = 111\ \mu\text{F}$$

(b) Equivalent high frequency circuit model is redrawn in Fig. 5.34 (b) with values of elements.

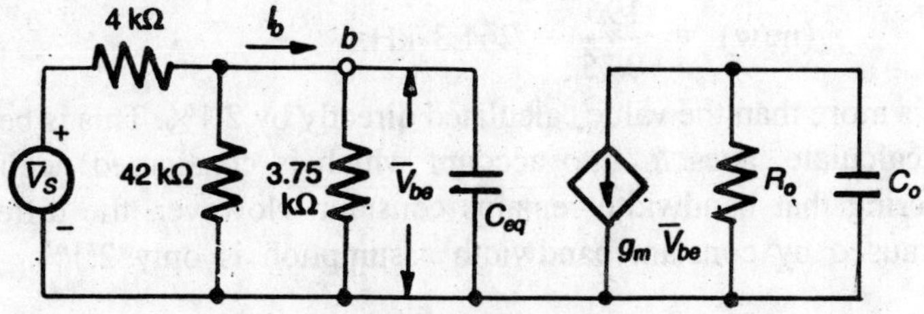

Fig. P. 5.34 (b)

$$C_{eq} = C_\pi + C_\mu\ (1 + g_m R_0) \qquad \text{...(i)}$$
$$C_o = (C_w + C_L) + C_\mu\ (1 + 1/g_m R_o) \qquad \text{...(ii)}$$
$$g_m = \beta/r_\pi = 100/3.75 = 27.7 \text{ mS}$$
$$R_o = R_L \parallel R_C = 10 \parallel 4 = 2.86 \text{ k}\Omega$$
$$g_m R_o = 27.7 \times 2.86 \approx 79.2$$

Substituting values in Eqs. (i) and (ii)

$$C_{eq} = 20 + 5\ (1 + 79.2) = 421 \text{ pF}$$
$$C_o = 50 + 5\ (1 + 1/79.2) \approx 55.1 \text{ pF}$$

Input circuit

$$\omega_{21} = \frac{1}{C_{eq}(r_\pi \parallel r_s)};\ R_B \gg r_\pi$$
$$= \frac{10^{12}}{421 \times 1.93 \times 10^3} = 1.23 \text{ Mrad/s or } 0.196 \text{ MHz}$$
$$\omega_{22} = 1/C_o R_o$$
$$= \frac{10^{12}}{55.1 \times 2.86 \times 10^3} = 6.36 \text{ Mrad/s or } 1.01 \text{ MHz}$$

Hence $\quad f_H = 196$ kHz

(c) Gain is reduced to 75% of original value by changing suitably external circuit elements.

So $\quad g_m R_o \rightarrow 79.2 \times 0.75 \approx 59.4$

$$C_{eq} \text{ (new)} = 20 + 5\ (1 + 59.4) = 322 \text{ pF}$$
$$\omega_L \text{ (new)} = \omega_{21} = \frac{10^{12}}{322 \times 1.93 \times 10^3} = 1.61 \text{ Mrad/s}$$
$$f_L \text{ (new)} = 1.61/2\pi = 0.256 \text{ MHz or } 256 \text{ kHz}$$

Remarks: Taking f_L to be proportional to gain (gain bandwidth product being constant)

$$f_L(\text{new}) = \frac{196}{0.75} = 261.3 \text{ kHz}$$

which is more than the value calculated directly by 2.1%. This is because direct calculate takes R_s into account which is considered negligible, considering that bandwidth remains constant. However, the difference in f_L caused by constant bandwidth assumption is only 2.1%.

Problem 5.35:

Consider the BJT amplifier of Fig. 5.35

(a) Find the quiescent value of I_C and V_o.

(b) Assuming C_E, to be very large, determine the value of C_{C1} such that low-frequency cut-off, f_L = 20 Hz.

(c) Assuming C_{C1} to be very large, determine the value of C_E for f_L = 20 Hz.

(d) Choose the value of C_{C1} and C_E for the total capacitance to be minimum ensuring that f_L = 20 Hz, while the corner frequency which is lower than this (nondominant) is less than 1/10 of f_L.

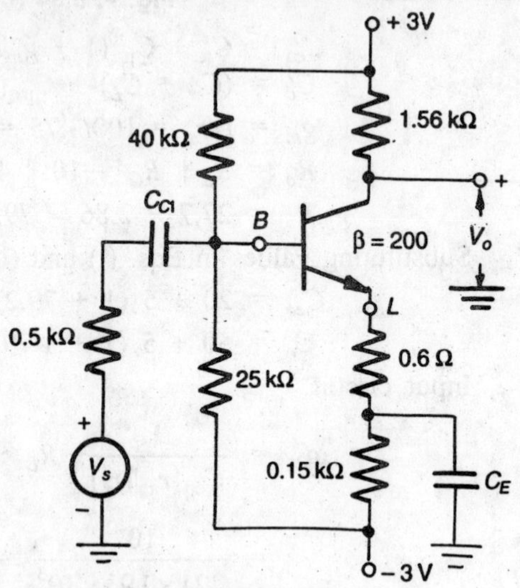

Fig. P. 5.35

Solution:

(a) DC analysis (C_{C1} acts as open circuit)

To find Thevenin equivalent at B (wrt ground), consider the circuit loop from + 3 V to –3 V, battery supply.

$$V_{Th} = 3 - \frac{3-(-3)}{(40+25)} \times 40 = -0.69 \text{ V}$$

$$R_{Th} = 40 \parallel 25 = 15.38 \text{ k}\Omega$$

dc Analysis from the circuit in Fig. 5.35 (a). Round the BE loop,

$$-0.69 - 15.38\ I_B - 0.7$$
$$- 0.75 \times 201\ I_B + 3 = 0$$

or $I_B = 9.69 \times 10^{-3}$ mA

$$I_C = \beta\ I_B = 200 \times 9.69 \times 10^{-3} = 1.94 \text{ mA}$$

$$V_o = 3 - 1.94 \times 1.56 = 0.023 \text{ V} \approx 0 \text{ V}$$

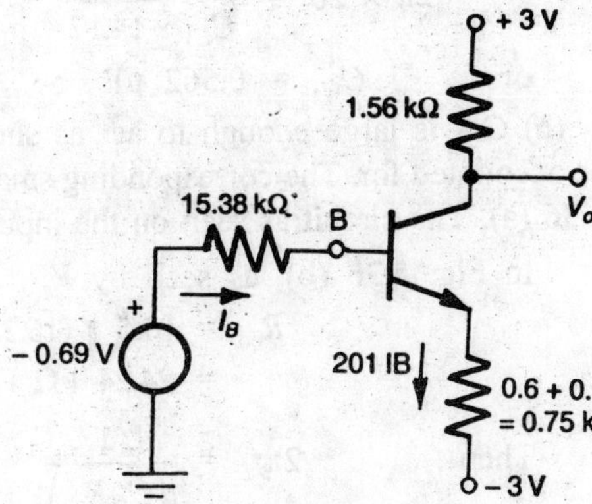

Fig. P. 5.35 (a)

(b) Small-signal model Fig. 5.35 (b)

$$r_\pi = \frac{25 \times 200}{1.94} \times 10^{-3} \approx 2.6 \text{ k}\Omega$$

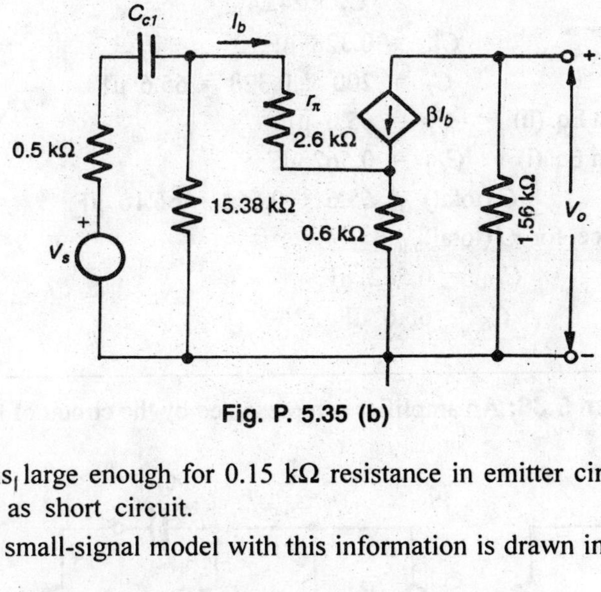

Fig. P. 5.35 (b)

C_E is large enough for 0.15 kΩ resistance in emitter circuit to be regarded as short circuit.

The small-signal model with this information is drawn in Fig. 5.35 (b).

0.6 kΩ resistance in emitter circuit acts as

$0.6 \times (200 + 1) = 120.6$ kΩ

for I_b circuit.

Effective resistance in series with C_{C1} is computed below

$$R_{eq} = 0.5 + 15.38 \parallel (2.6 + 120.6) = 14.17 \text{ k}\Omega$$

Then $\quad 2\pi f_L = 1/(C_{C1}\ R_{eq}) \qquad \ldots\text{(i)}$

$$2\pi \times 20 = \frac{10^{-3}}{C_{C1} \times 14.17}$$

or $\quad C_{C1} = 0.562\ \mu F$

(c) C_{C1} is large enough to act as short circuit but R_E and C_E must be accounted for. The corresponding small-signal model is drawn in Fig. 5.36 (a). The circuit as seen on the input side is drawn in Fig. 5.36 (b)

In Fig. 5.36 (b), as seen by V_s

$$R_{eq} = [0.5 \parallel (15.38) + (2.6 + 120.6)] \parallel 30.15$$
$$= 24.24\ k\Omega$$

Then
$$2\pi f_L = \frac{1}{(C_E R_{eq})}$$

$$2\pi f_L = \frac{10^{-3}}{C'_E \times 24.24} \quad f_L = 20\ Hz \qquad \text{...(ii)}$$

or
$$C'_E = 0.328\ \mu F$$
$$C_E = 200 \times 0.328 = 65.6\ \mu F$$

From Eq. (ii) $\quad C_E = 65.6\ \mu F$

From Eq. (i) $\quad C_{C1} = 0.562\ \mu F$

$$C\text{ (total)} = 65.6 + 0.562 = 66.16\ \mu F$$

Hence for $C\text{ (total)}|_{min}$

$$C_{C1} = 0.562\ \mu F,$$
$$C_E = 65.6\ \mu F$$

Problem 5.36: An amplifier is represented by the circuit of Fig. 5.36

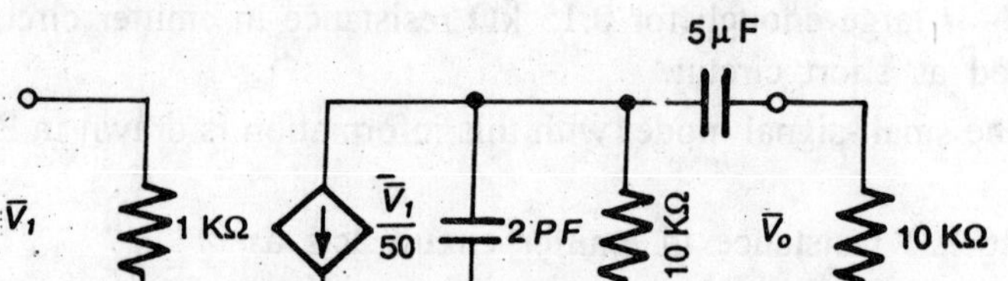

Fig. P. 5.36

(a) Predict the mid-frequency gain.
(b) Predict the upper and lower cut-off frequencies.

Solution:

(a) *Mid-frequency gain:* Shunt capacitor acts as open circuited and series capacitor acts as short circuit.

$$A_{Vo} = -(1/50) \times (10 \parallel 10) \times 1000 = -100$$

(b) *Upper-frequency region:* Series capacitor is shorted, while shunt capacitor has to be considered.

Take $\quad V_1 = 1$ V (for convenience)

$$V_o = \frac{1}{50} \times \frac{\left(5\times10^3\right)\times\dfrac{10^{12}}{j\omega\times2}}{\left(5\times10^3\right)+\dfrac{10^{12}}{j\omega\times2}}$$

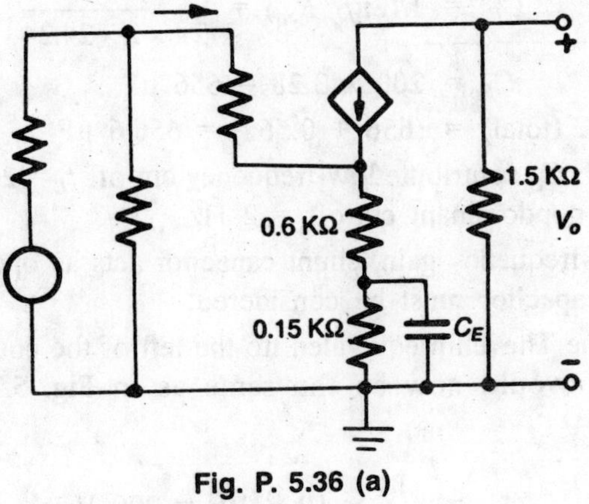

Fig. P. 5.36 (a)

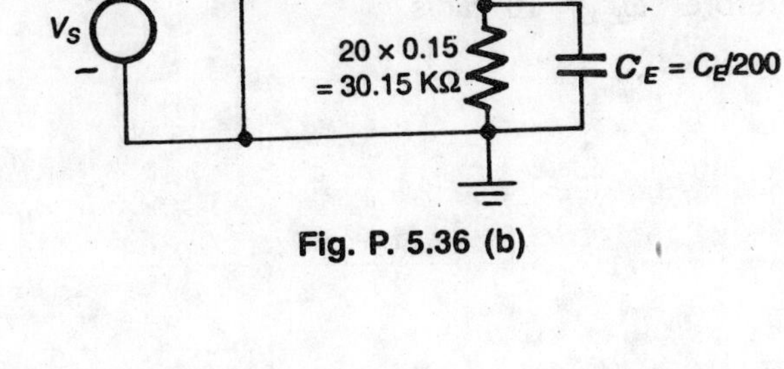

Fig. P. 5.36 (b)

$$A_{VH} = \overline{V}_o = \frac{1}{50} \times \left(\frac{5 \times 10^{15}}{j10^4 \omega + 10^{12}} \right) = \frac{100}{1 + j(\omega / 10^8)}$$

Therefore $\omega_H = 10^8$ rad/s or 100 M rad/s

(d) Case I: C_{C1} contributes low-frequency, $f_L = 20$ Hz, while C'_E contributes the nondominant cut-off = 20/10 = 2 Hz.

From eq. (i)

$$C_{C1} = 1/(2\pi f_L R_{eq}) = \frac{10^{-3}}{2\pi \times 20 \times 14.17} = 0.562\ \mu F$$

(same as already calculated in part (b))

From Eq. (ii)

$$C'_E = 1/(2\pi f_L R_{eq}) = \frac{10^{-3}}{2\pi \times 2 \times 24.24} = 3.28\ \mu F$$

$$C_E = 200 \times 3.28 = 656\ \mu F$$

$$C\ (\text{total}) = (656 + 0.562) = 656.6\ \mu F$$

Case II: C_E contribute low-frequency cut-off $f_L = 20$ Hz, while C_{C1} contributes nondominant cut-off = 2 Hz.

In low-frequency gain, shunt capacitor acts as open-circuit, while the series capacitor must be considered.

Take the Thevenin equivalen tto the left of the coupling capacitor. The circuit would now be the same as in Fig. 5.36 (b). Assume $v_1 = 1$ V.

$$V_{TH} = \frac{1}{50} \times 10 \times 10^3 = 200\ V$$

Then $$\overline{A}_{VL} = V_o = \frac{200 \times 10 \times 10^3}{\left(20 \times 10^3 + \frac{10^6}{j5\omega}\right)} = \frac{100}{(1 - j10/\omega)}$$

Therefore $\omega_L = 10$ rad/s.

Chapter 6

Transistor Biasing and Thermal Stabilization

Problem 6.1: (a) Determine the quiescent currents and the collector-to-emitter voltage for a silicon transistor with $\beta = 50$ in the self-biasing arrangement of Fig. 6.1. The circuit component values are $V_{CC} = 20\ V$, $R_C = 2$ K, $R_C = 0.1$ K, $R_1 = 100$ K, and $R_2 = 5$ K.

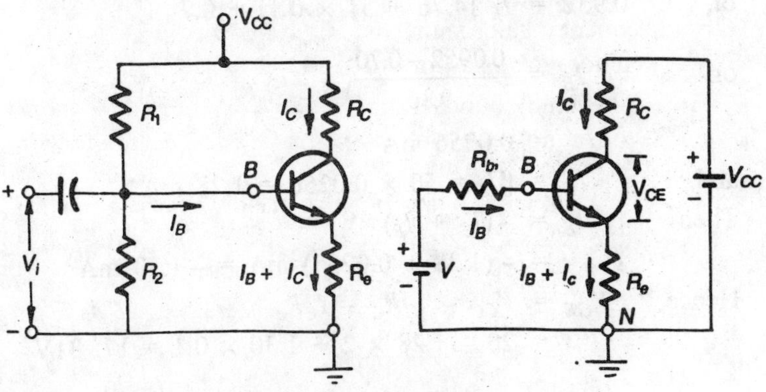

Fig. P. 6.1 (a) and (b)

Solution:

(a) Drawing the equivalent circuit using Thenenin's theorm to circuit left to base terminal.

Here, $$V = \frac{Vcc\ R_2}{R_1 + R_2} = \frac{5 \times 20}{100 + 5} = 0.952\ \text{V}$$

and, $R_b = R_1 \parallel R_2 = \dfrac{R_1 R_2}{R_1 + R_2} = \dfrac{5 \times 100}{105} = 4.76 \text{ K}$

Now, using KVL in the base to emitter loop, we have

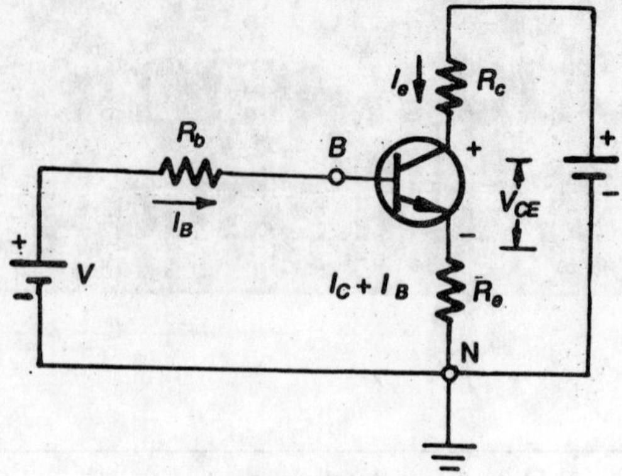

Fig. P. 6.1 (c)

$$V - R_b I_B - V_{BE} - (I_B + I_C) R_e = 0$$

or, $$V = R_b I_B + V_{BE} + (I_B + I_C) R_e$$

or, $$0.952 = I_B [4.76 + 51 \times 0.1] + 0.7$$

or, $$I_B = \frac{0.0952 - 0.70}{9.86}$$

$$= 0.0256 \text{ mA}$$

and, $$I_C = \beta I_B = 50 \times 0.0256 = 1.28 \text{ mA}$$

Then $$I_E = -(I_C + I_B)$$

$$= -(1.28 + 0.0256) \text{ mA} = -1.30 \text{ mA}$$

Hence, $$V_{CE} = V_{CC} - I_C R_C - I_E R_e$$

$$= 20 - 1.28 \times 2 + 1.30 \times 0.1 = 17.31\text{V}$$

Problem 6.2: A *p-n-p* germanium transistor is used in the self-biasing arrangement of Fig. 6.1. The circuit component values are $V_{cc} = 4.5$ V, $R_C = 1.5$ K, $R_e = 0.27$ K, $R_2 = 2.7$ K, and $R_1 = 27$ K. If $\beta = 44$

(a) Find the quiescent point.

(b) Recalculate these values if the base-spreading resistance of 690 Ω is taken into account.

Solution:

(a) Simplifying the given circuit using Thenvenin's theorem, we have circuit as shown,

here, $V = \dfrac{Vcc\ R_2}{R_1 + R_2}$

$= \dfrac{4.5 \times 2.7}{27 + 2.7}$

$= 0.409$ V

and $R = \dfrac{R_1\ R_2}{R_1 + R_2}$

$= \dfrac{2.7 \times 27}{2.7 + 27} = 2.45$ KΩ

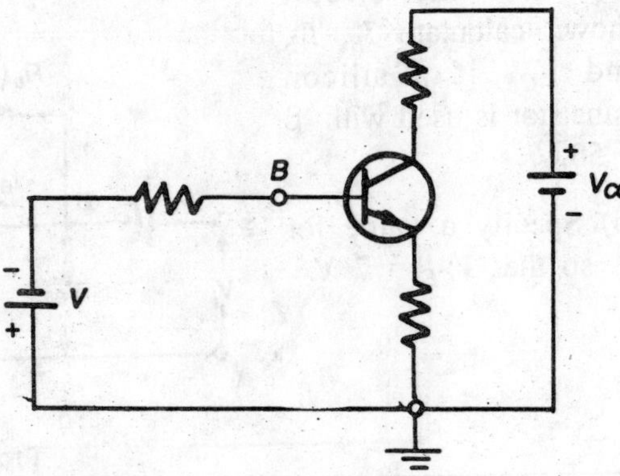

Fig. P. 6.2

For *Ge*, $|V_{BE}| = 0.2$ V

Using KVL in the base emitter circuit, we have

$-V - R_b I_B - V_{BE} - (1 + \beta) I_B R_e = 0$

or, $-0.409 = I_B\ [51 \times 0.27 + 2.45] - 0.2$

or, $I_B = \dfrac{-0.409 + 0.2}{14.6} = -14.3$ mA

and, $I_c = \beta I_B = 44 \times (-14.3) = -0.63$ mA

Then $V_{CE} = V_{CC} - I_C R_C - (I_B + I_C)\ R_e$

$= -4.5 + 0.63 \times 1.5 + 0.644 \times 0.27$

$= -3.38$ V

(b) Now, if R_b increased by 690 Ω, then

$R_b' = 2.46 + 0.69 = 3.15$ K

Then KVL around base-emitter loop gives.

$-0.409 = I_B\ (15.29) - 0.2$

or, $I_B = \dfrac{-0.209}{15.29} = -13.7$ mA

Then $I_C = -13.7 \times 44$ mA $= -0.60$ mA

$\therefore$ $V_{CC} = V_{CC} - I_C R_C - (I_B + I_C)\ R_e$

$= -4.5 + 0.6 \times 1.5 + 0.602 \times 0.27 = -3.43$ V

Problem 6.3:

(a) From the circuit shown calculate I_B, I_C, and V_{CE} if a silicon transistor is used with $\beta = 50$.

(b) Specify a value for R_b so that $V_{CE} = 7$ V.

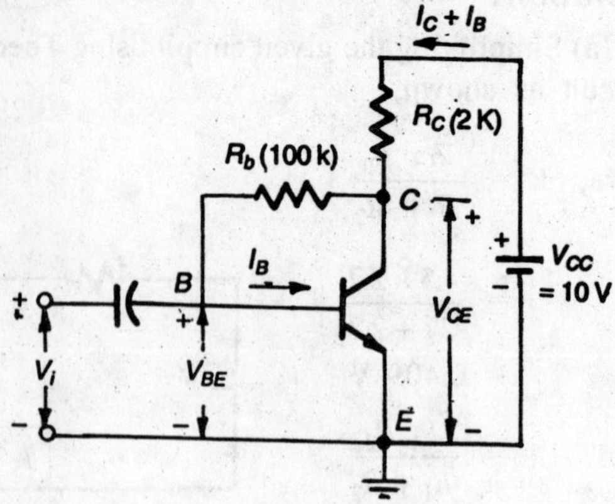

Fig. P. 6.3

Solution:

(a) Now, applying KVL around the loop $V_{cc} - C - B - E$, we have

$$10 - (I_C + I_B) R_C - I_B R_b - V_{BE} = 0 \quad \ldots\text{(i)}$$

Since, transistor is in active region, therefore

$V_{BE} = 0.7$ V

From equation (i),

$10 - (I_B + \beta I_B)$

$R_C - I_B R_b - 0.7 = 0$

or, $9.3 = [R_C (1 + \beta) + R_b] I_B$

or, $I_B = \dfrac{9.3}{202} = 0.046$ mA

$I_C = \beta I_B = 50 \times 0.046$

$= 2.30$ mA

and

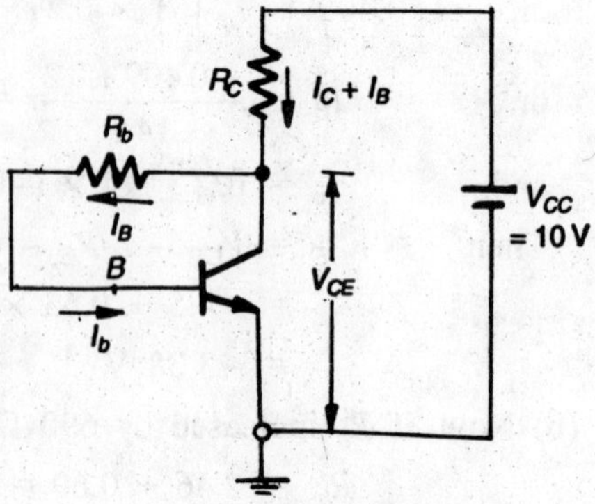

Fig. P. 6.3 (a)

$V_{CE} = 10 - R_c (I_C + I_B) = 10 - 2 \times 51 \times 0.046 = 5.20$ V

(b) Given $V_{CE} = 7$ V.

Applying *KVL* in the collector-emitter circuit, we have

$V_{CC} - 2\ (I_C + I_B) R_C - V_{CE} = 0$

or, $\dfrac{V_{CC} - V_{CE}}{2(1+\beta)\ R_c} = I_B$

or, $I_B = \dfrac{10-7}{2 \times 51 \times 2} = 0.0294$ mA

Also, $$V_{CE} = V_{BE} + R_b \cdot I_B$$

or, $$R_b = \frac{V_{CE} - V_{BE}}{I_B} = \frac{7 - 0.7}{0.0294 \times 10^{-3}}$$

$$= \frac{6.3 \times 10^3}{0.0294} = 214 \text{ K}\Omega$$

Problem 6.4:

For the two-battery transistor circuit shown, prove that the stabilization factor S is given by

$$S = \frac{1+\beta}{1+\beta R_e/(R_e + R_b)}$$

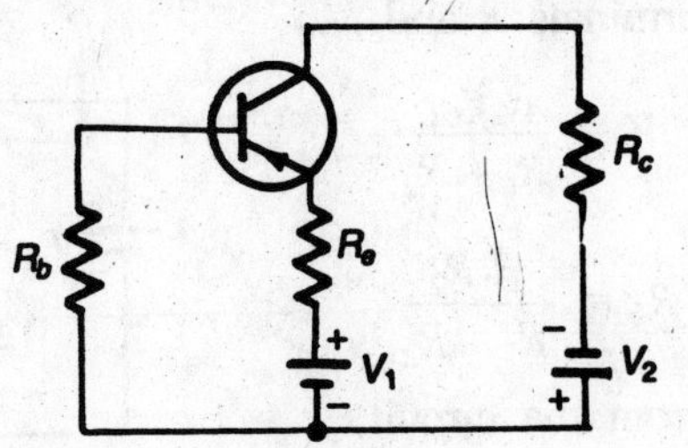

Fig. P. 6.4

Solution:

Neglecting V_{BE}, we obtain from the circuit using *KVL*,

$$V_1 - R_C I_E \div I_B R_b = 0 \quad \ldots\text{(i)}$$

But, $$I_E = -(I_B + I_C)$$

$$\therefore \quad V_1 = -(I_B + I_C)R_e - I_B R_b$$

or $$I_B = \frac{I_C R_e + V_1}{R_e + R_b} \quad \ldots\text{(ii)}$$

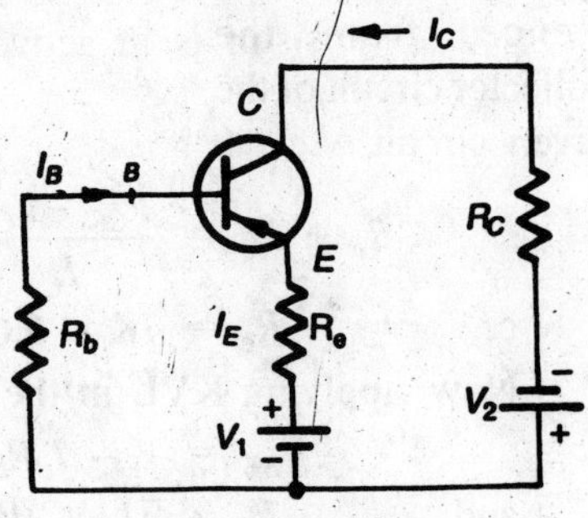

Fig. P. 6.4 (a)

Also, $$I_C = \beta I_B + (1 + \beta)\, I_{CO}$$

Using eq.(ii), $$= (1 + \beta)\, I_{CO} - \frac{\beta}{R_C + R_b} \times (I_C R_e + V_1)$$

$$I_C\left(1 + \frac{\beta R_e}{R_e + R_b}\right) = (1 + \beta)\ I_{CO}$$

Hence, stability factor,

$$S = \frac{\partial I_C}{\partial I_{CO}} = \frac{1+\beta}{1 + \dfrac{\beta R_e}{R_e + R_b}}$$

Problem 6.5: Assume that a silicon transistor with $\beta = 50$, $V_{BE\,\text{active}} = 0.7$, $V_{CC} = 22.5$ V, and $R_C = 5.6$ K is used in Fig. 6.5. It is desired to establish a Q point at $V_{CE} = 12$ V. $I_C = 1.5$ mA, and stability factor $S \leq 3$. Find R_e, R_1, and R_2.

Solution:

By the Thenenin's equivalent across terminals B and N.

$$V = \frac{R_2 V_{CC}}{R_1 + R_2}$$

$$R_b = \frac{R_1 R_2}{R_1 + R_2}$$

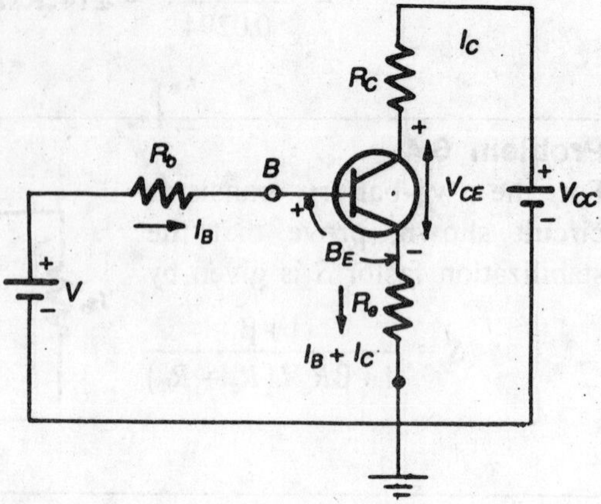

Fig. P. 6.5

From the circuit, current in R_e is $I_C + I_B \approx I_C$

Hence, from the collector circuit of the given circuit,

$$R_e + R_c = \frac{V_{CC} - V_{CE}}{I_C} = \frac{22.5 - 12}{1.5} = 7\ \text{K}\Omega$$

or, $R_e = 7\text{K} - 5.6\ \text{K} = 1.4\ \text{K}\Omega$

Now, applying KVL in the base-emitter loop, we have

$$V_{BE} = V - I_B R_b - (I_B + I_C) R_e$$

and $I_C = \beta I_B + (\beta + 1) I_{CO}$

Substituting the expression for I_B in eq. (ii), we have

$$V_{BE} = V + (R_B + R_C) \frac{\beta + 1}{\beta} I_{CO} - \frac{R_b + R_C(1+\beta)}{\beta} I_C \quad \text{...(iv)}$$

Differentiating with respect to I_{CO}, we get,
Stability factor,

$$S = \frac{\partial I_C}{\partial I_{CO}} = (\beta + 1) \frac{1 + \frac{R_b}{R_e}}{1 + \beta + \frac{R_b}{R_e}}$$

Given, $S \leq 3$

$$51 \times \left(\frac{1 + \frac{R_b}{R_e}}{51 + \frac{R_b}{R_e}} \right) \leq 3$$

or, $$\frac{R_b}{R_e} = 2.12$$

or, $$R_b = 2.12 \times 1.4 = 2.96 \text{ K}\Omega$$

From the relation,

$$V = \frac{R_2 V_{cc}}{R_1 + R_2}$$

and $$R_b = R_1 \parallel R_2 = \frac{R_1 R_2}{R_1 + R_2}$$

Solving these equation for R_1 and R_2, We get,

$$R_1 = R_b \left(\frac{V_{cc}}{V} \right); \quad R_2 = \left(\frac{R_1 V}{V_{cc} - V} \right)$$

From eq (ii), $V = V_{BE} + I_B R_b + (I_B + I_C) R_e$

$$= 0.7 + 0.03 \times 2.96 + (0.03 + 1.5) \times 1.4 = 2.93 \text{ V}$$

Then, $$R_1 = \frac{2.96 \text{ V} \times 22.5}{2.93 \text{ V}} = 22.8$$

and $$R_2 = \frac{22.8 \times 2.93}{22.5 - 2.93} = 3.4 \text{ K}\Omega$$

Problem 6.6:
Determine the stability factor S for the circuit shown.

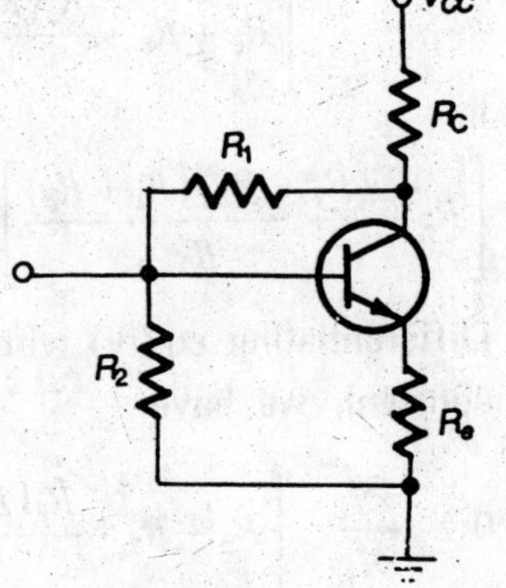

Fig. P. 6.6

Solution:

Redrawing the circuit

Applying KVL around the collection base-emitter loop gives,

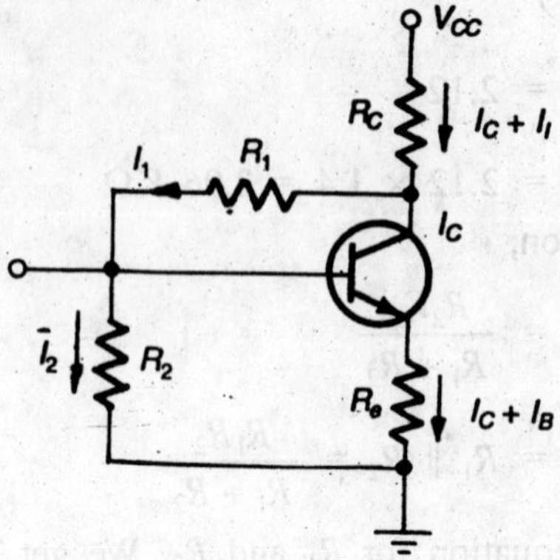

Fig. P. 6.6 (a)

$$V_{CC} = (I_C + I_1)\ R_C + (I_C + I_B)\ R_e + I_1R_1 + V_{BE} \quad \text{...(i)}$$

$$I_2 = \frac{V_{BE} + (I_B + I_c)\ R_e}{R_2} \quad \text{...(ii)}$$

$$I_1 = I_B + I_2 \quad \text{...(iii)}$$

$$I_1 = \frac{I_B R_2 + V_{BE} + (I_B + I_c) R_e}{R_2}$$

Substituting expression for I_1, in eq. (i), we have

$$V_{CC} = I_C \left[R_c + R_e + \frac{R_e(R_1 + R_c)}{R_2}\right] + \frac{I_c - (1+\beta)I_{co}}{\beta} \times$$

$$\left[R_e + \frac{(R_e + R_2)(R_1 + R_c)}{R_2}\right] + V_{BE}\left(1 + \frac{R_1 + R_c}{2}\right) \quad \text{...(iv)}$$

Differentiating eq.(iv) with respect to I_{CO} assuming V_{BE} constant, we have

$$0 = \frac{\partial I_c}{\partial I_{co}}\left[R_c + R_e + \frac{R_e(R_1 + R_c)}{R_2}\right] + \frac{\partial I_c}{\partial I_{co}}$$

$$\times \left[R_e + \frac{(R_e + R_2)(R_1 + R_c)}{R_2}\right] \times \frac{I}{B} - \frac{I+\beta}{\beta}$$

$$\times \left[R_e + \frac{(R_e + R_2)(R_1 + R_c)}{R_2}\right]$$

$$\text{or, } S \text{ (stability factor)} = \frac{(\beta+1)\left[R_2(R_c + R_1) + R_e(R_1 + R_2 + R_c)\right]}{R_1 R_2 + (\beta+1)\left[R_2 R_c + R_e(R_1 + R_2 + R_c)\right]}$$

Problem 6.7:
In the circuit shown, V_{CC} = 24 V, R_C = 10 K, and R_e = 270 Ω. If a silicon transistor is used with β = 45 and if under quiescent conditions V_{CE} = 5 V determine (a) R, (b) the stability factor S.

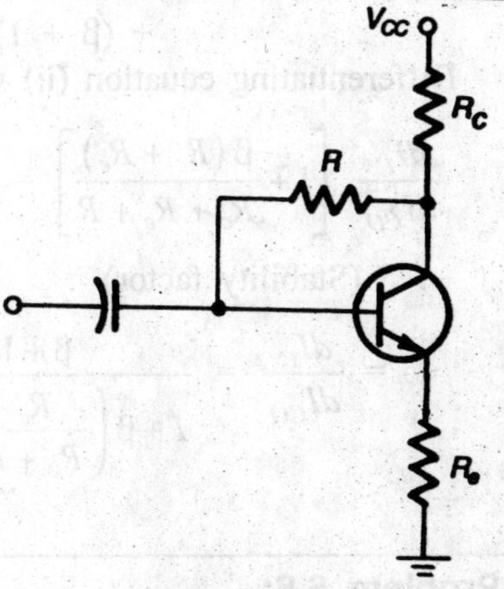

Fig. P. 6.7

Solution:

(i) Applying *KVL* to collector-emitter circuit,

$$24 = 10\ (I_c + I_B) + 5 + 0.27\ (I_c + I_B)$$

$$\text{or, } I_c + I_B = \frac{19}{10.27}$$

$$= 1.85 \text{ mA}$$

Since, $I_c = 45 I_B$

$$\therefore \quad I_B = \frac{1.85}{46}$$

$$= 0.043 \text{ mA}$$

and, I_c = 1.81 mA

Form the figure,

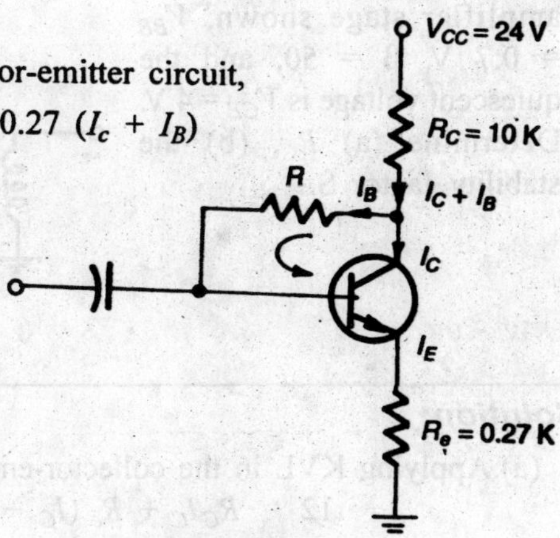

Fig. P. 6.7 (a)

$$V_{CE} = V_{BE} + I_B R$$

$$R = \frac{V_{CE} - V_{BE}}{I_B} = \frac{5-0.7}{0.0403} = 107 \text{ K}$$

(ii) From collector-base circuit,

$$V_{CC} = (I_B + I_c)(R_c + R_e) + I_{BR} + V_{BE}$$

or, $$I_B = \frac{V_{cc} - V_{BE} - I_c(R_c + R_e)}{R_c + R_e + R} \qquad \text{...(i)}$$

As $$I_C = \beta I_B + (\beta + 1) I_{CO}$$

From eq (i), $$I_C = \beta \left[\frac{V_{CC} - V_{BE}}{R_c + R_e + R}\right] - \frac{\beta(R_c + R_e)}{R_c + R_e + R} + (\beta + 1) I_{CO} \qquad \text{...(ii)}$$

Differentiating equation (ii) with respect to I_{CO}, we get

$$\frac{dI_C}{dI_{CO}}\left[1 + \frac{\beta(R_c + R_e)}{R_c + R_e + R}\right] = (\beta + 1)$$

$\therefore$ S (Stability factor)

$$= \frac{dI_C}{dI_{CO}} = \frac{\beta + 1}{I + \beta\left(\frac{R_c + R_e}{R_c + R_e + R}\right)} = \frac{46}{1 + 45\frac{10.27}{120.27}} = 9.5$$

Problem 6.8:
In the transformer-coupled amplifier stage shown, V_{BE} = 0.7 V, β = 50, and the quiescent voltage is V_{CE} = 4 V. Determine (a) R_C, (b) the stability factor S.

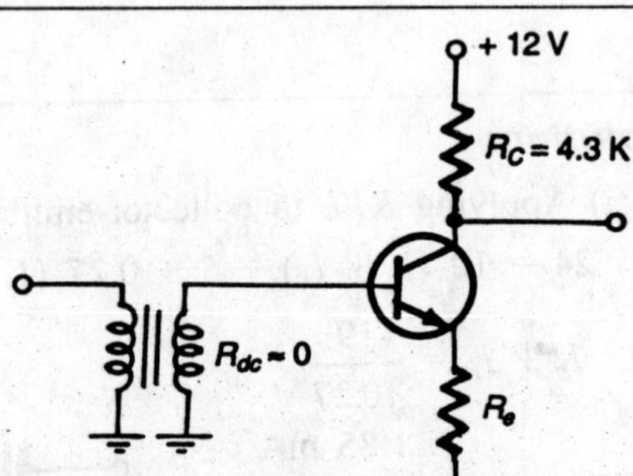

Fig. P. 6.8

Solution:

(a) Applying KVL in the collector-emitter circuit, we get

$$12 = R_C I_C + R_e (I_C + I_B) + V_{CE} + V_{EE} \qquad \text{...(i)}$$

or, $12 - V_{EE} = R_C I_C + R_e \left(I_C + \frac{I_C}{\beta}\right) + (+4)$

or, $$14 = \left[R_C + R_e\left(1+\frac{1}{\beta}\right)\right] I_C \quad \text{...(ii)}$$

Applying same in base emitter circuit, we get

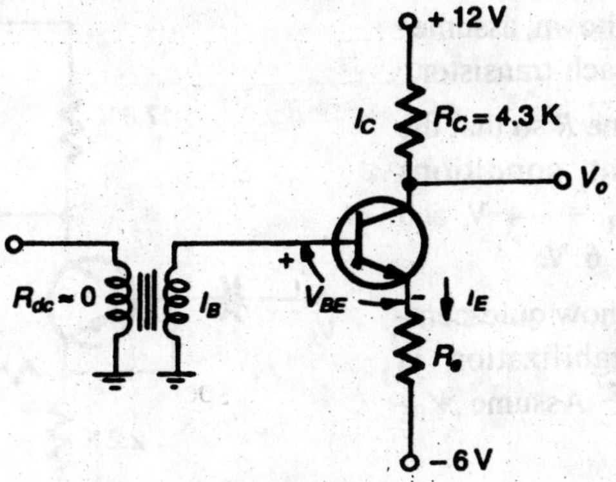

Fig. P. 6.8 (a)

$$V_{EE} + (I_C + I_B) R_e + V_{BE} = 0$$

or $$(I_C + I_B) R_e + V_{BE} = - V_{EE} \quad \text{...(iii)}$$

or $$I_C \left(1+\frac{1}{\beta}\right) R_e = 6 - 0.7 = 5.3 \quad \text{...(iv)}$$

Solving eq. (i) and (iv), we get

$$I_C = 2 \text{ mA}$$

$$R_C = 2.74 \text{ K}\Omega$$

(b) From eq (iii)

$$V_{EE} + (I_C + I_B) R_C + V_{BE} = 0$$

Using $I_C = \beta_{IB} + (1 + P) I_{CO}$

in this equation, we have

$$V_{EE} + \left[I_C + \frac{I_C - (1+\beta)I_{CD}}{\beta}\right] R_e + V_{DE} = 0$$

Assuming V_{BE} constant and differentiating eq. (v) with respect to I_{CO}, we get

$$\frac{\partial I_C}{\partial I_{CD}}\left[R_C+\frac{R_C}{\beta}\right]=\frac{(1+\beta)}{\beta}R_C$$

or, S (Stability factor) $= \frac{\partial I_C}{\partial I_{CD}} = 1$

Problem 6.9: In the two-stage circuit shown, assume $\beta = 100$ for each transistor.

(a) Determine R so that the quiescent conditions are $V_{CE1} = -4$ V and $V_{CE2} = -6$ V.

(b) Explain how quiescent-point stabilization is obtained. Assume $V_{BE} = 0.2$ V.

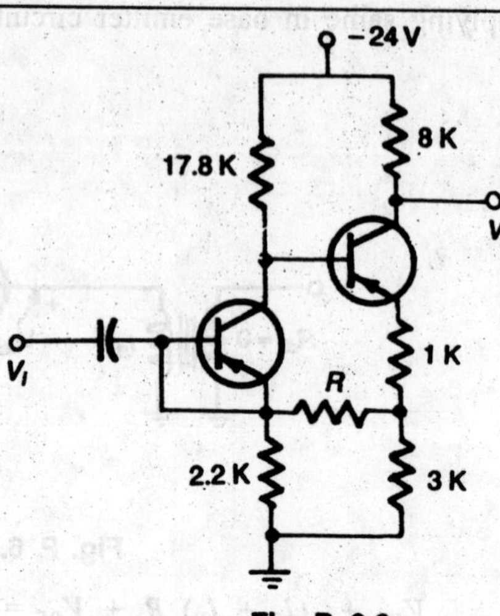

Fig. P. 6.9

Solution:

Redrawing the given circuit

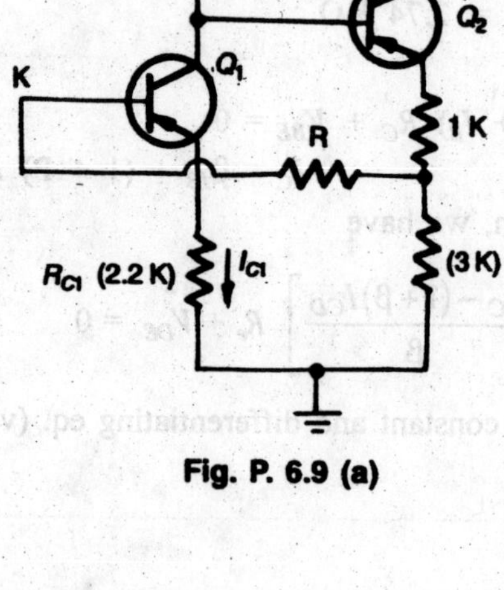

Fig. P. 6.9 (a)

KVL around collector-emitter circuit of Q_1,

$$V_{CC} = R_{C1}I_{C1} + R_{e1}I_{C1} + V_{CE1} \quad \text{...(i)}$$

or, $V_{CC} - V_{CE1} = (17.8 + 2.2)\ I_{C1}$

or, $$I_{C1} = -\frac{20\text{V}}{20\ \text{K}\Omega} = -1\ \text{mA}$$

Similarly for transistor Q_2,

$$24 - 6 = -(1 + 3 + 8)\ I_{C2}$$

or, $$I_{C2} = -\frac{18}{12\ \text{K}} = -1.5\ \text{mA}$$

Now, voltage at point P is,

$$V_P = -3 \times 1.5 = -4.5\ \text{V}$$

Voltage at point K is,

$$V_K = -2.2 \times 1 - 0.2 = -2.4\ \text{V (since } V_{BE} = 0.2\ \text{V)}$$

Then, voltage across R is,

$$V_R = -2.4 - (-4.5) = 2.1\ \text{V}$$

Now, $$I_{B1} = \frac{I_{C1}}{\beta} = \frac{-1}{100}\ \text{mA}$$

Then, $$R = \frac{V_R}{I_{B1}} = \frac{2.5}{10 \times 10^{-6}} = 210\ \text{K}\Omega$$

(b) In this circuit, bias stabilization is done by negative dc feed back through R and through the emitter resistances. For example, as I_{C1}, increases V_{CE1} decreases (in magnitude) and thus reducing I_{B2}, I_{E2}, and consequently the voltage at point P. Thus, I_{B1} decreases which offset the initial increase in I_{C1}.

Problem 6.10:
In the Darlington stage shown,

$V_{cc} = 24$ V
$\beta_1 = 24$
$\beta_2 = 39$
$V_{BE} = 0.7$
$R_C = 330\ \Omega$
$R_e = 120\ \Omega$.

If at the quiescent point V_{CE2} = 6 V, determine (a) R, (b) the stability factor defined as $S \equiv dI_c/dI_{CO1}$

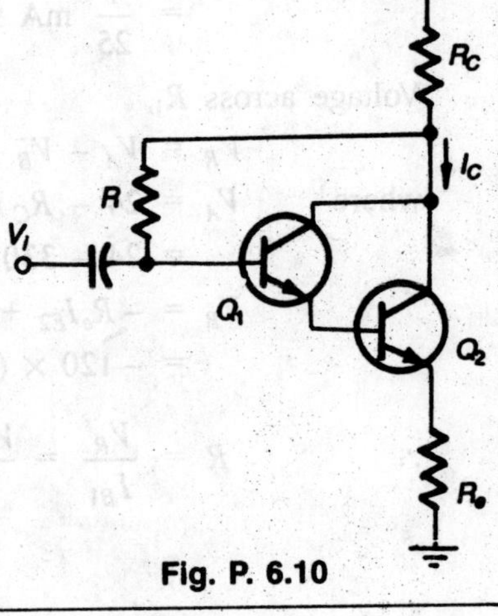

Fig. P. 6.10

Solution:

Given, $R_C = 300\ \Omega$

$R_e = 120\ \Omega$

Redrawing the given circuit,

Now applying KVL in collection-emitter circuit of Q_2

$$V_{CC} = (I_{B1} + I_{C1} + I_{C2}) \times (R_C + R_e) + V_{CE2} \quad ...(1)$$

or, $I_{B1} + I_{C1} + I_{C2}$

$$= \frac{V_{CC} - V_{CE2}}{R_C + R_e}$$

$$= -I_{E2}$$

or, $-I_{E2} = \frac{24 - 6}{0.45\ \text{K}}$

$= 40$ mA

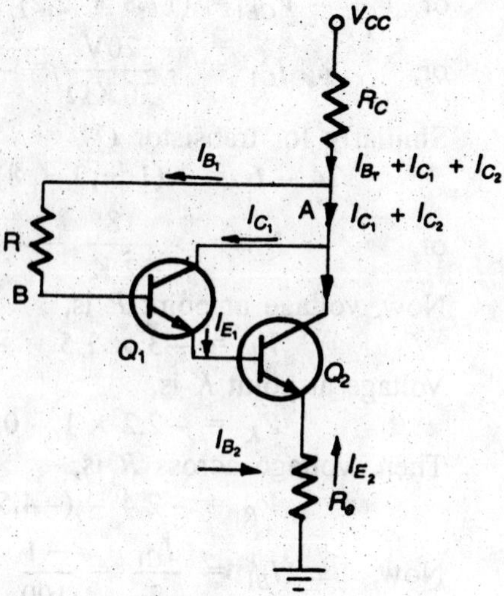

Fig. P. 6.10 (a)

Now, $I_{B2} = -I_{E1}$

$$= -\frac{I_{E2}}{1+\beta_2} = \frac{40}{40}\ \text{mA} = 1\ \text{mA}$$

Then, $I_{B1} = \dfrac{-I_{E1}}{1+\beta_1}$

$$= \frac{1}{25}\ \text{mA} = 40\ \mu\text{A}$$

Voltage across R_1,

$V_R = V_A - V_B$

where $V_A = 24 - R_C\,(I_{C1} + I_{C2} + I_{B1})$

$= 24 - 330 \times 40 \times 10^{-3} = 10.8$ V

$V_B = -R_e I_{E2} + V_{BE1} + V_{BE2}$

$= -120 \times (-40) \times 10^{-3} + 1.4 = 6.2$ V

$$\therefore \quad R = \frac{V_R}{I_{B1}} = \frac{V_A - V_B}{I_{B1}} = \frac{4.6}{40 \times 10^{-6}} = 115\ \text{K}\Omega$$

Problem 6.11: For the bias arrangement given in Prob. 6.3 prove that

$$S' = \frac{-\beta S}{1+\beta} \times \frac{1}{R_c + R_b}$$

$$S'' = \frac{I_C}{\beta} \times \frac{S}{\beta + 1}$$

Solution:

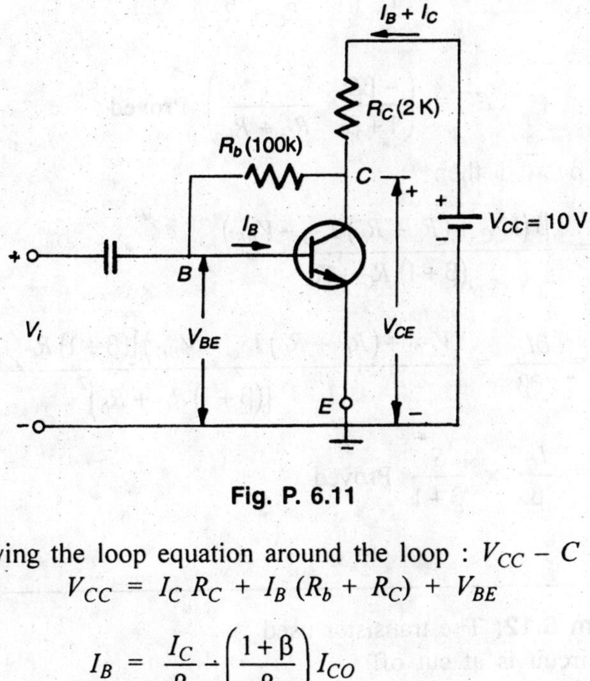

Fig. P. 6.11

Applying the loop equation around the loop : $V_{CC} - C - B - E$

$$V_{CC} = I_C R_C + I_B (R_b + R_C) + V_{BE} \qquad \text{...(i)}$$

$$I_B = \frac{I_C}{\beta} - \left(\frac{1+\beta}{\beta}\right) I_{CO} \qquad \text{...(ii)}$$

from eq. (i) and (ii), we have

$$V_{BE} = V_{CC} - I_C \left(R_C + \frac{R_c + R_b}{\beta}\right) + \frac{1+\beta}{\beta} (R_C + R_b) I_{CO}$$

Then $\quad S' = \dfrac{\partial I_C}{\partial V_{BE}}$

$$= \frac{-\beta}{(1+\beta)\,R_C + R_b}$$

$$= \frac{-\beta}{1+\beta} \times \frac{1}{R_C + R_b} \times \frac{\beta+1}{1+\beta\dfrac{R_C}{R_C+R_b}}$$

But S is found as $S = \left(\dfrac{\beta+1}{1+\dfrac{\beta R_C}{R_C+R_b}}\right)$

$$\therefore \qquad S' = \left(\frac{-\beta S}{1+\beta} \times \frac{1}{R_C+R_b}\right) \quad \text{Proved}$$

(b) If $\beta \gg 1$ then

$$I_C = \frac{\beta\left(V_{CC} + (R_e + R_b)\,I_{CO} - V_{BE}\right)}{(\beta+1)\,R_C + R_b}$$

$$S'' = \frac{\partial I_C}{\partial \beta} = \frac{\left(V_{CC} + (R_C + R_b)\,I_{CO} - V_{BE}\right)\left[(\beta+1)\,R_C + R_b + \beta R_c\right]}{\left[(\beta+1)\,R_C + R_b\right]^2}$$

$$= \frac{I_C}{\beta} \times \frac{S}{\beta+1} \quad \text{Proved}$$

Problem 6.12: The transistor used in the circuit is at cut off.

(a) Show that runaway will occur for values of I_{CO} in the range

$$\frac{V_{CC} - \sqrt{V_{CC^2} - 8R_C/0.07\theta}}{4R_C} \le I_{CO}$$

$$\le \frac{V_{CC} + \sqrt{V_{CC^2} - 8R_C/0.07\theta}}{4R_C}$$

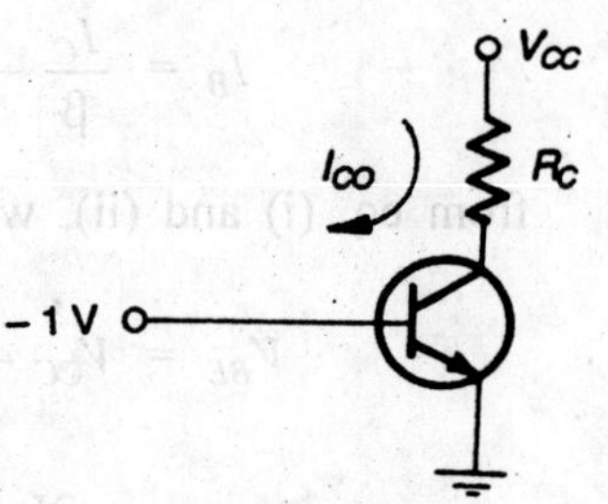

Fig. P. 6.12

(b) Show that if runaway is not destructive, the collector current I_{CO} after runaway can never $I_{CO} = V_{CC}/2\,R_c$.

Solution:

The following equation should helds good to avoid the therminal runaway:

$$\frac{\partial P_C}{\partial T_i} < \frac{1}{\theta}$$

$$P_C = I_C \cdot V_{CE} = I_{CO}\, V_{cc} - I^2{}_{CO}\, R_c$$

[$\therefore$ the transistor is at cut-off)

$$\text{Hence,} \quad \frac{\partial P_C}{\partial T_i} = \frac{\partial P_C}{\partial I_{C0}} \times \frac{\partial I_{CO}}{\partial T_i}$$

$$= (V_{CC} - 2\, I_{CO} R_C) \times 0.070\, I_{CO} < \frac{1}{\theta}$$

$$\text{or,} \quad 14\, R_C\, I_{CO}^2 - 0.07\, V_{CC} I_{CO} + \frac{1}{\theta} > 0$$

From the roots of this equation of I_{CO} (quadratic form) we find:

$$\frac{V_{CC} - \sqrt{V_{CC^2} - 8R_C / 0.07\theta}}{4R_C} \le I_{CO} \le \frac{V_{CC} + \sqrt{V_{CC^2} - 8R_C / 0.07\theta}}{4R_C}$$

(b) As I_{CO} increases due to thermal runsway, it will finally become,

$$I_{CO} > \frac{V_{CC}}{2R_C};$$

but then, $\quad V_{CC} - 2\, I_{CO}\, R_C < 0$

Only thermal runaway will stop

$$\text{when} \quad I_{CO} = \left(\frac{V_{CC}}{2R_C}\right)$$

Chapter 7

Field-Effect Transistors

Problem 7.1: For a *p*-channel silicon FET with $a = 2 \times 10^{-4}$ cm and channel resistivity $\rho = 10\ \Omega$-cm.
(a) Find the pinch-off voltage.
(b) Repeat (a) for a *p*-channel germanium FET with $\rho = 2\ \Omega$-cm.

Solution: Here the FET is a *p*-channel silicon FET.

Hence, $\mu_p = 500\ \text{cm}^2/V\text{-sec}$

$$\sigma = N_A \mu_p q$$

or $$qN_A = \frac{1}{\rho \mu_p}$$

Here, $\rho = 10\ \Omega\text{-cm}$

$$\therefore \quad qN_A = \frac{1}{10 \times 500} = 2 \times 10^{-4}$$

and $\in = 12\varepsilon_o$

∴ Pinch off-voltage

$$= \frac{qN_A}{2\in} \times a^2 = \frac{2 \times 10^{-4} \times \left(2 \times 10^{-4}\right)^2}{2 \times 12 \times \left(36\pi \times 10^{-11}\right)^{-1}} \cong 3.78 \text{ V}$$

(b) For Ge-FET,

$$\in = 16\in_o$$

and $\mu_p = 1800\ \text{cm}^2/\text{V-sec}$

$$\therefore \qquad V_p = \frac{qN_A}{2\in} \times a^2$$

$$\text{and} \qquad qN_A = \frac{1}{\mu_p \times \rho} = \frac{1}{1800 \times 2}$$

$$\therefore \qquad V_{P(Ge)} = 3.78 \times \frac{10 \times 12 \times 500}{2 \times 16 \times 1800} \cong 3.94 \text{ V}$$

Problem 7.2: (a) Plot the transfer characteristic curve of an FET with I_{DSS} = 10 mA and V_P = – 4 V.
(b) The magnitude of the slope of this curve at V_{GS} = 0 is g_{mo}. If the slope is extended as a tangent, show that it intersects the V_{GS} axis at the point $V_{GS} = V_P/2$.

Solution: From the equation relating with I_{DS} and V_{gs},

$$I_{DS} = I_{DSS}\left(1 - \frac{V_{GS}}{V_p}\right)^2$$

V_{GS} (V)	0	1	2	3
I_{DS} (MA)	10	5.62	2.5	0.625

and from this table, we obtain the curve,

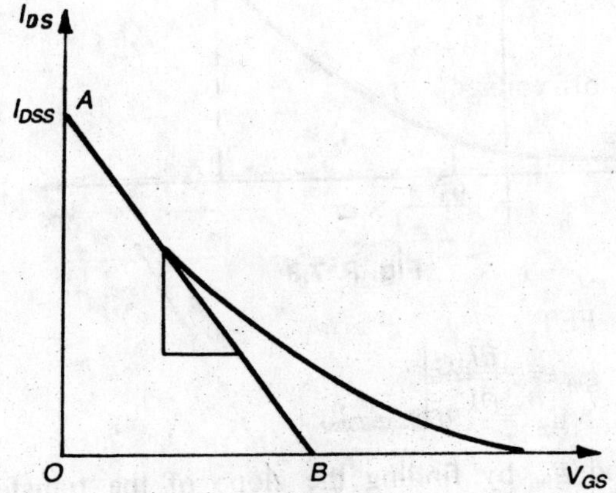

Fig. P. 7.2

(b) For the magnitude of the slope of the line $\overline{AB}$

$$g_{mo} = \frac{2I_{DSS}}{|V_P|}$$

$$OB = \frac{I_{DSS}}{g_{mo}}$$

$$= \frac{I_{DSS}}{2I_{DSS}/|V_p|} = \left(\frac{|V_P|}{2}\right)$$

Problem 7.3: Given the transfer characteristic of an FET, explain clearly how to determine g_m at a specified quiescent point.

Solution: The transfer function characteristic of an FET.

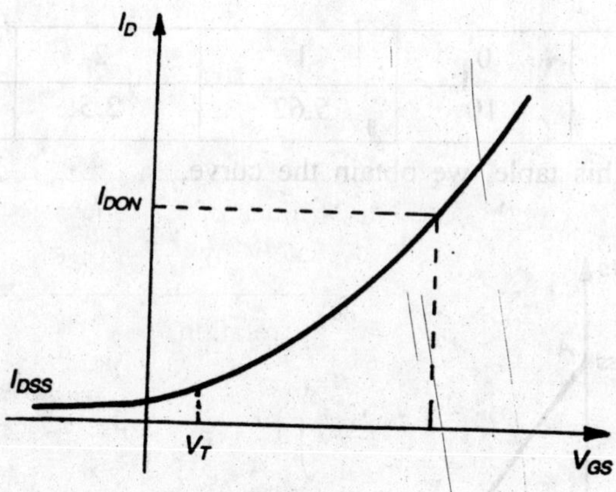

Fig. P. 7.3

Since, $$g_m = \left.\frac{\partial I_{DS}}{\partial V_{Gs}}\right|_{V_{DS}=const.}$$

We can find g_m by finding the slope of the transfer function characteristic at the given operating point.

Problem 7.4: (a) Show that the transconductance g_m of a JFET is related to the drain current I_{DS} by

$$g_m = \frac{2}{|V_P|} \sqrt{I_{DSS}\, I_{DS}}$$

Solution: The transconductance g_m of JFET given by expression.

$$g_m = g_{mo}\left(1 - \frac{V_{GS}}{V_P}\right) \qquad \ldots(i)$$

But, $$g_{mo} = \frac{-2I_{DSS}}{V_P}$$

$\therefore$ $$g_m = \frac{-2I_{DSS}}{V_P}\left(1 - \frac{V_{GS}}{V_P}\right) \qquad \ldots(ii)$$

Since, $$I_{DS} = I_{DSS}\left(1 - \frac{V_{GS}}{V_P}\right)^2$$

$\therefore$ $$g_m = \frac{-2I_{DSS}}{V_P}\left(\sqrt{\frac{I_{DS}}{I_{DSS}}}\right) = \frac{2}{|V_P|}\sqrt{I_{DSS} \times I_{DS}} \quad \textbf{Ans}$$

Problem 7.5: Show that for small values of V_{GS} compared with V_P, the drain current is given approximately by $I_D \approx I_{DSS} + g_{mo}V_{GS}$.

Solution: Drain current I_{DS} given by relation,

$$I_{DS} = I_{DSS}\left(1 - \frac{V_{GS}}{V_P}\right)^2 \qquad \ldots(i)$$

$$= I_{DSS}\left[1 - \frac{2V_{GS}}{V_P} + \frac{V^2{}_{GS}}{V_{P^2}}\right]$$

or, $$I_{DS} \approx I_{DSS}\left(1 - \frac{2V_{GS}}{V_P}\right)$$

Since $\quad V_{GS} \ll V_P$

or, $$I_{DS} \approx I_{DSS}\left[\frac{-2I_{DSS} \times V_{GS}}{V_P}\right]$$

Using $$g_{mo} = \frac{-2I_{DSS}}{V_P}$$

Equation (ii) becomes,

$$I_{DS} \approx I_{DSS} + g_{mo}V_{GS}$$

Problem 7.6: (a) Verify Eq. $\mu = r_d g_m$

(b) Starting with the definitions of g_m and r_d, show that if two identical FETs are connected in parallel, g_m is doubled and r_d is halved. Since $\mu = r_d g_m$, then μ remains unchanged.

(c) If the two FETs are not identical, show that

$$\frac{1}{r_d} = \frac{1}{r_{d1}} + \frac{1}{r_{d2}}$$

and that $$\mu = \frac{\mu_1 r_{d2} + \mu_2 r_{d1}}{r_{d1} + r_{d2}}$$

Solution: (a) The drain current i_D is a function of the gate voltage V_{GS} and drain voltage V_{DS} as expressed by

$$i_D = f(V_{GS}, V_{DS})$$

If both the gate and drain voltage are varied then, change in drain current, using Taylor's series expansion

$$\Delta i_D = \left.\frac{\partial i_D}{\partial V_{GS}}\right|_{V_{DS}} \Delta V_{GS} + \left.\frac{\partial i_D}{\partial V_{DS}}\right|_{V_{GS}} \times \Delta V_{DS} \qquad \ldots\text{(i)}$$

If $\quad \Delta i_D = i_d$, $\Delta V_{GS} = V_{gs}$ and $\Delta V_{DS} = V_{ds}$

then eq. (i) becomes,

$$i_D = g_m \times V_{gs} + \left(\frac{1}{r_d}\right) \times V_{ds}$$

Where $$g_m = \left.\frac{\partial i_D}{\partial V_{GS}}\right|_{V_{DS}} \approx \left.\frac{\Delta i_D}{\Delta V_{GS}}\right|_{V_{DS}} = \left.\frac{i_d}{V_{gs}}\right|_{V_{DS}}$$

and $$r_d = \left.\frac{\partial V_{DS}}{\partial i_D}\right|_{V_{GS}} \approx \left.\frac{\Delta V_{DS}}{\Delta i_D}\right|_{V_{GS}} = \left.\frac{V_{dS}}{i_d}\right|_{V_{GS}}$$

If $i_d = 0$, then, equation (ii) becomes,

$$-g_m r_d = \left.\frac{V_{dS}}{V_{gs}}\right|_{i_d=0}$$

As, $$\mu = -\left.\frac{V_{dS}}{V_{gs}}\right|_{i_d=0}$$

Then, $$\mu = +g_m r_d$$

(c) If the two FET's are parallel, then the circuit charge is double that of a single FET for a given change in gate voltage. Hence,

$$g'_m = \left.\frac{I_{ds\ \text{total}}}{V_{gs}}\right|_{V_{DS}=0}$$

$$= \left.\frac{i_{ds1}}{V_{gs}}\right|_{Vds=0} + \left.\frac{i_{ds2}}{V_{gs}}\right|_{Vds=0}$$

$$= g_{m1} + g_{m2} \qquad \text{...(iii)}$$

$$g'_d = \left.\frac{i_{dS\ \text{total}}}{V_{DS}}\right|_{V_{gs}=0} = \left.\frac{i_{ds1}}{V_{DS}}\right|_{V_{gs}=0} + \left.\frac{i_{ds2}}{V_{DS}}\right|_{V_{gs}=0}$$

or, $$\frac{1}{r_{d'}} = \frac{1}{r_{d1}} + \frac{1}{r_{d2}} \qquad \text{...(iv)}$$

(b) From eq. (iii)

$$g'_m = g_{m1} + g_{m2}$$

When $$g_{m1} = g_{m2}$$

$$g'_m = 2g_m$$

From eq. (iv), When $r_{d1} = r_{d2} = r_d$

$$\frac{1}{r_{d'}} = \frac{2}{r_d}$$

Or, $$r'_d = r_{d/2}$$

Problem 7.7: (a) Calculate the voltage gain $A_v = V_o/V_i$ at 1 kHz for the circuit shown. The FET parameters are g_m = 2 mA/V and r_d = 10 K. Neglect capacitances.

(b) Repeat part *a* if the capacitance 0.003 μF is taken under consideration.

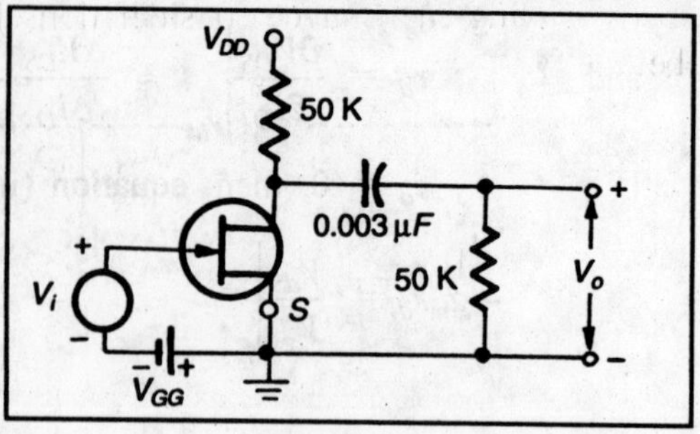

Fig. P. 7.7

Solution: The small-signal equivalent of the FET amplifier drawn above neglecting capacitance,

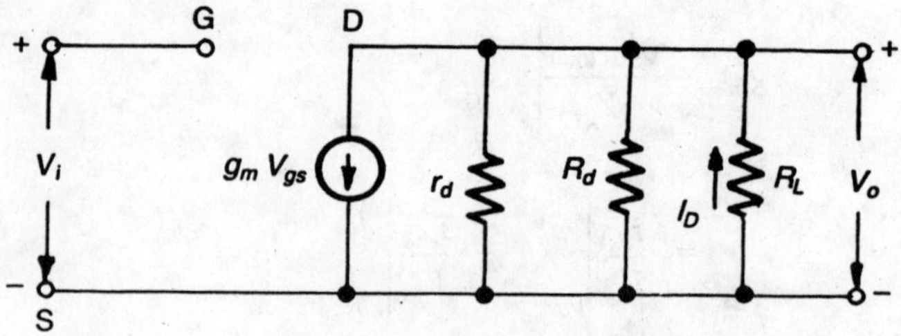

Fig. P. 7.7 (a)

From the circuit, we have,

$$-g_m V_{gs} = V_o \left[\frac{1}{R_L} + \frac{1}{r_d} + \frac{1}{R_d}\right]$$

$$\frac{V_o}{V_i} = \frac{V_o}{V_{gs}} = \frac{-g_m}{\left[\frac{1}{R_L} + \frac{1}{r_d} + \frac{1}{R_d}\right]}$$

Thus, voltage gain,

$$A_v = -g_m \times \frac{R_L \times r_d \times R_d}{R_L r_d + R_d R_L + R_d r_d}$$

$$= -2 \times 10^{-3} \times \frac{25000 \times 10^3}{2500 + 500 + 500} = \frac{-500}{35} = -14.28$$

(b) Taking capacitance consideration, the equivalent circuit is drawn below,

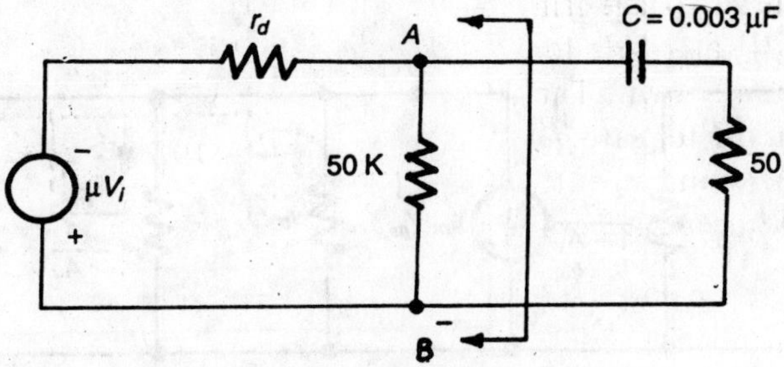

Fig. P. 7.7 (b)

Thevenin equivalent to the left of AB consists of

$$R_{Th} = 50\text{ K} \parallel r_d = 10\text{ K} \parallel 50\text{ K} = 8.33\text{ K}$$

and
$$V_{Th} = \mu V_i \frac{50\text{ K}}{10\text{ K} + 50\text{ K}} = \frac{20 \times 50}{60} = 16.67\ V_i$$

Equivalent circuit is drawn below

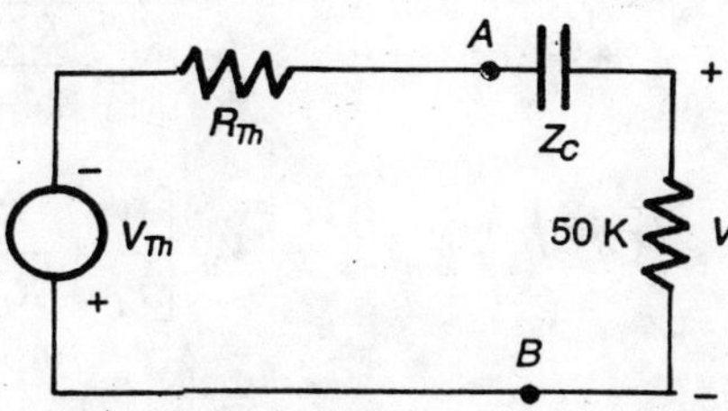

Fig. P. 7.7 (c)

$$Z_c = \frac{-1}{\omega c} j$$

$$= \frac{1}{2\pi \times 10^3 \times 0.003 \times 10^{-6}}$$

$$= -j153\text{ K}$$

$$\frac{V_o}{V_i} = \frac{-50\text{ K}(16.67)}{50\text{ K} + 8.33\text{ K} - 153\text{ K}j}$$

$$= \frac{-8.33}{58.33 - 153j} = 10.6\text{ L} - 137.7^\circ$$

Problem 7.8: If an input signal V_i is impressed between gate and ground, find the amplification $A_V = V_o/V_i$. Apply Miller's theorem to the 50-K resistor. The FET parameters are $\mu = 30$ and $r_d = 5$ K. Neglect capacitances.

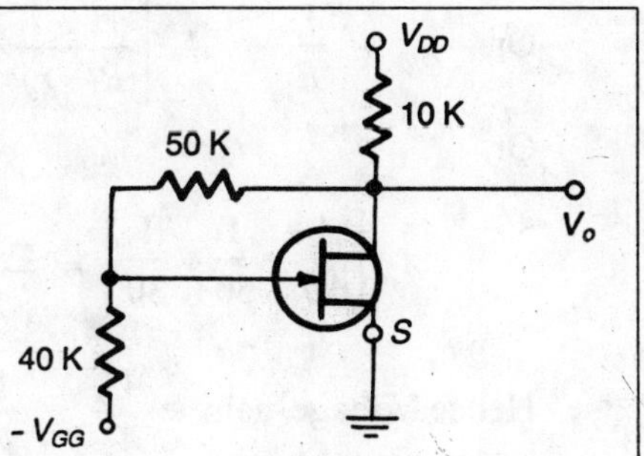

Fig. P. 7.8

Solution: Drawing the equivalent circuit of the given FET amplifier

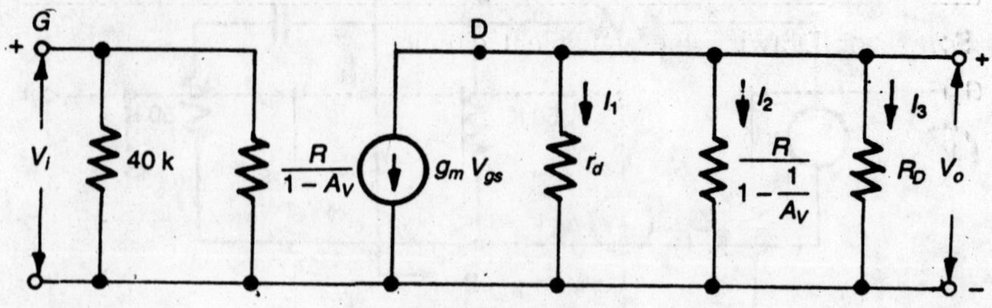

Fig. P. 7.8 (a)

Here $R_D = 10\ \text{K}\Omega$

$R = 50\ \text{K}\Omega$

From the circuit, using KCL at drain terminal, we have

$$-g_m V_{gs} = I_1 + I_2 + I_3$$

Or, $$-g_m V_{gs} = \frac{V_o}{r_d} + \frac{V_o}{\dfrac{RA_v}{A_v - 1}} + \frac{V_o}{R_D}$$

Or, $$-g_m V_{gs} = V_o\left[\frac{1}{r_d} + \frac{1}{R_D} + \frac{A_v - 1}{RA_v}\right]$$

Or, $$-g_m = \frac{V_o}{V_{gs}}\left[\frac{1}{r_d} + \frac{1}{R_D} + \frac{A_v - 1}{R \times A_v}\right]$$

$$= A_v\left[\frac{1}{r_d} + \frac{1}{R_D}\right] + \frac{(A_v - 1)}{R}$$

Or, $$-g_m + \frac{1}{R} = A_v\left[\frac{1}{R} + \frac{1}{r_d} + \frac{1}{R_D}\right]$$

Or,

$$A_v\left[\frac{1}{10} + \frac{1}{50} + \frac{1}{30}\right] = \frac{-30}{5} + \frac{1}{50} = \frac{-299}{50}$$

Hence, Voltage gain $= \dfrac{V_o}{V_i} = \dfrac{V_o}{V_{gs}} = A_v = \dfrac{-229}{16} = -18.7$

Problem 7.9: If in prob. 7.8 the signal V_i is impressed in series with the 40-K resistor (instead of from gate to ground), find $A_v = V_o/V_i$.

Solution: Drawing the equivalent circuit,

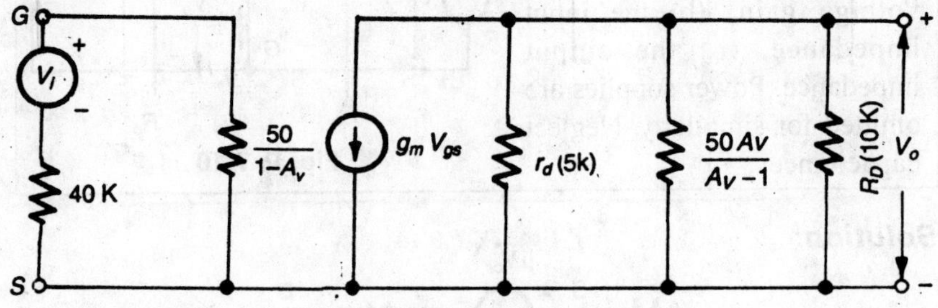

Fig. P. 7.9

We have, $$A_v = \frac{V_o}{V_{gs}} = -18.7$$

From the circuit,

$$\frac{V_{GS} - V_i}{40} = -\frac{V_{GS}}{R_i}$$

Where, $$R_i = \frac{50}{1 - A_v}$$

$$= \frac{50}{1 + 18.7} = 2.54 \text{ K}\Omega$$

$$\therefore\ V_{GS}\left[\frac{1}{R_i} + \frac{1}{40}\right] = \frac{V_i}{40}$$

$$\Rightarrow\ \frac{V_{GS}}{V_i} = \frac{R_i}{R_i + 40}$$

Then, Voltage gain,

$$A_v = \frac{V_o}{V_i} = \frac{V_o}{V_{gs}} \times \left(\frac{V_{gs}}{V_i}\right) = A_v \times \left(\frac{R_i}{R_i + 40}\right)$$

$$= -18.7\left(\frac{2.54}{254 \times 40}\right) = -1.11$$

Problem 7.10:
The circuit shown is called common-gate amplifier. For this circuit find (a) the voltage gain, (b) the input impedance, (c) the output impedance. Power supplies are omitted for simplicity. Neglect capacitances.

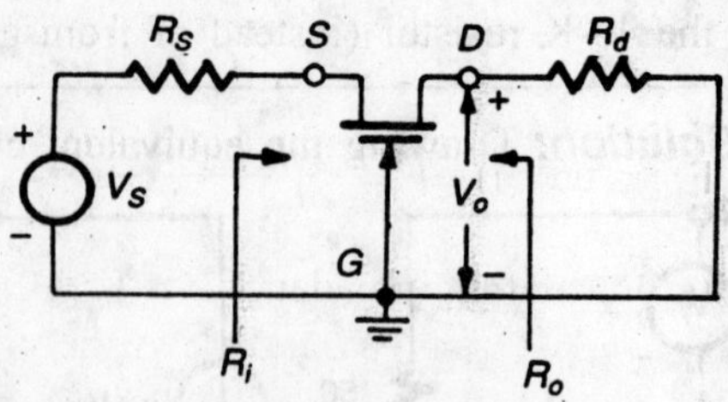

Fig. P. 7.10

Solution:

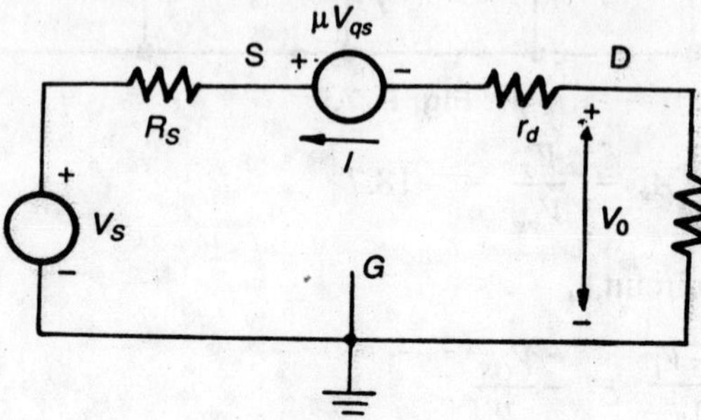

Fig. P. 7.10 (a)

From the circuit, we have,

$$V_{gs} = -V_s - R_s I \qquad \ldots\text{(i)}$$

Now, applying KVL around the loop, we get

$$-I\,(R_d + r_d + R_s) + \mu V_{qs} - V = 0 \qquad \ldots\text{(ii)}$$

Substituting the expression for V_{gs} from eq (i) in eq (ii), we get

Or, $$-I\,[R_d + r_d + (\mu + 1)\,R_s] = V_s\,(\mu + 1)$$

Or, $$I = -\frac{V_s(\mu+1)}{r_d + R_d + R_s(\mu+1)} \qquad \ldots\text{(iii)}$$

(a) Voltage gain,

$$A_v = \frac{V_o}{V_s} = \frac{-IR_d}{V_s}$$

From eq. (iii) $$A_v = \frac{(\mu+1)R_d}{r_d + R_d + R_s(\mu+1)}$$

(b) Input impedance,

$$R_i = -\frac{V_s}{I}$$

From eq. (i), $$R_i = \frac{r_d + R_d + R_s}{\mu + 1}$$

(c) For output impedance, set $V_s = 0$, and $R_d = \infty$ and supply a voltage V between D and ground, then the circuit will be as given below,

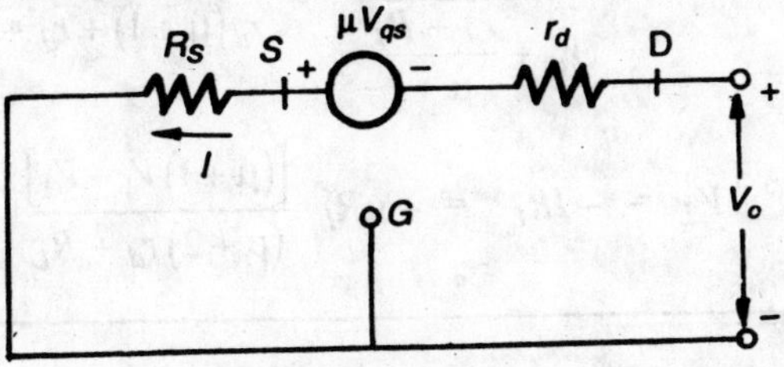

Fig. P. 7.10 (b)

Now applying KVL around the loop,

$$V + \mu V_{gs} = I\ (R_s + r_d)$$

Or, $V + \mu\,(-IR_s) = I\ (R_s + r_d)$

[Since $V_{gs} = -IR_s$ from eq. (i)]

Or, $$V = I\ [R_s\ (\mu + 1) + r_d]$$

Then, $\dfrac{V}{I}$ (Output impedance) $= r_d + (\mu + 1)\ R_s$

Problem 7.11:

Find an expression for the signal voltage across R_L. The two FETs are identical, with parameters μ, r_d and g_m.

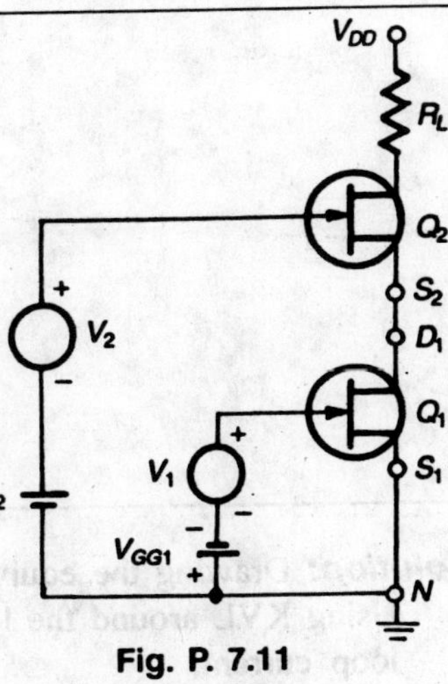

Fig. P. 7.11

Solution: Applying KVL around the loop, we have,

$$-\left[\mu V_1 + \frac{\mu V_2}{\mu+1}\right] + I\left[r_d + \frac{r_d + R_L}{\mu+1}\right] = 0$$

Solving for I, we obtain

$$I = \frac{\left[\mu V_1 + \dfrac{\mu V_2}{\mu+1}\right]}{r_d + \dfrac{r_d + R_L}{\mu+1}} = \frac{\mu V_1(\mu+1) + \mu V_2}{r_d(\mu+1) + r_d + R_L}$$

But, $$V_L = -IR_L = -\mu R_L \frac{[(\mu+1)V_1 + V_2]}{(\mu+2)r_d + R_L}$$

Problem 7.12:

Each FET shown has the parameters $r_d = 10$ K and $g_m = 2$ mA/V. Find the gain (a) v_o/v_1 if $v_2 = 0$, (b) v_o/v_2 if $v_1 = 0$.

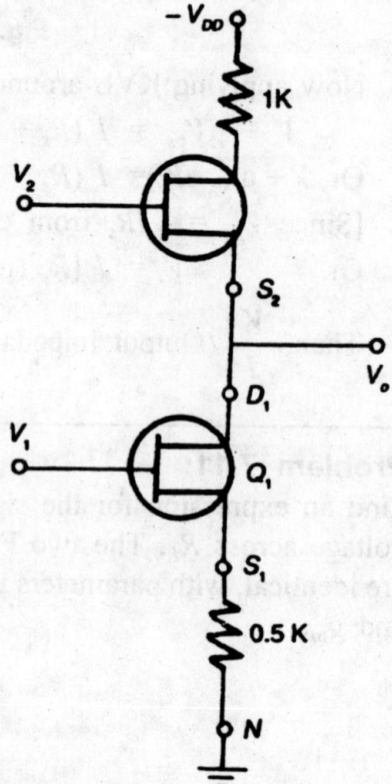

Fig. P. 7.12

Solution: Drawing the equivalent circuit

Using KVL around the loop, we have,

loop current,

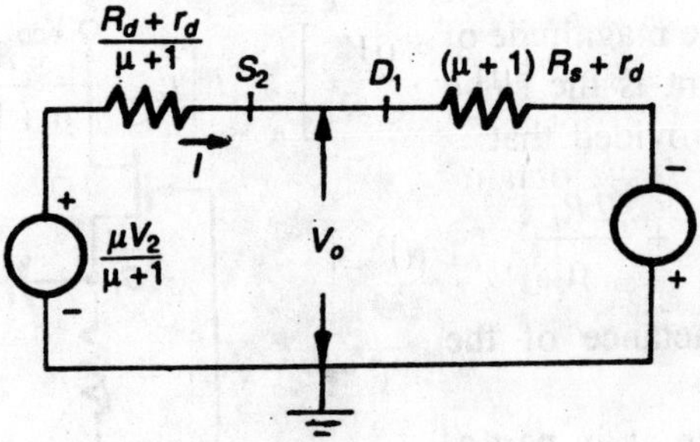

Fig. P. 7.12

$$I = \frac{\mu V_1 + \dfrac{\mu V_2}{\mu + 1}}{\dfrac{R_d + r_d}{\mu + 1} + (\mu + 1)R_s + r_d} = \frac{\mu V_1(\mu + 1) + \mu V_2}{R_d + r_d + (\mu + 1)[(\mu + 1)R_s + r_d]}$$

$$\text{Then, } V_o = \frac{\mu V_2}{\mu + 1} - I\left(\frac{R_d + r_d}{\mu + 1}\right)$$

$$= \frac{\dfrac{R_d + r_d}{\mu + 1} \times (-\mu V_1) + [(\mu + 1)\, R_s + R_d] \times \dfrac{\mu}{\mu + 1} \times V_2}{\dfrac{R_d + r_d}{\mu + 1} + (\mu + 1)\, R_s + r_d}$$

(a) When $V_2 = 0$

$$\frac{V_o}{V_1} = \frac{-\mu\,(R_d + r_d)}{R_d + (\mu + 1)^2 R_s + (\mu + 2)\, r_d}$$

$$= \frac{-20 \times 11\,\text{K}}{[1 \times 21^2 \times 0.5 + 22 \times 10]\,\text{K}} = \frac{-220}{441.5} = -0.49$$

(b) When $V_1 = 0$, $$\frac{V_o}{V_2} = \frac{\dfrac{\mu}{\mu + 1}[(\mu + 1)\, R_s + r_d]}{\dfrac{R_d + r_d}{\mu + 1} + (\mu + 1)\, R_s + r_d}$$

$$= \frac{20 \times 0.5 + \dfrac{20 \times 10}{21}}{\dfrac{11}{21} + 21 \times 0.5 + 10} = \frac{19.53}{21.02} = 0.93$$

Problem 7.13:

(a) Prove that the magnitude of the signal current is the same in both FETs provided that

$$r = \frac{1}{g_m} + \frac{2R_L}{\mu}$$

Neglect the reactance of the capacitors.

(b) If r is chosen as in part a, prove that the voltage gain is given by

$$A = \frac{-\mu^2}{\mu+1} \frac{R_L}{R_L + r_d/2}$$

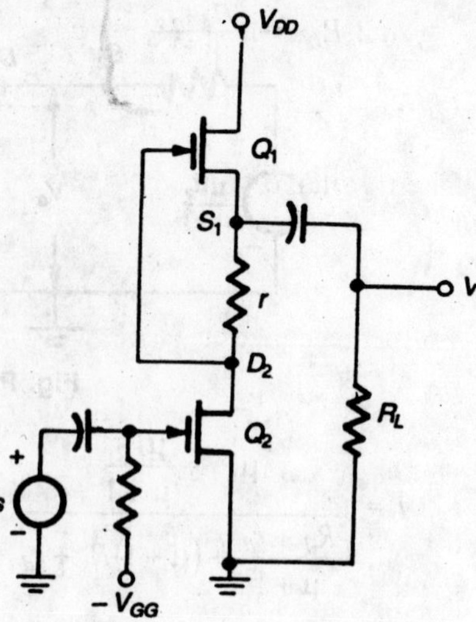

Fig. P. 7.13

Solution: (i) Drawing the equivalent circuit of the given circuit.

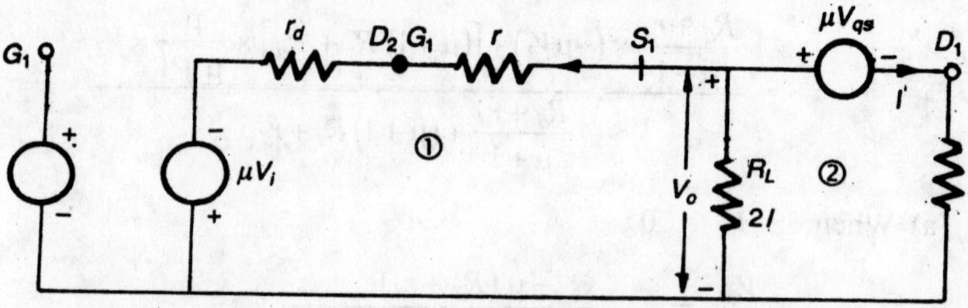

Fig. P. 7.13 (a)

Here we assumed that the signal current is same in both the FETs. Then to prove,

$$r_d = \frac{2R_L}{\mu} + \frac{1}{g_m}$$

Now, applying KVL rule to the loop (2), we have

$$2\,IR_L + \mu V_{gs1} + Ir_d = 0 \quad \ldots\text{(i)}$$

Where, $V_{gs1} = -Ir$

Then, $2\,IR_L - \mu Ir + Ir_d = 0$

Or, $$r = \frac{2R_L}{\mu} + \frac{r_d}{\mu} = \frac{2R_L}{\mu} + \frac{1}{g_m} \quad (\text{as } \mu = g_m r_d)$$

(b) Applying, KVL to loop (i), we have

$$2IR_L + Ir + Ir_d = \mu V_i \quad \text{...(ii)}$$

Or,
$$I = \frac{\mu V_i}{2R_L + r + r_d}$$

Then, voltage gain,

$$A = \frac{V_o}{V_i} = \frac{-2IR_L}{V_i}$$

From the equation (ii),

$$A = -2R_L\left(\frac{\mu}{2R_L + r + r_d}\right) \quad \text{...(iii)}$$

Putting
$$r = \frac{2R_L}{\mu} + \frac{1}{g_m}$$

in equation (iii), we get

$$A = \frac{-2R_L \times \mu^2}{(\mu+1)(2R_L + r_d)} = \frac{-\mu^2}{\mu+1}\left(\frac{R_L}{R_L + \frac{r_d}{2}}\right)$$

Problem 7.14:

(a) If $R_1 = R_2 = R$ and the two FETs have identical parameters, verify that the voltage amplification is $V_o/V_s = -\mu/2$ and the output impedance is $\frac{1}{2}[r_d + (\mu + 1) R]$.

(b) Given

$$r_d = 62 \text{ K}$$
$$\mu = 10$$
$$R_1 = 2 \text{ K}$$
and
$$R_2 = 1 \text{ K}$$

Find the voltage gain and the output impedance.

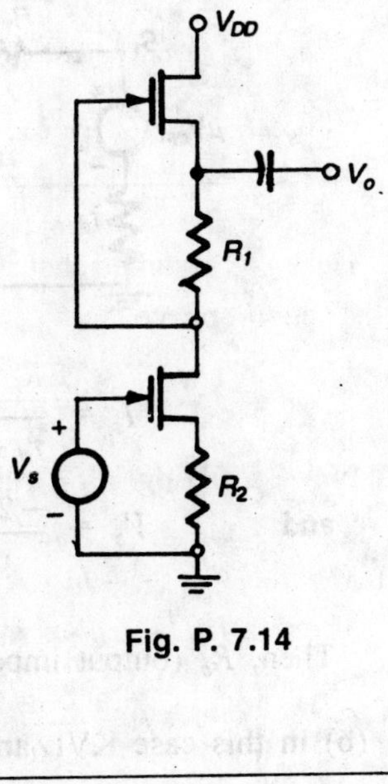

Fig. P. 7.14

Solution: Given,

The small-signal equivalent circuit given circuit is shown below,

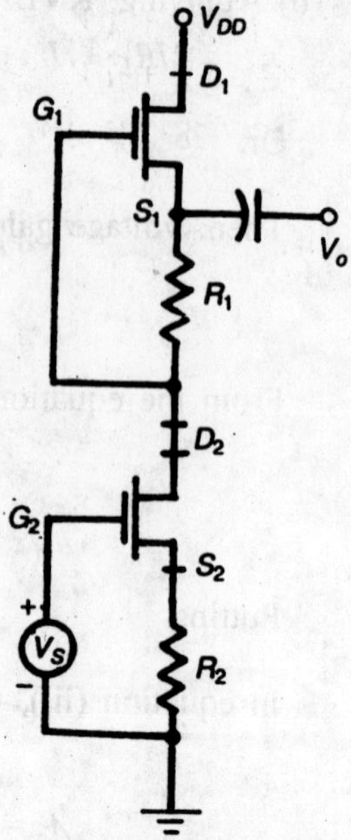

Fig. P. 7.14 (a)

Now, applying KVL around the loop, we have

$$\mu V_{gs1} + \mu V_s = [2r_d + (\mu + 2) R] I$$

Since $V_{gs1} = -IR$, then

$$-\mu RI + \mu V_s = I [2r_d + (\mu + 2)R]$$

$$\Rightarrow \quad I = \frac{\mu V_s}{2[r_d + (\mu + 1)R]}$$

Now, $V_o = -I (r_d + R) + \mu V_{gs1}$

$$= -I [r_d + R + \mu R]$$

$$= -\frac{\mu}{2} V_s$$

Hence, $\dfrac{V_o}{V_s} = \dfrac{-\mu}{2}$

To find the output impedance, we ground D_2 then we have

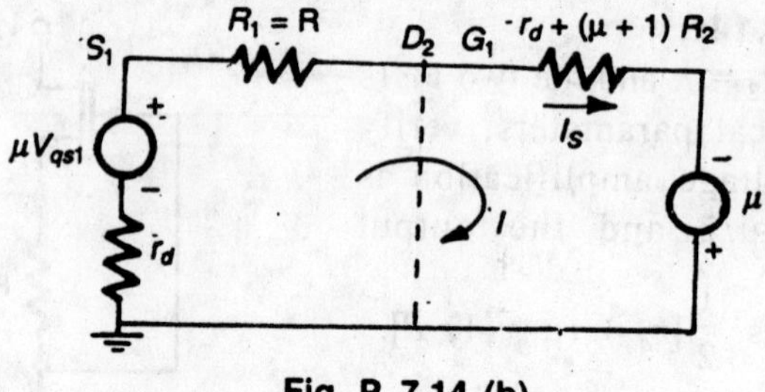

Fig. P. 7.14 (b)

$$I_s = \frac{-\mu V_s}{r_d + (\mu + 1) R}$$

and $\quad V_s = \dfrac{-2V_o}{\mu}$

Then, R_o (output impedance) $= \dfrac{V_o}{I_s} = \dfrac{1}{2} [r_d (\mu + 1) R]$

(b) In this case KVL around the loop gives,

$$\mu V_{gs1} + \mu V_s = I\,[2r_d + (\mu + 1)\,R_2 + R_1] \quad \text{...(i)}$$

$$-\mu R_1 I + \mu V_s = I\,[2r_d + (\mu + 1)\,R_2 + R_1] \quad \text{...(ii)}$$

Solving, we get,

$$I = \frac{\mu}{2r_d + (\mu+1)(R_1 + R_2)} \times V_s$$

and, $$V_o = -I\,(r_d + R_1) + \mu V_{gs1} = -I\,[r_d + (\mu + 1)\,R_1]$$

Then, $$A = \frac{V_o}{V_s} = \frac{-\mu\left[r_d + (\mu+1)\,R_1\right]}{2r_d + (\mu+1)(R_1 + R_2)}$$

$$= \frac{-10\,[62 + 22]}{124 + 3.3} = -5.35$$

In this case, $$I_s = \frac{-\mu V_s}{r_d + (\mu+1)\,R_2}$$

Then, output impedance

$$R_o = \frac{V_o}{I_s} = \left(\frac{V_D}{V_s}\right)\left(\frac{V_s}{I_s}\right)$$

$$= A \times \frac{r_d + (\mu+1)R_2}{-\mu} = 5.3 \times 7.3\ \text{K} = 39\ \text{K}$$

Problem 7.15: (a) If in the amplifier stage shown the positive supply voltage V_{DD} changes by $\Delta V_{DD} = v_a$, how much does the drain-to-ground voltage change?

(b) How much does the source-to-ground voltage change under the conditions in part a?

(c) Repeat parts *a* and *b* if V_{DD} is constant but V_{ss} changes by $\Delta V_{ss} = v_s$.

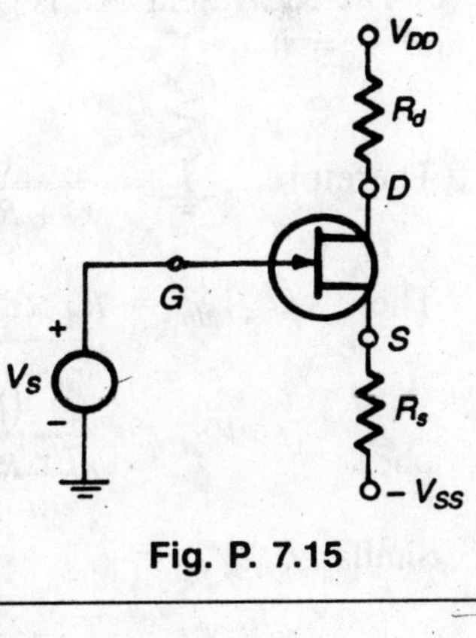

Fig. P. 7.15

Solution: The equivalent ckt. can be drawn as follows :

Applying KVL around the loop gives,

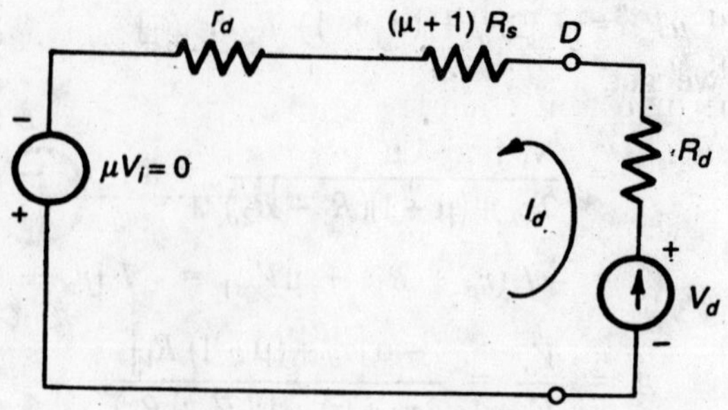

Fig. P. 7.15 (a)

$$I = \frac{V_d}{R_d + r_d + (\mu + 1)R_s}$$

Also, $V_{dn} = I\ (r_d + (\mu + 1)\ R_s)$

$\therefore$ $$V_{dn} = \frac{r_d + (\mu + 1)\ R_s}{r_d + R_d + (\mu + 1)\ R_s} \times V_d$$

(b) Also $V_{sn} = I\ (\mu + 1)\ R_s$

Hence, $$V_{sn} = \frac{(\mu + 1)\ R_s}{r_d + R_d + (\mu + 1)\ R_s}\ V_d$$

(c) The equivalent ckt. is given as below with a source in series with R_s and $v_i = 0$.

Therefore, $$I = \frac{(\mu + 1)\ V_s}{r_d + R_d + (\mu + 1)\ R_s}$$

Then, $V_{dn} = R_d \times I$

$\therefore$ $$V_{dn} = \frac{(\mu + 1)\ R_d}{r_d + R_d + (\mu + 1)\ R_s}\ V_s$$

Similarly, $V_{sn} = (r_d + R_d)$

$$I = \frac{(\mu + 1)(r_d + R_d)}{r_d + R_d(\mu + 1)\ R_s}\ V_s$$

Problem 7.16: The amplifier stage shown uses an *n*-channel FET having I_{DSS} = 1 mA, V_P = −1 V. If the quiescent drain-to-ground voltage is 10 V, find R_1.

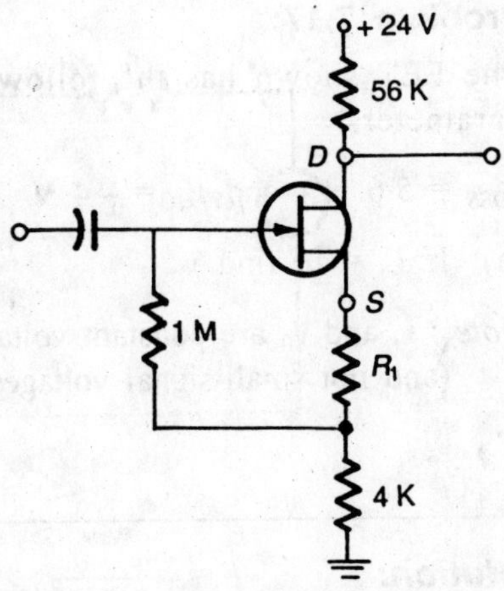

Fig. P. 7.16

Solution: From the given circuit,

$$I_D = \frac{V_{DD} - V_{DN}}{R_d} = \frac{24 - 10}{56} A = \frac{1}{4} \text{ mA}$$

As, $I_D = I_{DSS} \left(1 - \frac{V_{GS}}{V_p}\right)^2$

Or, $\frac{1}{4} = 1\left(1 - \frac{V_{GS}}{V_P}\right)^2$ (Given, I_{DSS} = 1 mA)

Or, $\frac{1}{\frac{V_{GS}}{V_P}} = \frac{1}{2}$

Or, $1 + \frac{V_{GS}}{1} = \frac{1}{2}$

Or, $V_{GS} = -0.5$ V

But, $V_{GS} = -I_D R_1$

then, $R_1 = -\frac{V_{Gs}}{I_D} = \frac{0.5}{\frac{1}{4} \times 10^{-3}} \Omega = 2 \text{ k}\Omega$

Problem 7.17:

The FET shown has the following parameters:

I_{DSS} = 5.6 mA and V_P = – 4 V.

(a) If V_i = 0, find v_o.

Note : v_i and v_0 are constant voltages (and not small-signal voltages).

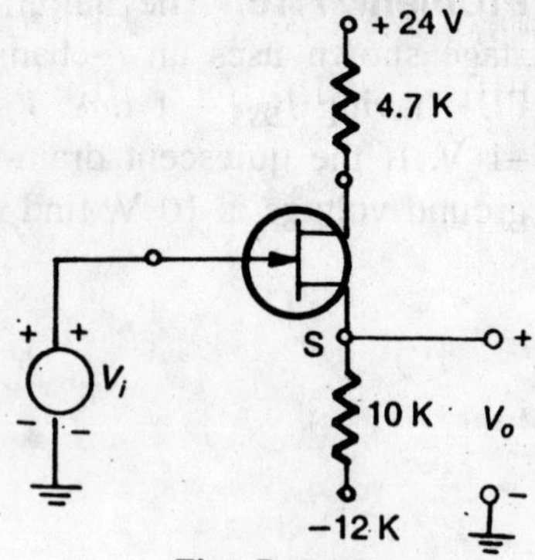

Fig. P. 7.17

Solution:

KVL in the $G - S$ loop gives,

$$V_i = V_{GS} + 10I_d - 12$$

$$\text{Or, } I_d = \frac{V_i + 12 - V_{GS}}{10} \quad \text{...(i)}$$

$$\text{(a) If } V_i = 0,\ I_d = \frac{12 - V_{GS}}{10}$$

$$= I_{DSS}\left(1 - \frac{V_{GS}}{V_P}\right)^2$$

$$\left[\because I_D = I_{DSS}\left(1 - \frac{V_{GS}}{V_P}\right)^2\right]$$

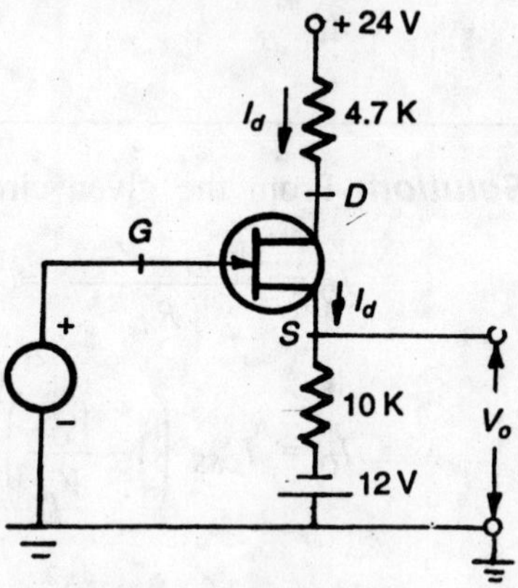

Fig. P. 7.17 (a)

$$\text{Or, } \frac{12 - V_{GS}}{10} = 5.6\left(1 + \frac{V_{GS}}{4}\right)^2$$

$$\text{Or, } 5.6\left[1 + \frac{V_{GS}}{2} + \frac{V_{GS^2}}{16}\right] = \frac{12 - V_{GS}}{10}$$

Solving we get, $V_{GS} = -2$ V

then, I_d = 1.4 mA

and, $V_D = 10\ I_d - 12 = 14 - 12 = 2$ V

Problem 7.18: If $|I_{DSS}| = 4$ mA, $V_P = 4$ V, calculate the quiescent values of I_D, V_{GS}, and V_{DS}.

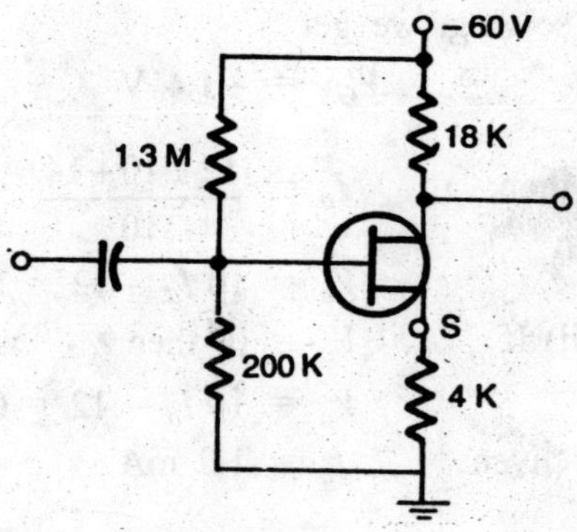

Fig. P. 7.18

Solution: If we find Thevenin's equivalent to the left of the gate of the circuit. So we have the resultant circuit as,

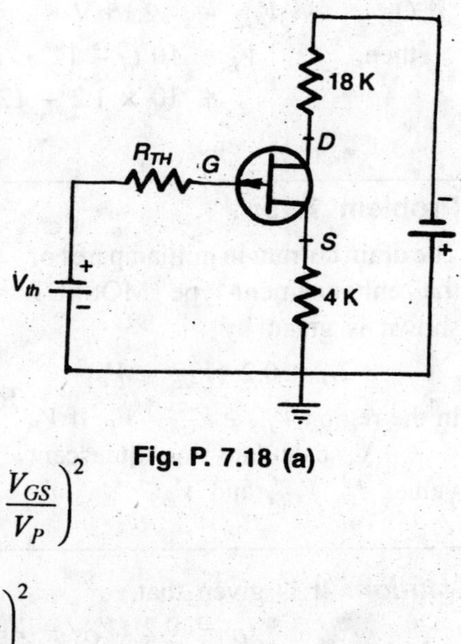

Fig. P. 7.18 (a)

R_{Th} = 1.3 M 11200 K = 173.5 K

$$\text{and } V_{Th} = \frac{200}{200+1300} \times -60$$

$$= -8 \text{ V}$$

KVL in the GS loop gives,

$$V_{GS} = -4\, I_d - 8$$

$$\text{or, } I_d = \frac{-V_{GS}-8}{4} = I_{DSS}\left(1-\frac{V_{GS}}{V_P}\right)^2$$

$$\text{or, } -\frac{V_{GS}}{4} - 2 = 4\left(1-\frac{V_{GS}}{4}\right)^2$$

$$\text{or } \frac{-V_{GS}}{4} - 2 = 4\left[1-\frac{V_{GS}}{2}+\frac{V_{Gs}^2}{16}\right]$$

Solving, we get $V_{GS} = 1$ V

and $I_D = -2.25$ mA

hence $V_{DS} = -60 + (18 + 4)\ 2.25 = -60 + 49.5 = -10.5$ V

solving, we get

$$V_{GS} = -1.4 \text{ V}$$

then, $$I_d = \frac{12+10+14}{10} = 2.34 \text{ mA}$$

$\therefore$ $$V_o = 10\,I_d - 12 = 11.4 \text{ V}$$

(iii) If $V_o = 0$, then

$$V_o = 10\,I_d - 12 = 0$$

Given; $$I_d = 1.2 \text{ mA}$$

then $$1.2 = 5.6\left(1+\frac{V_{GS}}{4}\right)^2$$

Or, $$V_{GS} = -2.15 \text{ V}$$

then, $$V_i = 10\,I_d - 12 + V_{GS}$$
$$= 10 \times 1.2 - 12 + (-2.15) = -2.15 \text{ V}$$

Problem 7.19:

The drain current in milliamperes of the enhancement-type MOSFET shown is given by

$$I_D = 0.2\,(V_{GS} - V_P)^2$$

in the region $V_{DS} \geq V_{GS} - V_P$. If $V_P = +3$ V, calculate the quiescent values I_D, V_{GS}, and V_{DS}.

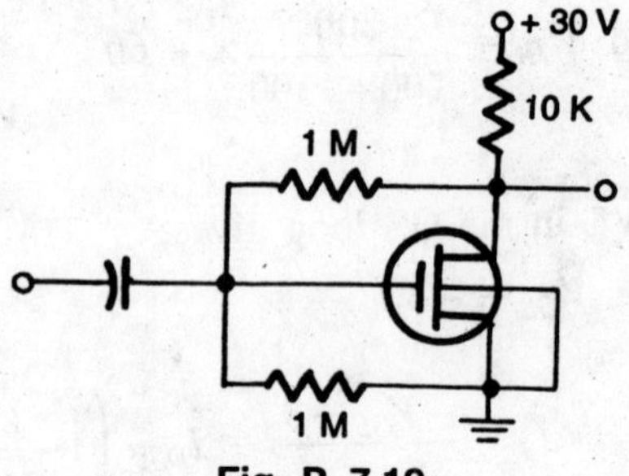

Fig. P. 7.19

Solution: It is given that,

$$I_D = 0.2\,(V_{GS} - V_P)^2 \quad \text{...(i)}$$

Now from the ckt.,

$$V_{GS} = \frac{1}{2}\,(30 - 10\,I_D) \quad \text{...(ii)}$$

From (i) and (ii),

$$I_D = 0.2\,(15 - 5\,I_D - 3) = 0.02\,(144 - 120\,I_D + 25\,I_D^2)$$

Or, $5\,I_D^2 - 25\,I_D + 28.8 = 0$

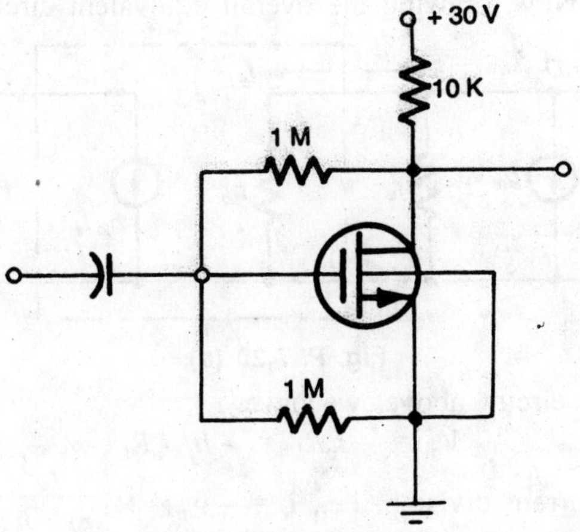

Fig. P. 7.19 (a)

Then, $I_D = \dfrac{25 \pm \sqrt{625 - 576}}{10} = \dfrac{25 \pm \sqrt{49}}{10}$

and $I_D = 1.8$ or 3.2 mA

The second value of 3.2 mA is rejected

Since, $V_{GS} = \dfrac{1}{2}(30 - 10 \times 1.8) = +6$ V

and $V_s = 30 - 10 \times 1.8 = +12$ V

Problem 7.20:
Show that if $R_L \ll 1/h_{ob2}$, the voltage gain of the hybrid cascode amplifier stage shown is given to a very good approximation by

$A_v = g_m h_{fb} R_L$

where g_m is the FET transconductance.

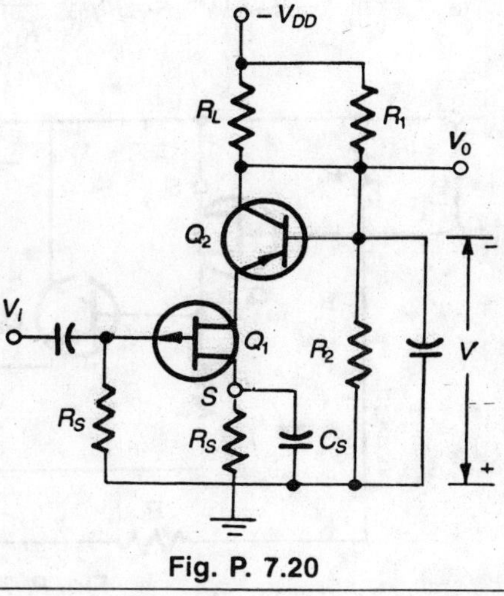

Fig. P. 7.20

Solution: $h_{ob2} \ll 1$, so we can draw simplified hybrid model for the transistor Q_2. Now, drawing the overall equivalent circuit.

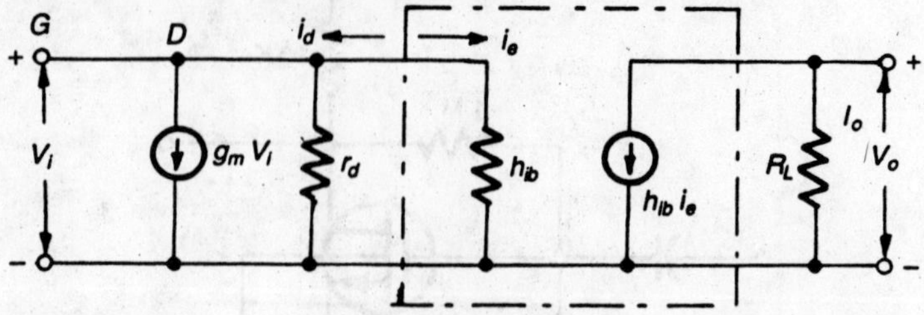

Fig. P. 7.20 (a)

From the circuit above, we have

$$V_o = -I_o R_L = -h_{fb} i_e R_L \quad \text{...(i)}$$

By the current division, i.e., $i_e = -g_m V_i \times \dfrac{r_d}{r_d + h_{ib}}$ ...(ii)

Voltage gain, $$A_V = \frac{V_o}{V_i} = -\frac{R_L h_{fb}\, i_e}{V_i}$$

Voltage eq. (2) $$A_v = \frac{+g_m h_{fb} r_d R_L}{r_d + h_{ib}}$$

Since drain resistance r_d much higher than transistor input resistance h_{ib} i,e $r_d \gg h_{ib}$

Hence, $$A_v \approx g_m h_{fb} R_L$$

Problem 7.21: If $r_d \gg R_1$, $R_2 \gg h_{ib3}$, $1/h_{oe2} \gg h_{oe3}$, $R' \gg R_3$, and $1/h_{ob3} \gg R_3$, show that voltage gain at low frequencies is given by

$$A_o = \frac{v_o}{v_i} = g_m (1 + h_{fe2}) h_{fb3} \frac{R_1 R_3}{R_1 + h_{ie2} + h_{ib3}(1 + h_{fe2})}$$

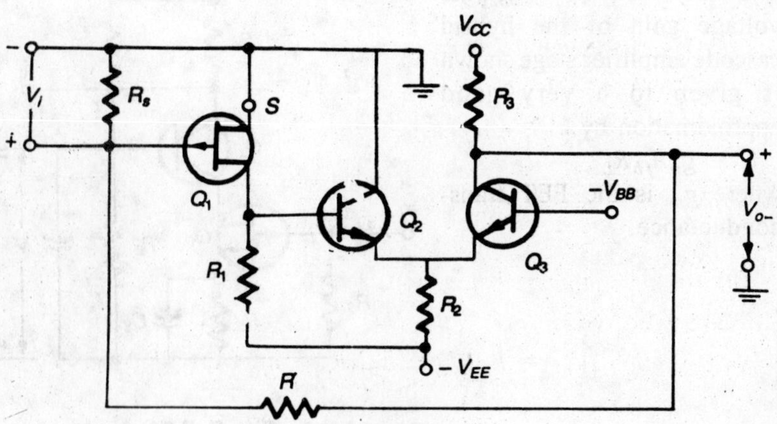

Fig. P. 7.21

Solution:

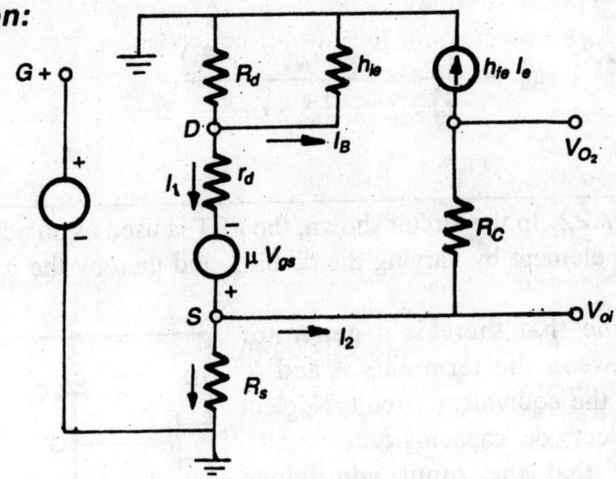

Fig. P. 7.21

From the circuit,

$$V_{gs} = V_i - IR_s = V_i - IR_s \qquad \text{...(i)}$$

Where, $I = I_1 - I_2 = I_1 - h_{fe}\, I_B$

since $h_{ie} \ll R_d$, then

$$I_B \approx -I_1$$

$$I = I_1 + h_{fe}I_1 = (1 + h_{fe})\, I_1 \qquad \text{...(ii)}$$

Using KVL in the FET loop gives,

$$[h_{ie} + r_d + R_s\,(1 + h_{fe})] = \mu V_{gs} = [\mu V_i - \mu(1 + h_{fe})\, R_s I_1]$$

Or,

$$I_1 = \frac{\mu V_i}{h_{ie} + r_d + (\mu+1)(1+h_{fe})R_s} \approx \frac{g_m V_i}{1 + g_m h_{fe} R_s} \qquad \text{...(iii)}$$

since, $r_d \gg h_{ie}$

Now, As

$h_{fe} \gg 1$ and $\mu \gg 1$

$$V_{o1} = IR_s = (1 + h_{fe})\, R_s I_1 \approx h_{fe} R_s I_1 \qquad \text{...(iv)}$$

(i)
$$A_{v1} = \frac{V_{o1}}{V_i} \approx \frac{g_m h_{fe} R_e}{1 + g_m h_{fe} R_s}$$

(ii)
$$V_{o2} = V_{o1} - h_{fe} R_s I_B R_c$$

From eq. (iv)

$$V_{o2} = h_{fe} R_s I_1 + h_{fe} I_1 R_c = h_{fe}\,(R_s + R_c)\, I_1$$

From eq. (iii),

$$A_{v2} = \frac{V_{o2}}{V_i} \approx \frac{g_m h_{fe}(R_s + R_c)}{1 + g_{m} {}_{fe} R_s}$$

Problem 7.22: In the circuit shown, the FET is used as an adjustable impedance element by varying the dc bias, and thereby the g_m of the FET.

(a) Assume that there is a generator V between the terminals A and B. Draw the equivalent circuit. Neglect interelectrode capacitances.

(b) Show that the input admittance between A and B is

$$Y_i = Y_d + (1 + g_m R) Y_{CR}.$$

where Y_d is the admittance corresponding to r_d, and Y_{CR} is the admittance corresponding to R and C in series.

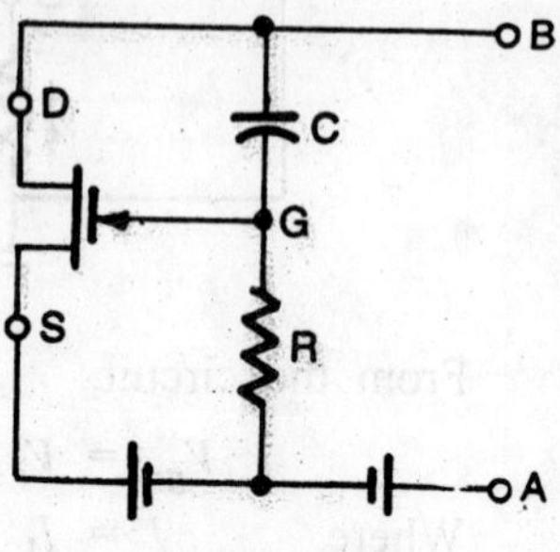

Fig. P. 7.22

(c) If $g_m R \gg 1$, show that the effective input capacitance is

$$C_i = \frac{g_m \alpha}{\omega\left(1+\alpha^2\right)}$$

and the effective input resistance is

$$R_i = \frac{\left(1+\alpha^2\right) r_d}{1+\alpha^2(1+\mu)}$$

where $\alpha \equiv \omega CR.$

(d) At a given frequency, show that the maximum value of C_i (as either C or R is varied) is obtained when $\alpha = 1$, and

$$(C_i)_{\max} = \frac{g_m}{2\omega}$$

Also show that the value of R_i corresponding to this C_i is

$$(R_i)_{\max} = \frac{2 r_d}{2 + \mu}$$

which, for $\mu \gg 2$, reduced to $(R_i)_{\max} = 2/g_m$.

Solution: (a) The equivalent ckt. is given as :

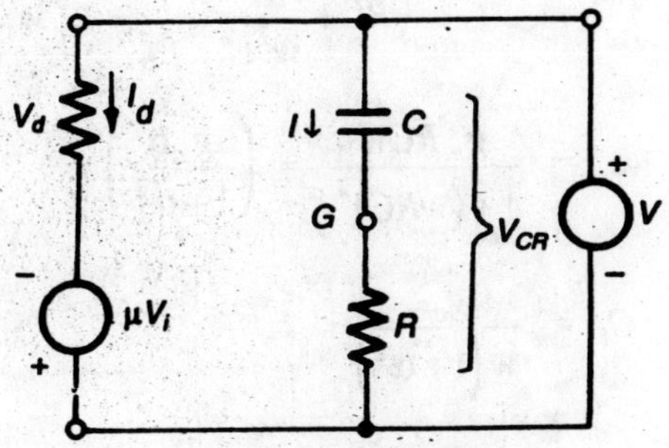

Fig. P. 7.22 (a)

(b) $I_d \left(\dfrac{V + \mu V_{gs}}{r_d}\right)$

but $\quad V_{gs} = IR$

and $\quad I = V \times Y_{CR}$

Then, $\quad V_{gs} = R \times (V \times Y_{CR})$

$\therefore \quad I_d = \left(\dfrac{1}{r_d} + g_m R\, Y_{CR}\right) \mathrm{V}$

but, $\quad Y_i\, V = I + I_d = \left(Y_{CR} + \dfrac{1}{r_d} + g_m R Y_{CR}\right) \mathrm{V}$

$\therefore \quad Y_i = Y_d + (1 + g_m R)\, Y_{CR}$

Where $\quad Y_d = \dfrac{1}{r_d}$

and $\quad Y_{CR} = \left(\dfrac{1}{R \| \dfrac{1}{\omega c}}\right)$

(c) Since, $g_m R \gg \mathrm{R}$

Hence we have,

$$Y_i = Y_d + R \times g_m \times \frac{1}{R - \dfrac{j}{wc}}$$

$$= \frac{1}{rd} + \frac{g_m R^2}{R^2 + \frac{1}{w^2c^2}} + i\,\frac{g_m(R/wc)}{R^2 + \frac{1}{w^2c^2}}$$

hence, $$wc_i = \frac{g_m RCw}{1+(wRC)^2} = \left(\frac{g_m\alpha}{1+\alpha^2}\right)$$

$\therefore$ $$C_i = \frac{g_{m\alpha}}{w\left(1+\alpha^2\right)}$$

Also $$C_i = \frac{1}{r_d} + \frac{g_m w^2 C^2 R^2}{1+(wCR)^2} = \frac{1}{r_d} + \frac{g_m\alpha^2}{1+\alpha^2}$$

$\therefore$ $$R_i = \frac{1}{C_i} = \left(\frac{\left(1+\alpha^2\right)r_d}{1+\alpha^2(1+\mu)}\right)$$

(d) For w = resonant C_i is a function of α, hence,

$$\frac{dC_i}{d\alpha} = \frac{g_m w\left(\alpha^2+1\right) - 2\alpha^2 g_m w^2}{w^2\left(1+\alpha^2\right)^2} = 0;$$

or, $\alpha^2 = 1;$

Or $\alpha = 1$

Since α is always greater than zero.

Also, $$\frac{d^2c_i}{d\alpha^2}$$

$$= -\frac{2g_m w^2\alpha\left(1+\alpha^2\right)^2 - g_m w^2\left(1-\alpha^2\right)2\left(1+\alpha^2\right)\times 2\alpha}{w^2\left(1+\alpha^2\right)^2} < 0$$

For $\alpha\sqrt{3}$ and for $\alpha = 1$,

$$(C_i)_{\max} = \frac{g_m}{2w} \text{ proved.}$$

and this follows that

$$(R_i)_{max} = \left(\frac{2r_d}{2+\mu}\right) \text{ proved}$$

Problem 7.23: Solve problem 7.22 if the capacitance C is replaced by an inductane L.

Solution: Here we replace the capacitance C by L in the above (last) problem.

We obtain, $Y_i = Y_d + (1 + g_m) Y_{LR}$

where $$Y_{LR} = \frac{1}{R + jwl}$$

Hence $$Y_i = G_i + jwC_i$$

$$= \frac{1}{r_d} + \frac{g_m R^2}{R_2 + w^2 L^2} - j\frac{w\, g_m\, RL}{R_2 + w^2 L^2}$$

$$jwc_i = -j\frac{w\, g_m RL}{R^2 + w^2 L^2}$$

But then Y_i is a parallel combination of a conductance G_i and inductance, then,

$$Y_i = G_i - j\,\frac{1}{wL_i}$$

or $$\frac{1}{wL_i} = \frac{W\, g_m RL}{R^2 + w^2 L^2}$$

or $$L_i = \frac{R^2 + w^2 L^2}{w^2 g_m R_L} = \frac{1 + w^2 L^2 / R^2}{w\, g_m\left(\frac{L}{R}\right)}$$

Therefore, $$G_i = \frac{r_d g_m R^2 + R^2 + w^2 L^2}{\left(R^2 + w^2 L^2\right) r_d} = \left(\frac{\mu + 1 + \alpha^L}{\left(1+\alpha^2\right) r_d}\right)$$

Where $\alpha = \dfrac{wL}{R}$

then $L_i = \left(\dfrac{1+\alpha^2}{wg_m\alpha}\right)$

For w constant, L_i is the function of α, and hence for $(L_i)_{max}$

$$\frac{dL_i}{d\alpha} = \frac{2wg_m\alpha^2 - wg_m(1+\alpha^2)}{w^2 g_m \alpha^2} = 0$$

then $\alpha = 1$

$$\therefore \quad (L_i)_{max} = \left(\frac{2}{wg_m}\right)$$

For $\alpha = 1$,

$$(R_i)_{max} = \left(\frac{2r_d}{\mu+2}\right)$$

Problem 7.24: Calculate the input admittance of an FET at 10^3 and 10^6 Hz when the total drain circuit impedance is (a) a resistance of 50 K, (b) a capacitive reactance of 50 K at each frequency. Take the interelectrode capacitances into consideration. The FET parameters are $\mu = 20$, $r_d = 10$ K, $g_m = 2.0$ mA/V, $C_{gs} = 3.0$ pF, $C_{ds} = 1.0$ pF, and $C_{gd} = 2.0$ pF. Express the results in terms of the input resistance and capacitance.

Solution: (a) From the eqn of A_v,

$$A_v = \frac{-g_m + y_{gd}}{Y_d + g_d + Y_{ds} + Y_{gd}}$$

and eqn of $Y_i = Y_{gs} + (1 - A_v)\, Y_{gd}$

(a) $R_d = 50$ K for $f = 10^3$ Hz

$$Y_{gs} = jwC_{gs} = j2\pi \times 10^3 \times 3.0 \times 10^{-12} = j1.88 \times 10^{-8}$$

$$Y_{ds} = jwC_{ds} = j2\pi \times 10^3 \times 1.0 \times 10^{-12} = j0.625 \times 10^{-8}$$

$$Y_{gd} = jwC_{gd} = j2\pi \times 10^3 \times 2.0 \times 10^{-12} = j1.26 \times 10^{-8}$$

$$Y_d = 2 \times 10^{-4},\ g_d = 10^{-4}$$

Hence $$A_v = \frac{-2\times10^{-3}\times j1.26\times10^{-8}}{12\times10^{-4}+j1.88\times10^{-8}}$$

$$\approx \frac{-2}{12}\times 10^2 \cong -16.7$$

and $$C_i = 3.0 + 17.7 \times 2 = 38.4 \text{ pF}$$

For $$f = 10^6 \text{ Hz}$$

$$Y_{gs} = j1.88 \times 10^{-5},$$

$$Y_{dss} = j0.628 \times 10^{-5}$$

$$Y_{gd} = j1.23 \times 10^{-5}$$

$$A_v = \frac{-2\times10^{-3}+j1.26\times10^{-5}}{1.2\times10^{-4}+j1.88\times10^{-5}}$$

$$= \left(\frac{-200+j1.26}{12+j1.88}\right) \cong (-16.3 + j1.97)$$

$$Y_i = \frac{1}{R_i} + jwC_i$$

or $$C_i = 37.6 \text{ pF}$$

$$R_i = 40.4 \text{ K}$$

(b) $$A_v = \frac{-g_m}{g_d+Y_L} = \frac{-2\times10^{-3}}{10^{-4}+j2\times10^{-5}} = \frac{-200}{10+j2}$$

$$\cong -19.3 + j3.84$$

$$Y_i = \frac{1}{Ri} + jwC_i$$

$$= jwC_{gs} + (20.2 - j3.84)\, jwC_{gd}$$

∴ $$C_i = 43.4 \text{ pF and } R_i = 20.8 \text{ M}$$

Problem 7.25: A MOSFET can be modelled as a voltage-controlled current device as shown in Fig. 7.25 wherein the controlled source is defined by

$$I_2 = 2.5\ V_1 \text{ mA}$$

(a) Sketch the output characteristic (I_2 *vs* V_2) with V_1 as a parameter (0 ≤ 2 V in 0.5 V steps).
(b) For V_i = 1 V, determine I_2 and V_2.
(c) At what value of V_i (i) V_2 = 0, and (ii) V_2 = 10 V.
(d) Sketch the input pulse for which the output pulse sequence is 10 V, 0 V, 10 V...
(e) For $V_i = 1.0 + 0.5 \sin \omega t$, sketch one cycle of V_2 and voltage across 2 kΩ resistance.

Solution:

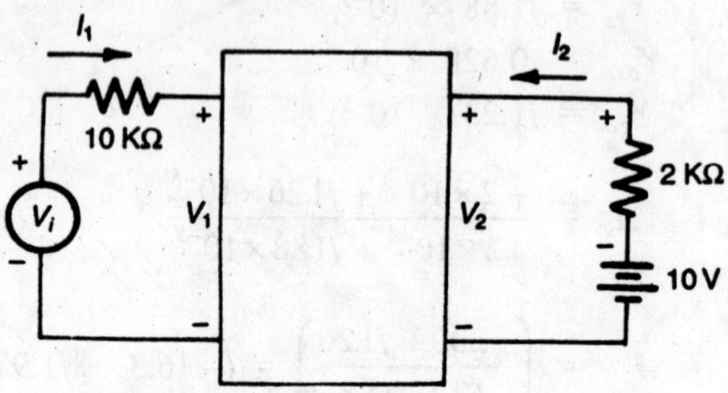

Fig. P. 7.25

(a) From above, we know

$$I_2 = 2.5\ V_1 \text{ mA} \qquad \text{...(i)}$$

$$V_2 = 10 - 2\ I_2$$

or

$$V_2 = 10 - 5\ V_1 \qquad \text{...(ii)}$$

Now, using eqs. (i) and (ii) V_1, I_2, V_2 and V_R (voltage across resistance) are calculated using table A and the output characteristic is depicted in Fig. 7.25 (a) against parameter (V_1) values marked thereon.

Table A

V_1(V)	I_2(mA)	V_2(V)	V_R(V)
0	0	10	0
0.5	1.25	7.5	2.5
1.0	2.5	5	5
1.5	3.75	2.5	7.5
2.0	5	0	10

(b)

$$V_i = 1 \text{ V}$$

$$V_1 = V_i = 1 \text{ V};$$

as $I_1 = 0$ in an ideal voltage-controlled current source

$$V_2 = 5 \text{ V}, I_2 = 2.5 \text{ mA}$$

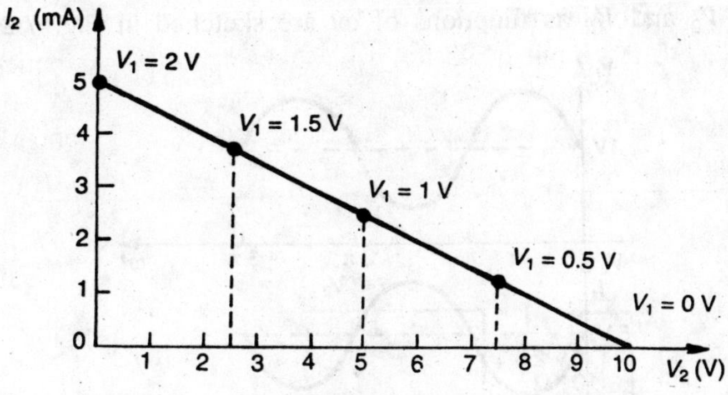

Fig. P. 7.25 (a)

(c) (i) $V_2 = 0$ V

for $V_i = 2$ V

(ii) $V_2 = 10$ V

for $V_i = 0$ V

(d) The output pulse sequence specified and the input pulse sequence necessary to produce this output are sketched in Fig. 7.25 (b) and (c).

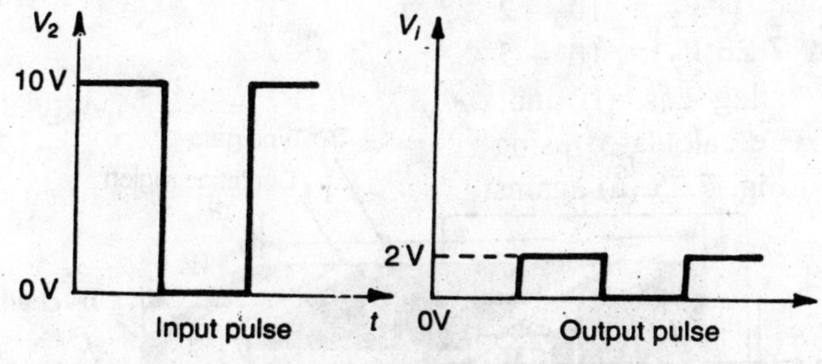

Fig. P. 7.25 (b) & (c)

(e) $V_1 = V_i = 1.0 + 0.5 \sin \omega t$

Linearity of device operation is obvious from Fig. 7.25 (a). So the output will be a sine wave imposed on dc value.

$$V_2 = 5 - 2.5 \sin \omega t$$

Note that the sinusoidal component of output is in phase opposition to the corresponding input component. The device acts as amplifier for this component with a voltage gain of

$$A_v = -2.5/.05 = -5$$

$$V_R = V_2 - 10$$

$$= -5 + 2.5 \sin \omega t$$

V_i, V_2 and V_R as functions of ωt are sketched in Fig. 7.25 (d)

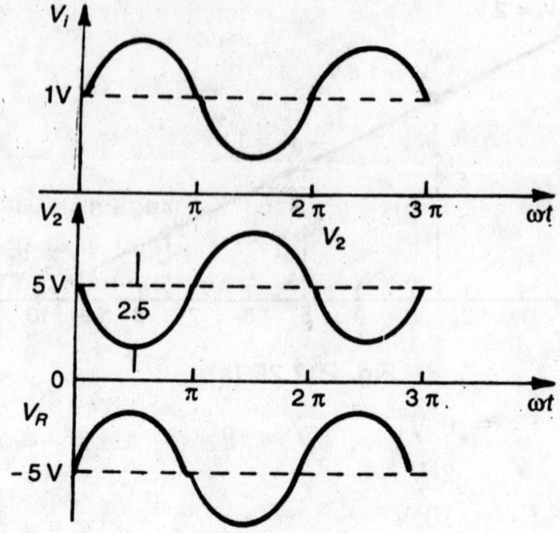

Fig. P. 7.25 (d)

Problem 7.26:

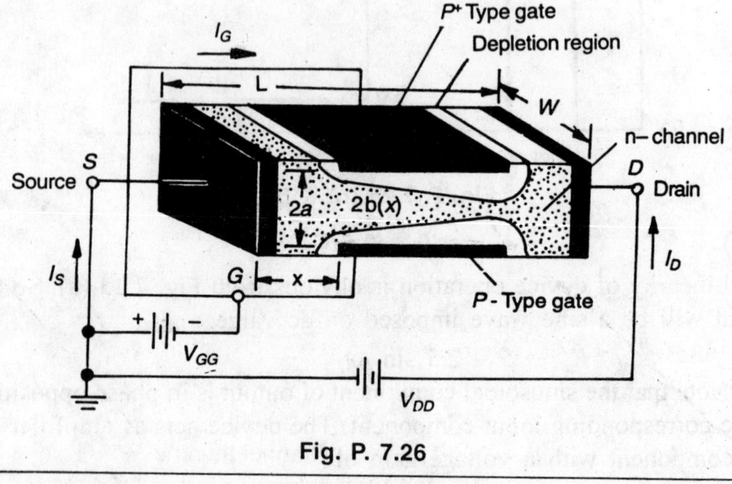

Fig. P. 7.26

Figure 7.26 shows the structure of an *n*-channel JFET showing the depletion region constricting the channel. Stating the assumptions made, derive an expression for I_D as function of V_{DS} for physical dimension as marked in the Fig. 7.26.
Find the expression for $r_{DS(ON)}$ which is defined as

$$R_{DS}\,(\text{ON}) = V_{DS} / I_D\big|_{V_{GS}=0}\quad 0$$

The term $r_{DS(ON)}$ is known as *drain resistance*.

Solution: We shall assume that the V_{DS} applied is small and so is I_D such that there is no pinch effect of V_{DS} and as a result the channel thickness $2b$ is uniform throughout the channel length L. Of course $b < a$ because of depletion caused by V_{GS} (negative).

Channel cross-sectional area, $A = 2\,bW$

Field intensity (uniform along channel length) $\in_x = \dfrac{V_{DS}}{L}$

Drift velocity of free electrons, $v_n = \mu_n \in_x$; μ_n = mobility of electrons

Since no current flows in the depletion region, one can then write the drain current as

$$I_D = AeN_D\mu_n\in_x;$$

N_D = density of donor atoms in the channel

or $$I_D = 2beN_D\mu_n(W/L)\,V_{DS} \qquad \text{...(i)}$$

When $V_{GS} = 0$. the depletion region is narrow and can be assumed uniform throughout channel length, i.e.,

$$b = a$$

Then $$I_D = 2aeN_D\mu_n(W/L)\,V_{DS} \qquad \text{...(ii)}$$

Equations (i) and (ii) both represent ohmic behaviour, i.e., constant channel resistance. The value of this resistance for $V_{GS} = 0$, called drain resistance $r_{DS(ON)}$, can be expressed as

$$r_{DS(ON)} = \frac{1}{2aeN_D\mu_n}\,(L/W)$$

This parameter is of importance in switching applications of JFETs and DEMOSFETs. For fast switching this resistance should be quite low (an ideal switch has zero resistance). $r_{DS(ON)}$ is governed by (W/L) ratio of the channel. As $\mu_n \gg \mu_p$ (hole mobility), this resistance is much lower in *n*-channel devices and hence their prevalence in switching applications.

As V_{GS} increases in magnitude, channel thickness $b(x)$ reduces along the channel length and $r_{DS(ON)}$ increases.

Problem 7.27: Consider an *n*-channel JFET with geometry of its structure as in Fig. 7.26. Here

N_D = concentration of donor atoms/m^3

N_A = concentration of acceptor atoms/m^3

$N_A \gg N_D$; and the gate-channel junction is assumed abrupt.

$$V_{DS} = 0$$

Assume junction contact potential to be much smaller than $|Vp|$.

$$|Vp| = \frac{eN_D}{2\,\epsilon_0\,\epsilon_r}\,a^2$$

where

ϵ_r = dielectric constant of the channel material

$2a$ = total channel thickness in m (special case when $b = a$)

Given : $N_D = 7 \times 10^{14}$ atom/m^3, $a = 2$ μm, $\epsilon_r = 12$

Solution: The following are the consequences of $V_{DS} = 0$

(a) Depletion region will have uniform thickness along channel length L and so,

(b) Channel thickness will be uniform even when it is constricted by application of V_{GS} (negative).

Channel will be fully pinched off and thus become non-conducting as $|V_{GS}|$ is increased to the extent that depletion region on each side of channel has a thickness 'a'.

Further, as the *p*-type region under the gate is heavily doped (p^+) and $N_A \gg N_D$, the thickness of depletion region on this side of the junction will be very small. Therefore, the immobile ions (negative) on this side act like a sheet charge.

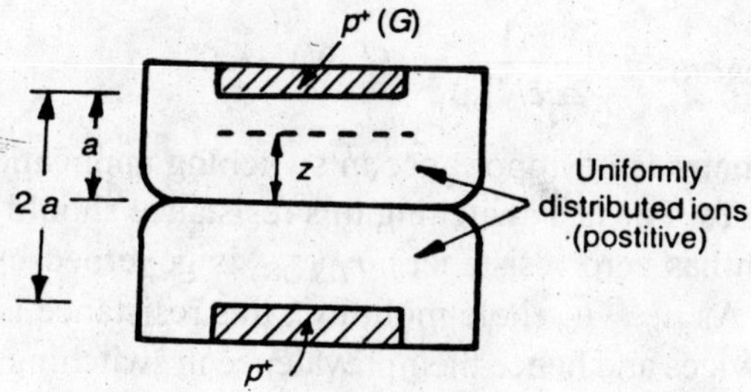

Fig. P. 7.27

When $|V_{GS}| = |V_p|$, as stated above the two depletion regions, above and below, just touch each other as shown in Fig. 7.27. Consider a sheet at height z above the meeting plane of the depletion regions.

Charge underneath this sheet = eN_Dz Coulomb/m^2

Electric displacement piercing the sheet = eN_Dz Coulumb/m^2

= Electric flux density

Electric field intensity, $\in_z = \left(\dfrac{eN_Dz}{\in_o \in_r}\right)$ V/m (= Potential gradient)

$$\text{Voltage applied} = \int_0^a \in_z \, dz$$

or
$$|V_P| = \frac{eN_D}{\in_o \in_r}\int_0^a z\,dz = \frac{eN_D}{2\in_o \in_r} \times a^2$$

Substituting values

$$|V_P| = \frac{1.6\times10^{-19}\times\left(7\times10^{14}\times10^{6}\right)}{2\times8.85\times10^{-12}\times12} \times (2 \times 10^{-6})^2$$

$$= 2.11 \text{ V}$$

Problem 7.28:

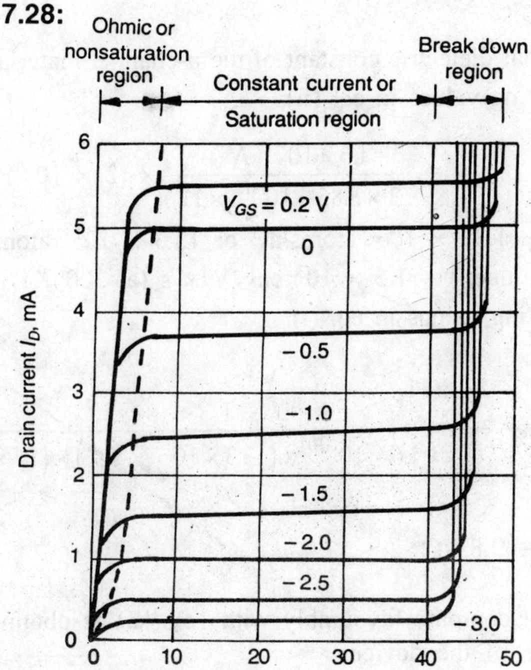

Fig. P. 7.28

(a) For the JFET whose characteristics are given in Fig. 7.28 evaluate $r_{DS(ON)}$ at $V_{GS} = 0$.
(b) If the above JFET has the structure of Fig. 7.26 evaluate $r_{DS(ON)}$ at $V_{GS} = 0$. Given : $L = 10\ \mu m$, $a = 2\ \mu m$, $W = 8\ \mu m$, $V_P = -4$ V.

Solution: (a) Reading from Fig. 7.28 in the initial straight line part of the characteristic corresponding to $V_{GS} = 0$, we have

$$I_D = 5 \text{ mA},$$
$$V_{DS} \approx 6.4 \text{ V}$$

Therefore,

$$r_{DS(ON)} = (6.4/5) = 1.28 \text{ k}\Omega$$

(b) We know drainage resistance $r_{DS(ON)}$ is given by

$$r_{DS(ON)} = \frac{1}{2aeN_D\mu_n}\left(\frac{L}{W}\right) \quad \text{...(i)}$$

In the data given, we do not have the value for N_D. So we use eq. for pinch-off voltage

$$|V_P| = \frac{eN_D}{2\,\epsilon_o\,\epsilon_r} \times a^2 \quad \text{...(ii)}$$

We assume that dielectric constant of the n-channel material is $\epsilon_r = 12$

Substituting values in eq. (ii)

$$4 = \frac{1.6\times 10^{-19}\,N_D}{2\times 8.85\times 10^{-12}\times 12} \times (2 \times 10^{-6})^2$$

Or $N_D = 1.33 \times 10^{21}$ atoms/m^3 or 13.3×10^{14} atoms/cm^3

μ_n (for silicon) = 1.5×10^3 cm^2/V × s (at 300 K)

Substituting values in eq. (i)

$r_{DS(0N)}$

$$= \frac{1}{2\times 2\times 10^{-6}\times 1.6\times 10^{-19}\times\left(13.3\times 10^{14}\times 10^{6}\right)\times\left(1.5\times 10^{3}/10^{4}\right)}$$

$$\left(\frac{10}{8}\right) = 9.8 \text{ k}\Omega$$

Does not compare favourably with 1.28 kΩ as obtained from the characteristics of the device.

Problem 7.29:
An NMOS transistor is used in the circuit of Fig. 7.29. The transistor has $k = 1$ mA/V^2, $W/L = 2$, $V_T = 4$ V. The circuit has $R_D = R_S = 2.5$ kΩ, $R_1 = 100$ kΩ and $R_2 = 200$ kΩ.
Determine :
(a) I_D and V_{DS}.
(b) Calculate the new value of R_S, if I_D has to be maintained at the value calculated in part (a), while the transistor is substituted with one having $W/L = 4$.

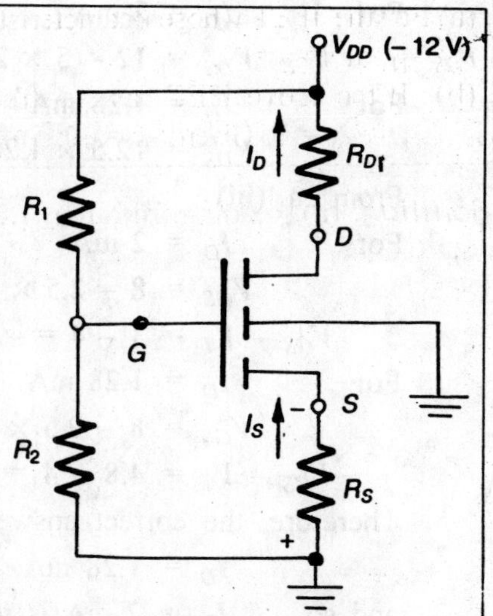

Fig. P. 7.29

Solution: As the current drawn by the gate terminal is almost zero

$$V_{GG} = \frac{R_2}{R_1 + R_2} \times V_{DD} = \frac{200}{100+200} \times 12 = 8 \text{ V}$$

(a) We shall assume $(V_{GS} - V_T) < V_{DS}$ in the saturation region, so

$$I_D = k\,(W/L)\,(V_{GS} - V_T)^2 \approx I_{DS}$$

Substituting values

$$I_D = 1 \times 2 \times (V_{GS} - 4)^2 = 2(V_{GS} - 4)^2 \qquad \text{...(i)}$$

Now $I_D = I_s$; as

$$I_G \approx 0$$

KVL for the DS loop

$$12 = 2 \times 2.5\, I_D + V_{DS} = 5\, I_D + V_{DS} \qquad \text{...(ii)}$$

KVL for the GS loop

$$V_{GS} = V_{GG} - I_S R_S = 8 - 2.5\, I_D \qquad \text{...(iii)}$$

Substituting V_{GS} from eq. (iii) in eq. (i), we get

$$I_D = 2(8 - 2.5\, I_D - 4)^2$$

or $I_D^2 - 3.28\, I_D + 2.56 = 0$

$$I_D = 2 \text{ mA, } 1.28 \text{ mA}$$

From eq. (ii)

$$V_{DS} = 12 - 5\, I_D$$

For $\quad I_D = 2$ mA

$V_{DS} = 12 - 5 \times 2\ V = 2$ V

For $\quad I_D = 1.28$ mA

$V_{DS} = 12.5 \times 1.28 = 5.6$ V

From eq. (iii)

For $\quad I_D = 2$ mA

$V_{GS} = 8 - 2.5 \times 2 = 3$ V;

$V_{GS} - V_T = 3 - 4 = -1$ V (not valid)

For $\quad I_D = 1.28$ mA

$V_{GS} = 8 - 2.5 \times 1.28 = 4.8$ V;

$V_{GS} - V_T = 4.8 - 4 = 0.8 \text{ V} < 5.6 \text{ V } (V_{DS})$

Therefore, the correct answer is

$I_D = 1.28$ mA,

and so $\quad I_D = 2$ mA is rejected. The answer then is

$I_D = 1.28$ mA,

$V_{DS} = 5.6$ V

(b) $\quad W/L = 4$

$I_D = 1.28$ mA (to be maintained at the value calculated in part (a)

Equation (i) now modifies to

$$I_D = 4\,(V_{GS} - 4)^2;$$

as W/L is increased by a factor of 2

$$1.28 = 4\,(V_{GS} - 4)^2$$

or $\quad V_{GS} = 4.57$ V

But $\quad V_{GS} = V_{GG} - I_D R_S$

$$4.57 = 8 - 1.28\,R_S \text{ (new)}$$

which gives

$$R_S \text{ (new)} = \frac{8 - 4.57}{1.28} = 2.68 \text{ k}\Omega$$

Problem 7.30: A PMOS transistor has the following parameters: $k = 2$ mA/V^2, $W/L = 1$, $V_T = -1$
It is used in the circuit of Fig. 7.29 with circuit element values as $V_{DD} = 21$ V, $R_1 = 300$ kΩ, $R_2 = 100$ kΩ
(a) Determine R_S so that $V_{GS} = -2$ V.

(b) Determine R_D to make $V_{DS} = -4$ V.
(c) Find the new value of R_2 to maintain I_D as in (a) and (b), while V_T is changed to – 1.5 V. Assume that all other parameters remain at the values given or computed in (a) and (b).

Solution: Figure 7.29 is redrawn for PMOS in Fig. 7.30

$$V_{GG} = \frac{100}{300+100} \times (-12) = -3 \text{ V}$$

(a) We shall assume $(V_{GS} - V_T) < V_{DS}$. So

$$I_D = k(W/L)\,(V_{GS} - V_T)^2$$

substituting values

$$I_D = 2\,(V_{GS} + 1)^2 \quad \text{...(i)}$$

$$V_{GS} = -2 \text{ V}$$

Thus $\quad I_D = 2\,(-2 + 1)^2 = 2$ mA

KVL for the GS loop

$$V_{GG} - V_{GS} + I_S R_S = 0 \quad \text{...(ii)}$$

or $\quad V_{GS} = V_{GG} + R_S I_S$

$$-2 = -3 + 2R_S$$

or $\quad R_S = 0.5$ kΩ

(b) KVL for the *DS* loop

$$-V_{DD} - R_D I_D + V_{DS} - R_S I_S = 0$$

Substituting values

$$12 - 2R_D - 4 - 0.5 \times 2 = 0$$

or $\quad R_D = 3.5$ kΩ

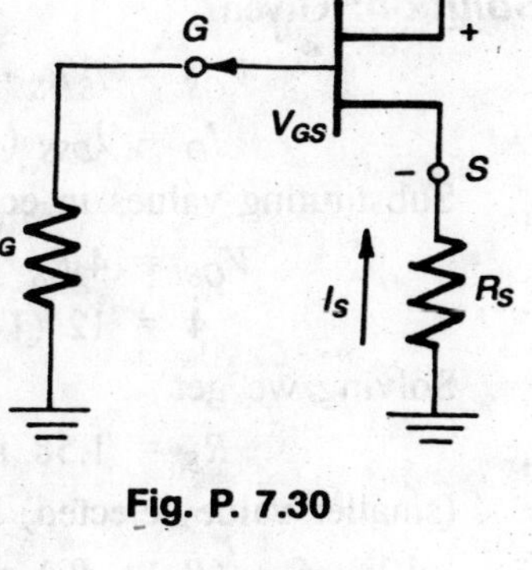

Fig. P. 7.30

(c) $\quad I_D = 2$ mA,

$R_D = 3.5$ kΩ,

$R_S = 0.5$ kΩ

V_T is now changed to – 1.5 V. This requires that V_{GG} be suitably changed, while R_2 is adjusted to a new value.

From eq. (i) $2 = 2(V_{GS} + 1.5)^2$

or $\quad V_{GG} = V_{GS} - I_S R_S = -0.5 - 2 \times 0.5 = -1.5$ V

But $$V_{GG} = \frac{R_2(new)}{R_1 + R_2(new)} \times (-12)$$

$$-1.5 = \frac{R_2(new)}{300 + R_2(new)} \times (-12)$$

which gives

$$R_2 \text{ (new)} = 42.86 \text{ k}\Omega$$

Problem 7.31:
A *p*-channel JFET has $V_P = 4$ V and $I_{DSS} = 12$ mA. It is used in the circuit of Fig. 7.31 with $V_{DD} = 12$ V. Determine R_D and R_S so that $I_D = 4$ mA and $V_{DS} = 6$ V.

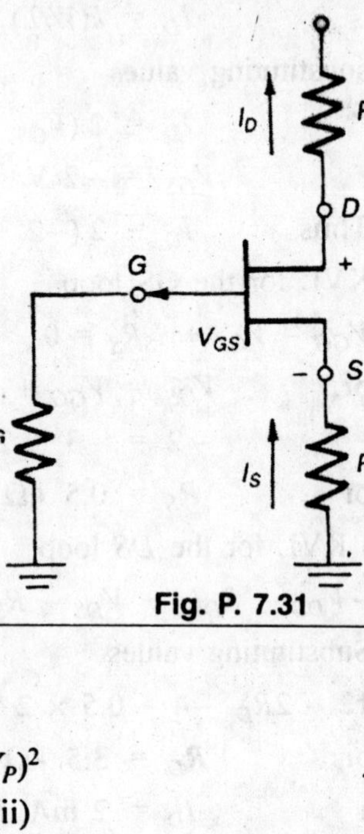

Fig. P. 7.31

Solution: Given

$$V_{GS} = I_S R_S \approx I_D R_S \quad \text{...(i)}$$

$$I_D = I_{DSS} (1 - V_{GS}/V_P)^2 \quad \text{...(ii)}$$

Substituting values in eq. (i) and (ii)

$$V_{GS} = 4 R_S$$

$$4 = 12 (1 - 4 R_S/4)^2 \quad \text{...(iii)}$$

Solving we get

$$R_S = 1.58 \text{ k}\Omega \text{ or } 0.42 \text{ k}\Omega$$

(smaller value selected) KVL for the *DS* loop

$$-12 - 6 + (R_D + R_S) = 0$$

$$(R_D + R_S) = 4.5 \text{ k}\Omega$$

$$R_S = 0.42 \text{ k}\Omega \text{ gives}$$

$$R_D = 4.08 \text{ k}\Omega$$

So the answer is

$$R_D = 4.08 \text{ k}\Omega,$$
$$R_S = 0.42 \text{ k}\Omega$$

Problem 7.32:
An *n*-channel JFET, having $V_p = -4$ V and $I_{DSS} = 10$ mA, is used in the circuit of Fig. 7.32. The parameter values are $V_{DD} = 18$ V, $R_S = 2$ kΩ, $R_D = 2$ kΩ, $R_1 = 450$ kΩ and $R_2 = 90$ kΩ. Determine I_D and V_{DS}.

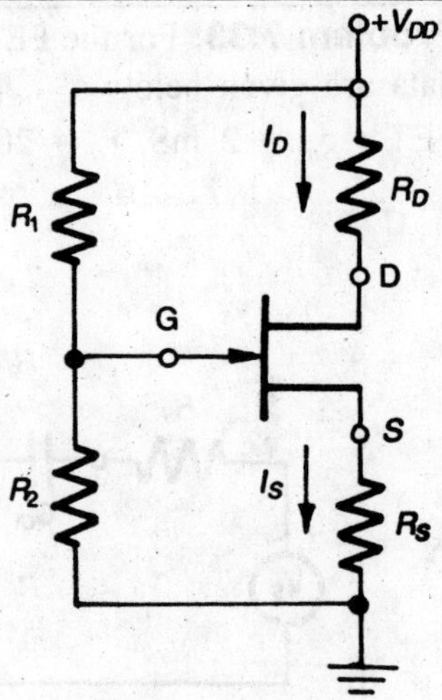

Fig. P. 7.32

Solution:

$$V_{GG} = \frac{90}{450+90} \times 18 = 3 \text{ V}$$

$$V_{GS} = V_{GG} - I_D R_S \qquad \text{...(i)}$$

$$I_D = I_{DSS}\,(1 - V_{GS}/V_P)^2 \qquad \text{...(ii)}$$

Substituting values in eqs. (i) and (ii)

$$V_{GS} = 3 - 2I_D \qquad \text{...(iii)}$$

$$I_D = 10\left[1 + \frac{3-2I_D}{4}\right]^2 \qquad \text{...(iv)}$$

Solving, we get

$$I_D = 4.9 \text{ mA},\ 2.5 \text{ mA}$$

For $I_D = 4.9$ mA

$$V_{GS} = 3 - 2 \times 4.9 = -6.8 \text{ V}$$

$$V_{DS} = 18 - (2 + 2) \times 4.9 = -1.6 \text{ V}$$

(rejected as $V_{DS} > 0$ for *n*-channel)

For $I_D = 2.5$ mA

$$V_{GS} = 3 - 2 \times 2.5 = -2 \text{ V}$$

$$V_{DS} = 18 - (2 + 2) \times 2.5 = 8 \text{ V} > (V_{GS} - V_P = 2 \text{ V})$$

Therefore, the answer is

$$I_D = 2.5 \text{ mA},$$
$$V_{DS} = 8 \text{ V}$$

Problem 7.33: For the FET amplifier circuit shown in Fig. 7.33 the data are given below :

FET : $g_m = 2$ mS, $r_d = 200$ kΩ, $C_{gs} = 10$ pF, $C_{gd} = 0$

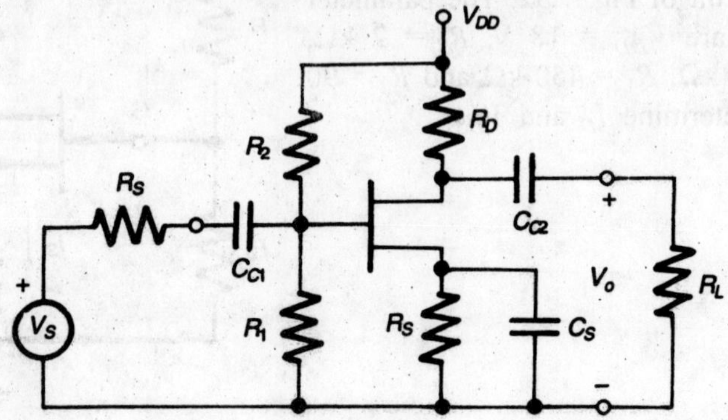

Fig. P. 7.33

External circuit elements :

$$R_s = 1 \text{ k}\Omega,\ R_1 \parallel R_2 = R_G = 1 \text{ M}\Omega$$
$$R_D = 5 \text{ k}\Omega$$
$$R_S = 2 \text{ k}\Omega$$
$$C_S = \text{very large}$$
$$R_L = \text{very large}$$
$$C_{C1} = C_{C2} = 0.1\ \mu\text{F}$$
$$C_W + C_L \approx 10 \text{ pF}$$

(a) Draw the mid-frequency circuit and compute the voltage gain.

(b) Draw the low-frequency circuit and compute the low-frequency cut-off.

(c) Draw the high-frequency circuit and compute the high frequency cut-off.

(d) Calculated the gain-bandwidth product.

(e) The external circuit element R_D is now reduced to 4 kΩ. Calculate the new gain-bandwidth product. Compare the result with that of part (d) and comment.

Solution: (a) Mid-frequency band–small-signal model (Fig. 7.33 (a))

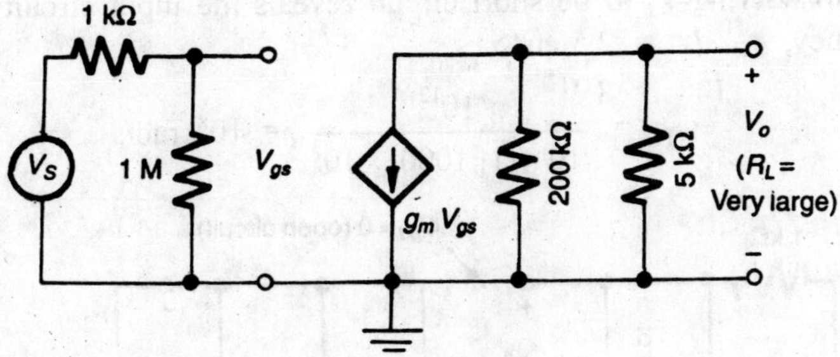

Fig. P. 7.33 (a)

$$\text{Take} \quad V_s = 1 \text{ V}$$

$$V_{gs} \approx V_s = 1; \; R_G \text{ very high}$$

$$g_m V_{gs} = 2 \times 1 = 2 \text{ mA}$$

$$200 \parallel 5 = 4.88 \text{ k}\Omega$$

$$A_{Vo} = V_2 = -2 \times 4.88 = -9.76$$

(b) Low-frequency band-small-signal model (Fig. 7.33 (b)) C_S being large it continues to short circuit R_S.

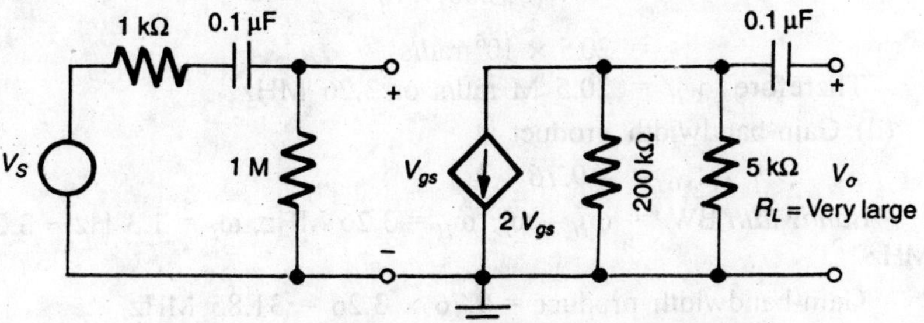

Fig. P. 7.33 (b)

It is easily seen from the circuit of Fig. 7.33 (b) that $C_{C2} = 0.1$ μF is not effective as $R_L = \infty$. The only corner frequency is that due to the input circuit.

Imaging V_s to be short circuit, we can calculate the corner frequency as

$$\omega_{l1} = \frac{1}{0.1 \times 10^{-6} \times (1000 + 1) \times 10^3} = 10 \text{ rad/s}$$

$$\omega_L = 10 \text{ rad/s or } 1.6 \text{ Hz}$$

(c) High-frequency band–Small-signal model (Fig. 7.33 (c)).

Considering V_s to be short circuit reveals the input circuit corner frequency,

$$w_{11} = \frac{10^{12}}{10 \times (1 \| 1000) \times 10^3} = 10^8 \text{ rad/s}$$

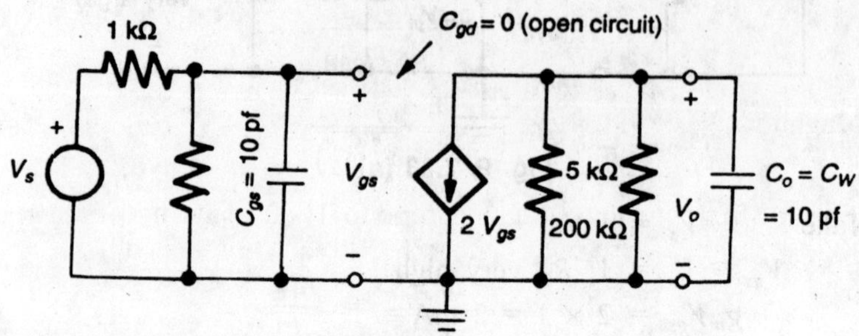

Fig. P. 7.33 (c)

Output circuit corner frequency is

$$w_{22} = \frac{10^{12}}{10 \times (5 \| 200) \times 10^3} = \frac{10^8}{4.88}$$

$$= 20.5 \times 10^6 \text{ rad/s}$$

Therefore $w_H = 20.5$ M rad/s or 3.26 MHz

(d) Gain-bandwidth product

$$A_{VO} = -9.76$$

Bandwidth BW $= \omega_H - \omega_L$; $\omega_H = 3.26$ MHz, $\omega_L = 1.3$ Hz ≈ 3.26 MHz

Gain-bandwidth product = 9.76 × 3.26 = 31.85 MHz

(e) Ref. Fig. 7.33 (a)

R_D reduced to 4 kΩ

200 ∥ 4 = 3.92

A_{Vo} (new) = − 2 × 3.92 = − 7.84

$$\text{BW (new)} = \omega_H \text{ (new)} = \frac{10^8}{3.92} = 25.5 \text{ M rad/s or 4.06 MHz}$$

Gain-bandwidth product (new) = 7.84 × 4.06 = 31.83 MHz

This is almost the same as the original value i.e. 31.8 = MHz.

So we have verified that the gain-bandwidth product remains constant as it depends only upon the device.

Problem 7.34: An RC-coupled amplifier has equal input and output circuit corner frequencies in the high-frequency band i.e. $\omega_{21} = \omega_{22} = \omega_2 = 200$ k rad/s
Calculate the cut-off frequency of this band.

Solution:

$$\bar{A}_{VH} = \frac{A_{VO}}{1+(j\omega/\omega_2)^2}$$

Magnitudewise $$A_{VH} = \frac{A_{VO}}{\sqrt{1+(\omega/\omega_2)}}$$

At cut-off frequency the gain drops to $1/\sqrt{2}$ of its mid-frequency gain (3 dB drop). Therefore,

$$A_{VH}(\omega_H) = \frac{A_{VO}}{\sqrt{1+(\omega_H/\omega_2)^2}} = \frac{A_{VO}}{\sqrt{2}}$$

Solving for ω_H, we have

$$\omega_H = \omega_2 = 200 \text{ k rad/s}$$

Problem 7.35: The *iv* characteristics of a JFET are drawn in Fig. 7.35 (a). It is used in the amplifier circuit of Fig. 7.35 (b). Corresponding to the circuit element values the load line is already drawn on the characteristics of the JFET.

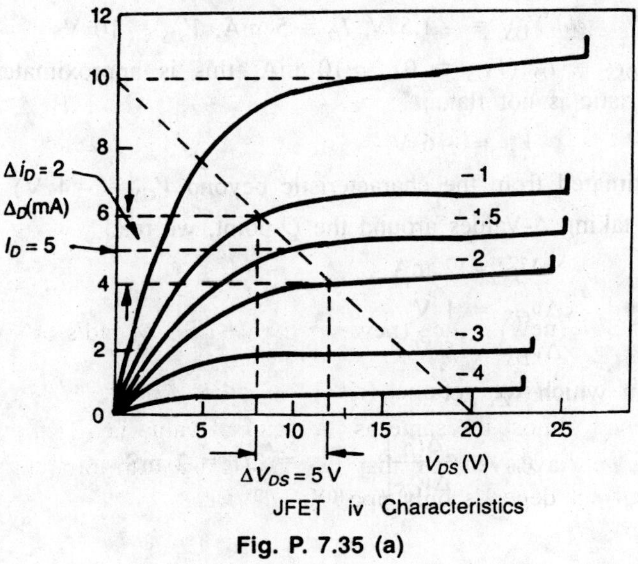

Fig. P. 7.35 (a)

(a) Estimate I_{DSS} and V_P from the characteristic. Also estimate from these the value of g_m and r_{ds}. Also find the value of g_m.

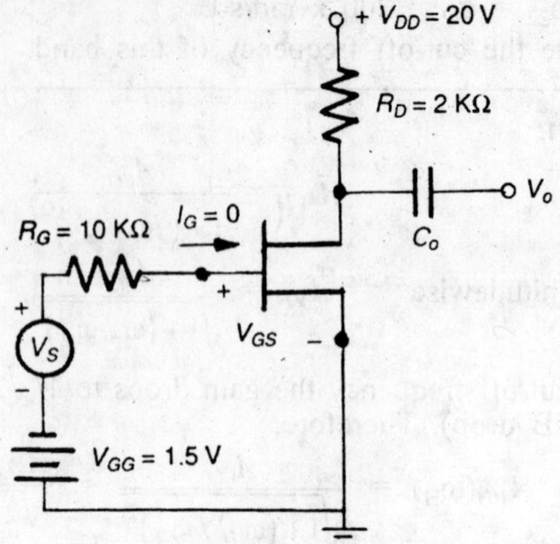

Fig. P. 7.35 (b)

(b) Using the value of g_m and r_{ds} found directly from the characteristics in part (a), find the small-signal amplifier gain.

Solution: As $I_G = 0$, $V_{GS} = -V_{GS} = -1.5$ V

Load line. Join $V_{DS} = 20$ V and $I_D = 10$ mA

The Q-point as identified on the interpolated characteristic for $V_{GS} = -1.5$ V is given by :

$$V_{GS} = -1.5 \text{ V}, \; I_D = 5 \text{ mA}, \; V_{DS} = 10 \text{ V}$$

(a) $I_{DSS} = i_D \; (V_{GS} = 0) \approx 10$ mA; this is approximate as the characteristic is not flat.

$$V_P \approx -6 \text{ V}$$

(estimated from the characteristic beyond $V_{GS} = -4$ V)

By taking Δ-Values around the Q-point, we read

$$\Delta i_D \approx 2 \text{ mA}$$

$$\Delta v_{GS} \approx 1 \text{ V}$$

$$\Delta v_{DS} \approx 5 \text{ V}$$

from which we get

$$g_m = \left. \frac{\Delta i_D}{\Delta v_{GS}} \right|_{V_{DS}=10\text{V}} = 2/1 = 2 \text{ mS}$$

$$r_{ds} = \left.\frac{\Delta v_{DS}}{\Delta i_D}\right|_{V_{GS}=1.5} = 5/2 = 2.5\ \text{k}\Omega$$

We have $$g_{mo} = -\frac{2I_{DSS}}{V_P} = \frac{2 \times 10}{6} = 10/3\ \text{mS}$$

$$g_m = g_{mo}\sqrt{(I_D / I_{DSS})} = (10/3)\ \sqrt{(5/10)} = 2.36\ \text{mS}$$

(b) Small-signal circuit model is drawn in Fig. 7.35 (c). As $i_g = 0$, $v_{gs} = v_s$

Gain, $$A_V = -g_m(4.5 \,\|\, 2)$$

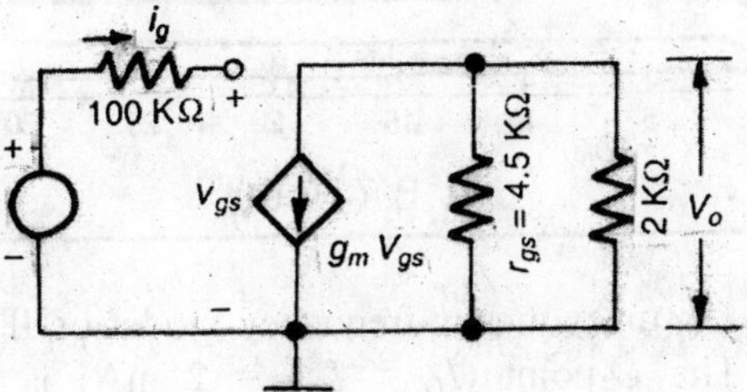

Fig. P. 7.35 (c)

Problem 7.36:

The JFET circuit given in Fig. 7.36(a) is devised to amplify the signal v_i.

The device characteristics are given in Fig. 7.36(b). Other data on this circuit is as under :

I_{DD} = 2.0 mA,

v_i = 2 sin ω*t* mV

R_D = 100 kΩ,

r_d = 100 kΩ

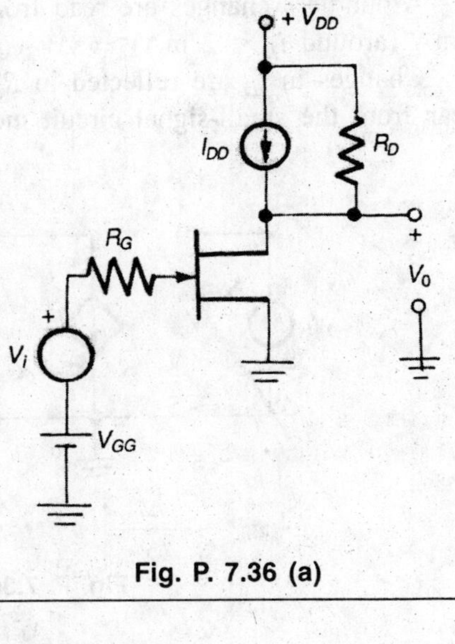

Fig. P. 7.36 (a)

Assume that R_D does not draw any dc current.

Using low-frequency model of JFET, find the signal component in v_o (output).

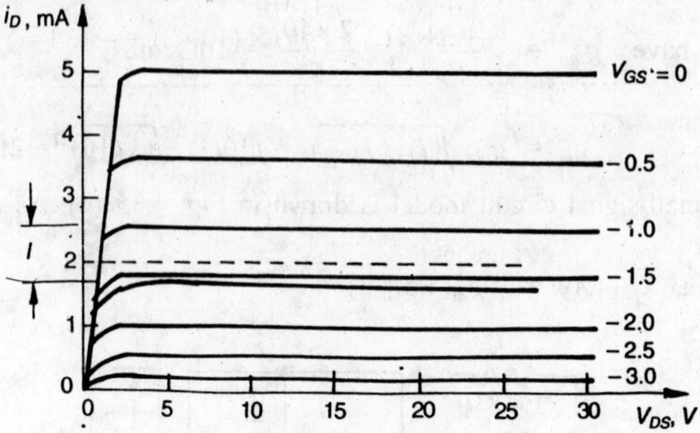

Fig. P. 7.36 (b)

Solution: The circuit with low-frequency model of JFET is drawn in Fig. 7.36 (c). The Q-point ($I_D = I_{DD} = 2$ mA) is located on the characteristics of Fig. 7.36 (b).

As per basic definition

$$g_m = \Delta i_D / \Delta v_{GS}$$

Around i_D, changes are read from the characteristics. For $\Delta i_D =$ 1 mA (around $i_D = 2$ mA) $\Rightarrow \Delta v_{GS} \approx (1.75 - 1) = 0.75$ V.

Changes in i_D are reflected in R_d and the device as will become clear from the small-signal circuit model of Fig. 3.6 (c). Therefore,

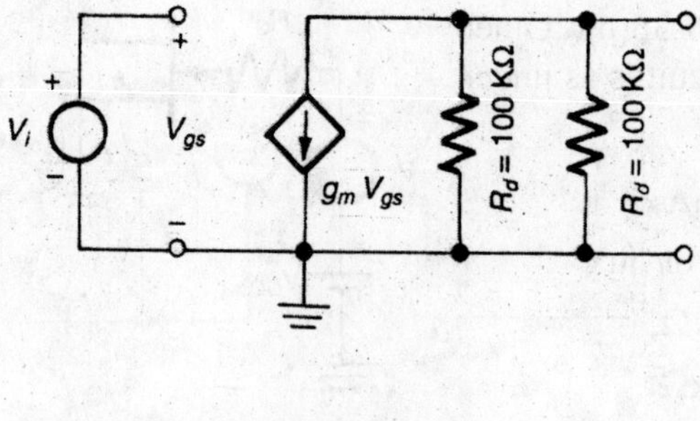

Fig. P. 7.36 (c)

$$g_m = \Delta i_D / \Delta v_{GS} = /0.75 = 1.33 \text{ mS}$$

For the circuit model of Fig. 7.36 (c)

$$v_{gs} = v_i = 2 \sin \omega t \text{ mV}$$

$$V_0 = -g_m v_{gs} \times (r_d \parallel R_D)$$

$$= -1.33 \times (2 \sin \omega t) \times (100/2) = -133 \sin \omega t \text{ mV}$$

Problem 7.37:

Consider the JFET circuit of Fig. 7.37

(a) Determine the resistance seen from terminal 2 and ground at low frequencies.

(b) Evaluate the resistance in part (a) for R_D = 4 kΩ, R_s = 2.5 kΩ, R_1 = 200 kΩ, R_2 = 100 kΩ, g_m = 2.5 mS and r_d = 60 kΩ.

(c) Repeat part (b) if R_D = 0

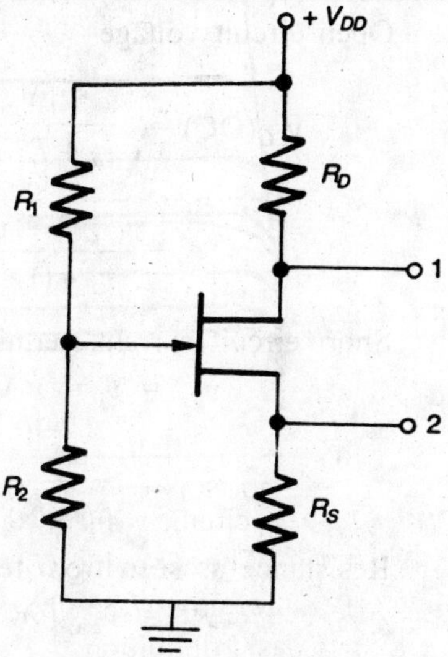

Fig. P. 7.37

Solution: (a) The small-signal model of the circuit is drawn in Fig. 7.37 (a). Use the Thevenin equivalent converts the circuit to that of Fig. 7.37(b)

Let us apply input signal of 1 V

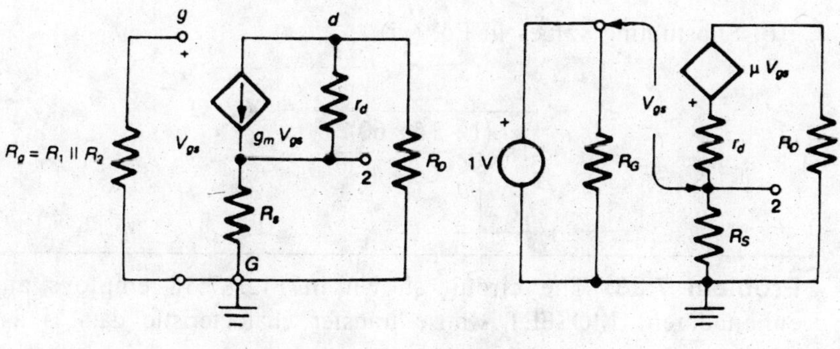

Fig. P. 7.37 (a) & (b)

$$v_{gs} = 1 - R_s\left(\frac{\mu v_{gs}}{r_d + R_s + R_D}\right) \quad \text{...(i)}$$

or $$v_{gs} = \frac{r_d + R_s + R_D}{r_d + (1+\mu)R_s + R_D} \quad \text{...(ii)}$$

Open circuit voltage

$$v_{2G}(\text{OC}) = \frac{\mu v_{gs}}{r_d + R_s + R_D} \times R_S \quad \text{...(iii)}$$

$$= \frac{\mu R_S}{r_d + (1+\mu)R_S + R_D} \quad \text{...(iv)}$$

Short circuit current (terminal 2 shorted to G)

$v_{gs} = v_i = 1$ V as R_s is shorted.

$$i_{2G\ (\text{SC})} = \frac{\mu v_{gs}}{r_d + R_D} = \frac{\mu}{r_d + R_D} \quad \text{...(v)}$$

Resistance as seen from terminal 2 and G

$$R_o = v_{2G}\ (\text{OC})/i_{2G}\ (\text{SC})$$

$$= \frac{\mu R_S}{r_d + (1+\mu R_S) + R_D} \times \frac{(r_d + R_d)}{\mu}$$

$$= \frac{R_S}{1 + \dfrac{(1+\mu)R_S}{r_d + R_D}}; \ \mu = g_m r_d \quad \text{...(vi)}$$

(b) Substituting values in Eq. (vi)

$$R_o = \frac{2.5}{1 + \dfrac{(1 + 2.5 \times 60)2.5}{60 + 4}} = 362\ \Omega$$

Problem 7.38: The circuit shown in Fig. 7.38 employs an enhancement MOSFET whose transfer characteristic data is as under

$V_T = 2$ V, $V_{GS(ON)} = 5$ V, $I_{DS(ON)} = 200$ μA

Find the change in bias current I_D, if V_{DD} increases by 0.3 V. Use the small-signal model and take $r_{ds} = 50$ kΩ.

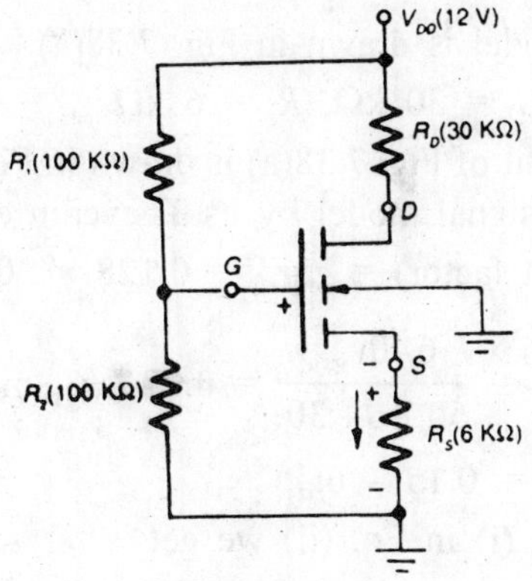

Fig. P. 7.38

Solution: $I_D = K(V_{GS} - V_T)^2$ ...(i)

Substituting values

$$200 \times 10^{-3} = K(5 - 2)^2 \text{ mA}$$

or $K = (0.2/9) = 1/45 \text{ mA/V}^2$

$\therefore$ $I_D = (1/45)\,(V_{GS} - 2)^2$ ...(ii)

As $I_G = 0$

$$V_{GG} = 12 \times \frac{100}{100+100} = 6 \text{ V}$$

Round the gate-source loop

$V_{GG} - V_{GS} - R_S I_S = 0$

or $6 - V_{GS} - 6\,I_D = 0$ ...(iii)

Also from eq. (ii) for dc conditions

$I_D = (1/45)\,(V_{GS} - 2)^2$ ...(iv)

Solving Eqs. (iii) and (iv) we get

$V_{GS} = 4.89$ V, $- 8.375$ V (rejected)

$I_D = (1/45)\,(4.89 - 2)^2 = 0.185$ mA

$V_{DS} = 12 - (30 + 6) \times 0.185 = 5.34$ V

We have $g_m = 2\sqrt{(KI_D)} = 2\sqrt{(0.185/45)} = 0.128$ mS

V_{DD} increased by 0.3 V. This can be regarded as small-signal which will cause a signal input ΔV_{GG} at G.

$$\Delta V_{GG} = 0.3 \times \frac{100}{100+100} = 0.15 = v_{gg}$$

Small-signal model is drawn in Fig. 7.38(a)

$r_{ds} = 50\ \text{k}\Omega,\ R_D = 30\ \text{k}\Omega,\ R_s = 6\ \text{k}\Omega$

Equivalent circuit of Fig. 7.38(a) is drawn in Fig. 7.38(b) replacing the MOSFET small-signal model by its Thevenin equivalent where

μ (amplification factor) $= g_m r_{ds} = 0.128 \times 50 = 6.40$

$$i_d = \frac{6.40 v_{gs}}{50+6+30} = 0.074\ V_{gs}\ \text{mA} \quad \text{...(i)}$$

$$v_{gs} = 0.15 - 6i_d \quad \text{...(ii)}$$

Substituting eq. (i) in Eq. (ii) we get

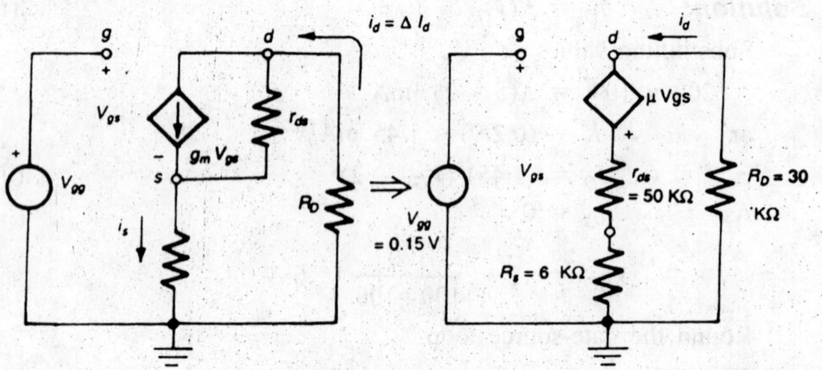

Fig. P. 7.38

$$v_{gs} = 0.105\ \text{V}$$

It then follows

$$i_d = 0.074 \times 0.105 \times 10^3 = 7.77\ \mu\text{A}$$

Chapter 8

The Transistor at High Frequencies

Problem 8.1: Show that at low frequencies the hybrid-II model with $r_{b'c}$ and r_{ce} taken as infinite reduces to the approximate CE h-parameter model.

Solution:

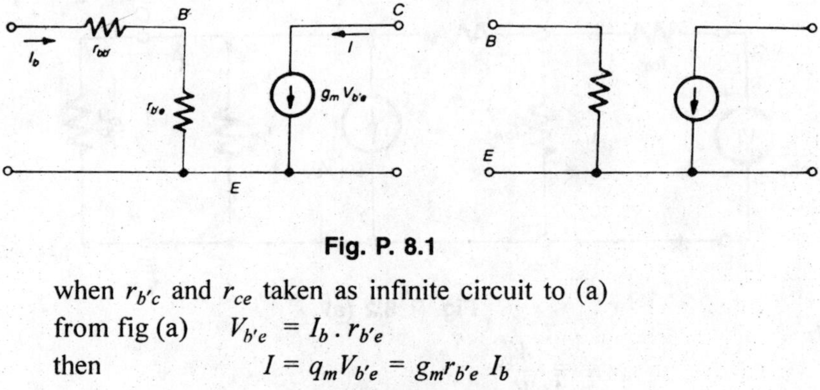

Fig. P. 8.1

when $r_{b'c}$ and r_{ce} taken as infinite circuit to (a)

from fig (a) $\quad V_{b'e} = I_b \cdot r_{b'e}$

then $\quad I = q_m V_{b'e} = g_m r_{b'e} I_b$

$\quad = h_{fe} I_b$; where $g_m r_{b'e} = h_{fe}$

hence, $\quad I = h_{fe} I_b$

Also, $\quad h_{ie} = r_{bb'} + r_{b'e}$

Hence this gives the approximate CE-h-parameter model.

Problem 8.2: (a) Consider the hybrid-II circuit at low-frequencies, so that C_e and C_c may be neglected. Omit none of the other elements in the circuit. If the load resistance is $R_L = 1/gL$, prove that

$$K = \frac{V_{ce}}{V_{b'e}} = \frac{-g_m + g_{b'c}}{g_{b'c} + g_{ce} + g_L}$$

(b) Using Miller's theorem, draw the equivalent circuit between C and E. Applying KCL to this network, show that the above value of K is obtained.

(c) Using Miller's theorem, draw the equivalent circuit between B and E. Prove that the current gain under load is

$$A_I = \frac{g_L}{(g_{b'c} + g_{b'e}) / K - g_{b'c}}$$

(d) Using the results of parts a and c and the relationships between the hybrid-II and the h-parameters, prove that

$$A_I = \frac{-h_{fe}}{1 + h_{oe} R_L}$$

Solution: (a)

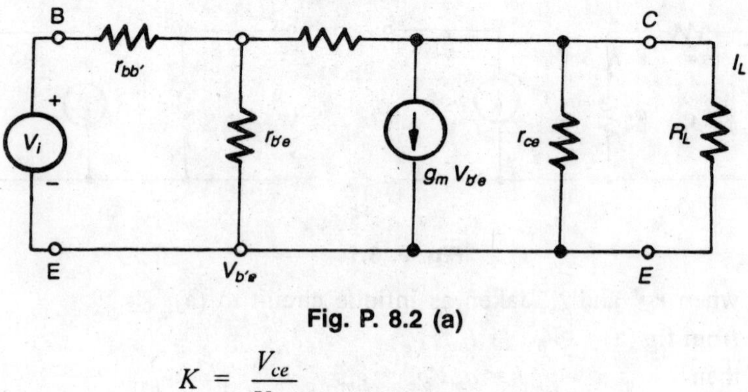

Fig. P. 8.2 (a)

$$K = \frac{V_{ce}}{V_{b'e}}$$

where $V_{ce} = I_{sc} Z_{ce}$ with $R_L = 0$

$$I_L = I_{sc} = V_{b'e}\, g_{b'c} - g_m V_{b'e} = (g_{b'c} - g_m) V_{b'e}$$

Z_{ce} is the impedance seen between $C - E$ Thus

$$Z_{ce} = (r_{b'c} \parallel r_{ce} \parallel R_L) \text{ or, } Z_{ce} = \frac{1}{g_{b'c} + g_{ce} + g_L}$$

Hence, $$\frac{V_{ce}}{V_{b'e}} = K = \left(\frac{-g_m + g_{b'c}}{g_{b'c} + g_{ce} + g_L}\right)$$

(b) From ckt we have,

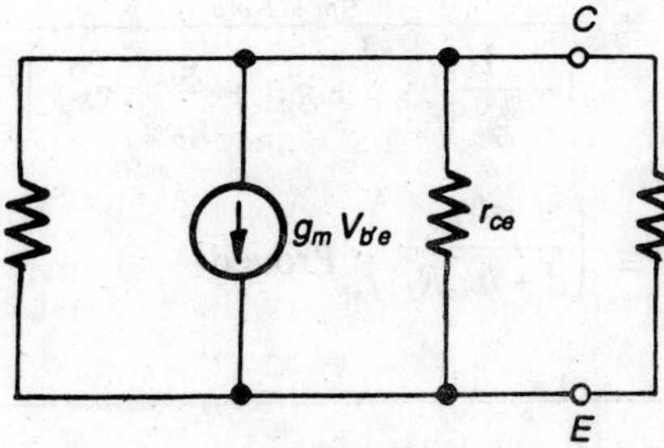

Fig. P. 8.2 (b)

$$V_{ce} = -g_m V_{b'e} \frac{1}{g_L + g_{ce} + \frac{K-1}{K} g_{b'c}}$$

$$= \frac{-K g_m V_{b'e}}{K(g_L + g_{ce} + g_{b'c}) - g_{b'c}} = KV_{b'e}$$

or, $k\ (g_{ce} + g_{b'c} + g_{ce}) - g_{b'c} = -g_m$

$$\therefore \quad k = \left(\frac{-g_m + g_{b'c}}{g_{b'c} + g_{ce} + g_L}\right) \text{Proved}$$

(c) $$A_I = \frac{I_L}{I_i}$$

where $$I_L = g_L V_{ce}$$

$$= g_L\ kV_{b'e} = \frac{g_L K I_i}{g_{b'e} + (1-K) g_{b'c}}$$

Hence, $$A_I = \frac{g_L K}{g_{b'c} + g_{b'e} - K g_{b'c}} = \left(\frac{g_L}{g_{b'c} + g_{b'e} / k - g_{b'c}}\right)$$

(d) It is to be note that, $g_m \gg g_{b'c}$ and $g_{b'e} \gg g_{b'c}$

Thus form part (a) and (c), we have,

$$K \cong \left(\frac{-g_m}{g_{b'c} + g_{ce} + g_L}\right) \text{and } A_3 \cong \frac{g_L K}{g_{b'e} - K g_{b'c}}$$

$$\therefore \quad A_I = \left(\frac{-g_m g_L}{g_{b'e}(g_{b'c} + g_{ce} + g_L) + g_m \cdot g_{b'c}}\right)$$

Hence
$$A_I \cong \frac{-g_m / g_{b'e}}{1 + \frac{1}{g_L}\left(g_{b'c} + g_{ce} + \frac{g_m}{g_{b'e}} . g_{b'c}\right)}$$

$$\cong \left(\frac{-h_{fe}}{1 + h_{oe} R_L}\right) \text{ Proved}$$

Problem 8.3: The following low-frequency parameters are known for a given transistor at I_c = 10 mA, V_{CE} = 10 V, and at room temperature.

$$h_{ie} = 500\ \Omega$$
$$h_{oe} = 10^{-5}\ \text{A/V}$$
$$h_{fe} = 100$$
$$h_{re} = 10^{-4}$$

At the same operating point, f_T = 50 MHz and C_{ob} = 3 *pF*, compute the values of all the hybrid-II parameters.

Solution:

$$g_m = \frac{|I_e|}{V_T} = \frac{10 \text{ mA}}{26 \text{ mV}} = 0.385 \frac{\text{mA}}{\text{V}}$$

$$r_{b'e} = \frac{h_{fe}}{g_m} = \frac{100}{0.385} \cong 260\ \Omega$$

$\therefore$
$$r_{bb'} = h_{ie} - r_{b'e}$$
$$= 500 - 260 = 240\ \Omega$$

$$r_{b'c} = \frac{r_{b'e}}{h_{re}} = 260 \times 10^4\ \Omega = 2.6 \text{ M}$$

$$g_{ce} = h_{oe} - (1 + h_{fe})\, g_{b'c} \cong 0.12 \times 10^{-6}$$

Hence
$$r_{ce} = \frac{1}{g_{ce}} \cong 833 \text{ K}$$

$$C_e = \frac{g_m}{2\pi f_T} = \frac{385 \times 10^{-3}}{2\pi \times 50 \times 10^6}\ 1224 \text{ pF}$$

$$C_c = 3 \text{ pF}$$

Problem 8.4: Given the following transistor measurements made at $I_c = 5$ mA, $V_{CE} = 10$ V, and at room temperature:

$$h_{fe} = 100$$
$$h_{ie} = 600 \ \Omega$$
$$[A_{ie}] = 10 \text{ at } 10 \text{ MHz}$$
$$C_c = 3 \text{ pF}$$

Find f_β, f_T, C_e, $r_{b'e}$ and $r_{bb'}$.

Solution:

From eq of A_i,

$$|A_i| = \frac{h_{fe}}{\left[1+(f/f_B)^2\right]^{\frac{1}{2}}} = 10$$

or,
$$\frac{100}{\left[1+(f/f_B)^2\right]^{\frac{1}{2}}} = 10 \ ; \text{ or, } 1 + \left(\frac{f}{f_m}\right)^2 = 100$$

and
$$\frac{f}{f_B} = \sqrt{99} = 9.95$$

Here
$$f = 10 \text{ MHz},$$

then
$$f_\beta = \frac{10}{9.95} = 1.005 \text{ MHz}$$

and
$$f_T = h_{fe} f_\beta = 100.5 \text{ MHz}$$

$\therefore$
$$C_e = \frac{g_m}{2\pi f_T} = \frac{1 I_c 1}{V_T} \ 304 \text{ pF}$$

$$r_{b'e} = \frac{h_{fe}}{g_m} = \frac{100}{5/26} = 520 \ \Omega$$

$\therefore$
$$r_{bb'} = h_{ie} - r_{b'e}$$
$$= 600 - 520 = 80 \ \Omega$$

Problem 8.5: A silicon *p-n-p* transistor has an $f_T = 400$ MHz. What is the base thickness?

Solution:

$$f_T = 400 \text{ MHz}$$

$$f_T = h_{fe}\, f_\beta \cong \frac{g_m}{2\pi c_e} \quad \text{...(1)}$$

and $$C_e = C_{De} = g_m \frac{w^2}{2D_B} \quad \text{...(2)}$$

D_B in the diffusion constant for minority carrier

w = width/base thickness

from eq. (1) & (2)

$$f_T = \frac{g_m}{2\pi\, g_m \dfrac{w^2}{2D_B}} = \frac{D_B}{\pi w^2}$$

or, $$w^2 = \frac{D_B}{\pi f_T}$$

But for a *p-n-p* transistor $D_B = D_P = 13 \text{ cm}^2/\text{sec}$

Then, $$w^2 = \frac{13}{3.14 \times 400 \times 10^6} = 1.035 \times 10^{-8}$$

Hence, $$w = 1.03 \times 10^{-4} \text{ cm} = 1.03 \ \mu\text{m}$$

Problem 8.6: Given a germanium *p-n-p* transistor whose base width is 10^{-4} cm. At room temperature and for a dc emitter current of 2 mA, find (a) the emitter diffusion capacitance, (b) f_T.

Solution:

(a) We have the equation

$$C_e = C_{De}$$

$$= g_m \frac{w^2}{2D_B}$$

but $$g_m = \left(\frac{1 I_c 1}{V_T}\right) \cong \left|\frac{I_E}{V_T}\right|$$

$$\therefore \quad C_e = \frac{I_E}{V_T} \times \frac{w^2}{2D_B} = \frac{2 \times 10^3}{26 \times 10^{-3}} \times \frac{10^{-8}}{2 \times 47}$$

$$[D_B = D_P = 47 \text{ cm}^2\text{/sec for Ge}]$$

$$\therefore \quad C_e = 8.2 \text{ pF.}$$

(b)
$$f_T = \frac{g_m}{2\pi C_e} = \frac{I_c}{V_T \times 2\pi \times C_e}$$

$$\frac{2 \times 10^{-3}}{26 \times 10^{-3} \times 2\pi \times 8.2 \times 10^{-12}} = \frac{I_c}{V_T . 2\pi \times C_e}$$

$$= \frac{2 \times 10^{-3}}{26 \times 10^{-3} \times 2\pi \times 8.2 \times 10^{-12}} = 1500 \text{ MHz}$$

Problem 8.7: The hybrid-II parameter of the transistor used in the circuit shown are given below. Using Miller's theorem and the approximate analysis, compute

(a) The upper 3-dB frequency of the current gain $A_I = I_L/I_i$.

(b) The magnitude of the voltage gain $A_{vs} = V_o/V_s$ at the frequency of part a.

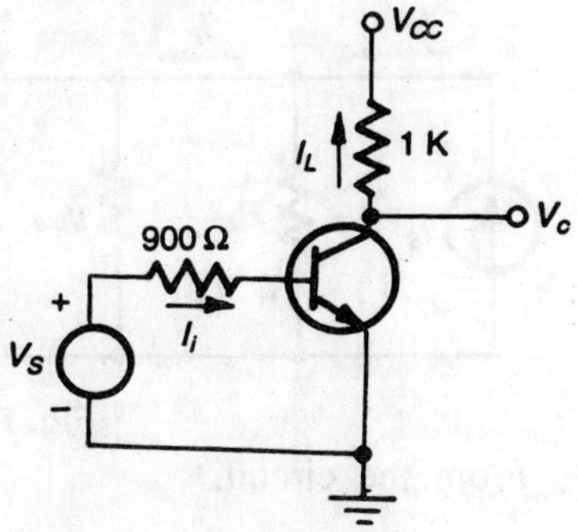

Fig. P. 8.7

Solution: *Given* :

$g_m = 50$ mA/V $\qquad r_{b'e} = 1000\ \Omega$

$r_{bb'} = 100\ \Omega$ $\qquad r_{b'c} = 4$ M

$C_c = 3$ pF $\qquad C_e = 100\ pF$

$r_{ce} = 80$ k

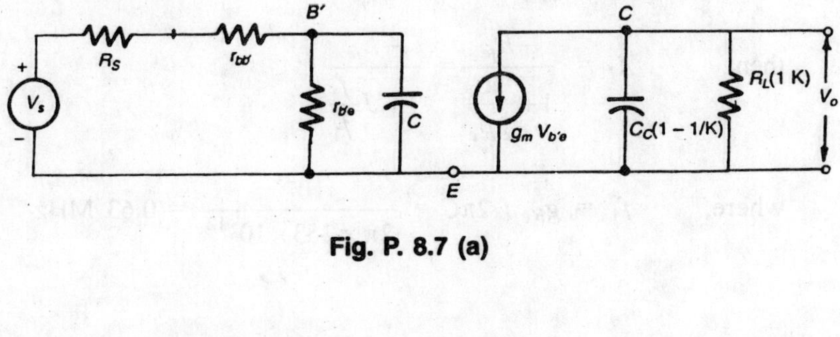

Fig. P. 8.7 (a)

(a) Assume the internal voltage gain $|k| = \left|\dfrac{V_{ce}}{V_{b'e}}\right| \gg 1$ and using miller's theorm we have the following equivalent circuit;

Now, Neglecting $C_{c,} \left(1 - \dfrac{1}{K}\right)$ we have

$$K = -g_m R_L = -(50 \text{ mA/V}) \times 1 \text{ K} = -50$$

then input capacitance,

$$C = C_e + C_c\ (1 - k) = C_e\ C_c\ (1 + g_m R_L)$$
$$= 100 + 3 \times 51 = 253 \text{ pF}$$

Now, using Norton's theorm, redrawing the equivalent circuit for the above Here $R_{s'} = R_s\ r_{bb'}$

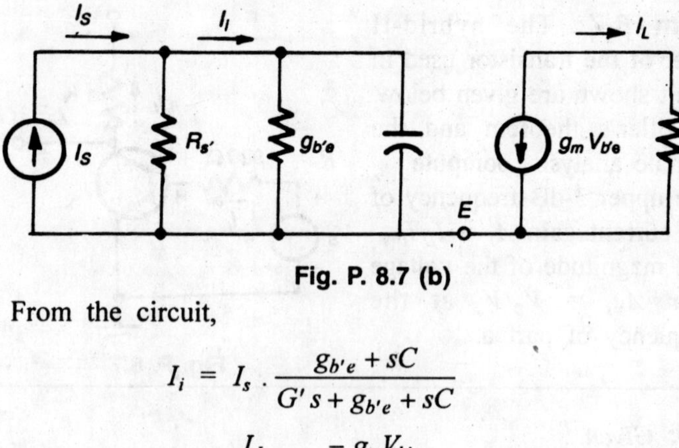

Fig. P. 8.7 (b)

From the circuit,

$$I_i = I_s \cdot \frac{g_{b'e} + sC}{G' s + g_{b'e} + sC}$$

and, $\quad A_I = \dfrac{I_L}{I_1} = \dfrac{-g_m V_{b'e}}{I_i}$

but, $\quad I_i = V_{b'e}\ (g_{b'e} + SC)$

$\therefore \quad A_I = -g_m / g_{b'e} + SC$

Now since, $g_m = g_{b'e}\ h_{fe}$

then $\quad A_I = \dfrac{-h_{fe}}{1 + \dfrac{sC}{g_{b'e}}} = \dfrac{-h_{fe}}{1 + J\dfrac{f}{f_1}}$

where, $\quad f_1 = g_{b'e} / 2\pi C = \dfrac{10^{-3}}{2\pi \times 253 \times 10^{-12}} = 0.63 \text{ MHz}$

(b) Input loop resistance,

$$R \equiv R_{s'} \parallel r_{b'e}$$

$$V_o = -g_m R_L V_{b'e}$$

and, $$V_{b'e} = \frac{(r_{b'e} \parallel SC)V_s}{R_{s'} + (r_{b'e} \parallel SC)}$$

Then, voltage gain,

$$A_{vs} = \frac{V_o}{V_s} = \frac{-g_m R_L}{1 + R_{s'}(G_{b'e} + SC)} = \frac{-g_m R_L G_{s'}}{(G_{s'} + g_{b'e}) + SC}$$

which is in form $A_{Vs} = K_2/(S - S_1)$

where $$S_1 = \frac{-G_{s'} + g_{b'e}}{C} = \frac{-1}{RC}$$

Then, upper- $3dB$ frequency,

$$f_H = \frac{|S_1|}{2\pi} = \frac{1}{2\pi RC}$$

as, $C = 253\ pF$ and $R = 1/(Gs' + g_{b'c})$

$$R = \frac{1}{\dfrac{1}{R_{s'}} + \dfrac{1}{r_{b'e}}} = \frac{1}{\dfrac{1}{10^3} + \dfrac{1}{10^3}} = 500\ \Omega$$

Then, $$f_H = \frac{1}{2\pi \times 500 \times 253} \times 10^{-12} = 1.26\ \text{MHz}$$

Now, we have,

$$A_{vs} = \frac{-g_m R_L G_{s1}/(G_{s'} + g_{b'e})}{1 + j\dfrac{f_c}{G_{s'} + g_{b'e}}}$$

$$= \frac{-g_m R_L G_{s'}/(G_{s'} + g_{b'e})}{1 + j\dfrac{f}{f_H}}$$

or, $$|A_{vs}| = \frac{g_m R_L G_{s'}}{G_{s'} + g_{b'e}} \left(\sqrt{1 + \frac{f^2}{f^2 H}}\right)^{-1}$$

where $\quad G_s = \dfrac{1}{R_s + r_{bb'}} = 10^{-3}$ V

then $\quad \dfrac{g_m R_L G_{s'}}{G_{s'} + g_{b'e}} = \dfrac{50 \times 10^{-3} \times 10^3 \times 10^{-3}}{10^{-3} \times 10^{-3}} = 25$

Hence, $\quad |A_{vs}| = 25 \times \dfrac{1}{\sqrt{1+\left(\dfrac{f_1}{f_H}\right)^2}}$ (at $f = f_1$)

$$25 \times \frac{1}{\sqrt{1+\left(\dfrac{0.63}{1.26}\right)^2}} = 25 \times \frac{1}{\sqrt{1.25}} = 22.32 \textbf{ Ans}$$

Problem.8.8: Consider a single-stage CE transistor amplifier with the load resistor R_L shunted by a capacitance C_L.

(a) Prove that the internal voltage gain $K = V_{ce}/V_{b'e}$ is

$$K \approx \frac{-g_m R_L}{1 + j\omega(C_c + C_L)R_L}$$

(b) Prove that the 3-dB frequency is given by

$$f_H \approx \frac{1}{2\pi(C_C + C_L)R_L}$$

provided that the following condition is satisfied:
$g_{b'e} R_L (C_c + C_L) \gg C_e + C_c (1 + g_m R_L)$

Solution:

Using miller's theorm, the equivalent circuit for single stage transistor amplifier shown below :

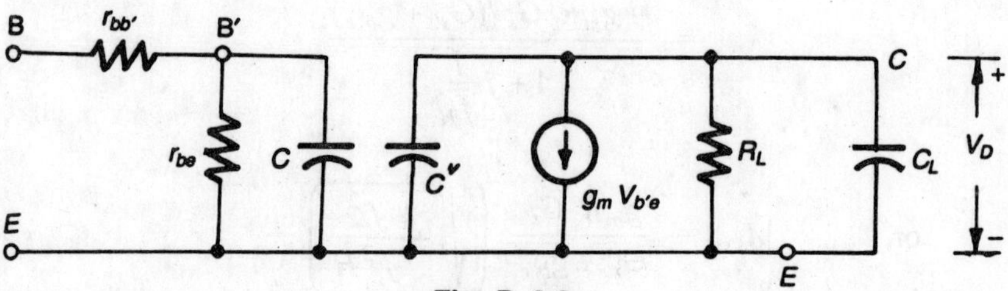

Fig. P. 8.8

Here

$$C = C_e + C_c\,(1 - K)$$
$$C' = C_c\,(K - 1)/K$$
$$C' \approx C_c \qquad \text{(since } |K| \gg 1)$$

(a) Now integral voltage gain,

$$K = \frac{V_{ce}}{V_{b'e}} = \frac{1}{V_{b'e}}\left[\frac{-g_m V_{b'e} R_L}{1 + jwR_L(c_c + c_L)}\right]$$

$$= \left[\frac{-g_m R_L}{1 + jw(c_e + c_L)R_L}\right]$$

(b) It can be inferred from the circuit, that there is two time constants, $R_L\,(C_c + C_L)$ and $r_{b'e}\,[C_e + C_c\,(1 - k)]$

Now, if, $R_L\,(C_c + C_L) \gg \dfrac{1}{r_{b'e}}\,[C_e + C_c\,(1 - k)]$

or, $R_L\,(C_c + C_L) \gg r_{b'e}\,[C_e + C_c\,(1 + g_m R_L)]$

As $\qquad k = -\,g_m\,R_L$

then, f_H is determined by larger time constant and from part (i)

Hence, $\qquad f_H = \dfrac{w_H}{2\pi} \approx \dfrac{1}{2\pi(C_c + C_L)R_L}$

Problem 8.9: For a single stage CE-transistor amplifier where hybrid II parameters have the values

$$g_m = 50 \text{ mA/V},$$
$$r_{bb'} = 100\ \Omega$$
$$r_{b'e} = 1\text{ K}$$
$$r_{b'c} = 4\text{ M}$$
$$r_{ce} = 80\text{ K}$$
$$C_c = 3\text{ pF}$$
$$C_e = 100\text{ pF}$$

What are the source resistance R_s will give a 3-dB frequency f_H which is (a) half the value for $R_s = 0$ (b) twice the value for $R_s \to \infty$? Do these values of R_s depend upon the magnitude of the load R_L.

Solution:

As we know,

$$f_H = \frac{1}{2\pi\, Rc}$$

where $$R = \frac{R_s + r_{bb'}}{R_s + h_{fe}}\, r_{b'e}$$

and $$C = C_e + C_c\,(1 + g_m R_L)$$

(a) If $$R_s = 0$$

$$f_H = \frac{1}{2\pi} \times \frac{1}{c_e + C_c(1 + g_m R_L)}$$

Here $$f'_H = \frac{1}{2}\, f_H$$

$$f'_H = \frac{1}{2\pi} \times \frac{h_{ie}}{(R_s + r_{bb'})\, r_{b'e}} \times \frac{1}{C_e + C_c(1 + g_m R_L)}$$

$$= \frac{1}{2}\left[\frac{1}{2\pi}\frac{h_{ie}}{r_{bb'}\, r_{b'e}} \times \frac{1}{C_e + C_c(1 + g_m R_L)}\right]$$

or

$$\frac{R_s + h_{ie}}{(R_s + r_{bb'})r_{b'e}} = \frac{1}{2}\;\frac{h_{ie}}{r_{bb'} + r_{b'e}}$$

$$\therefore \quad R_s = \frac{h_{ie}\, r_{bb'}}{h_{ie} - 2r_{bb'}}$$

Their $$R_s = \frac{(r_{bb'} + r_{b'e})r_{bb'}}{r_{b'e} - r_{bb'}} = \frac{1100 \times 100}{900} \cong 122\ \Omega$$

(b) If $$R_s = \infty$$

$$f_H = \frac{1}{2\pi}\;\frac{1}{r_{b'e}}\;\frac{1}{C_e + C_e(1 + g_m R_L)}$$

Here $$f'_H = 2\, f_H$$

or, $$\frac{f'_H}{f_H} = 2$$

or, $$\frac{(R_s + h_{ie})\, r_{b'e}}{(R_s + r_{bb'})\, r_{b'e}} = 2$$

or, $$2R_s + 2r_{bb'} = R_s + h_{ie} = R_s + r_{bb'} + r_{b'e}$$

or, $$R_s = r_{b'e} - r_{bb'} = 1000 - 100 = 900\ \Omega$$

Hence the above values of R_s does not depend upon R_L.

Problem 8.10: A single-stage CE amplifier is measured to have a voltage gain band width f_H of 5 MHz with R_L = 500 Ω. Assume h_{fe} = 100, g_m = 100 mA/V, $r_{bb'}$ = 100 Ω,
C_c = 1 pF, and f_T = 400 MHz

(a) Find the value of the source resistance that will give the required band width.

(b) With the value of R_S found in part *a*, find the midband voltage gain V_o/V_s.

Solution:

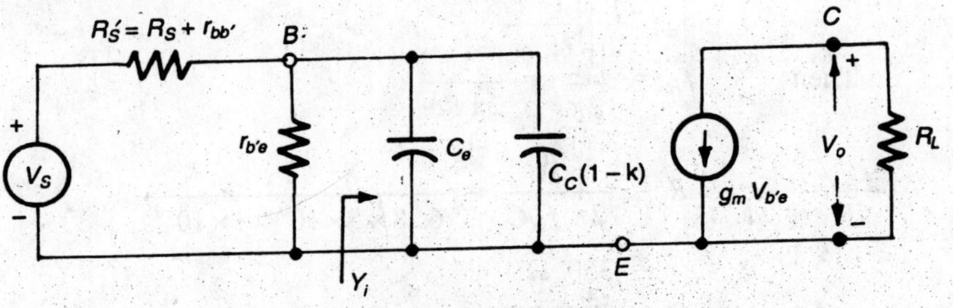

Fig. P. 8.10

From the circuit, internal voltage gain

$$k = \frac{V_{ce}}{V_{b'c}}$$

$$= -g_m R_L \qquad \text{...(i)}$$

and the input capacitance

$$C \equiv C_e + C_c(1 - k) \equiv C_e + C_c(1 + g_m R_L) \qquad \text{...(ii)}$$

$$= 1$$

Now, $$C_e = g_m/2\pi f_T$$

$$= \frac{100 \times 10^{-3}}{2\pi \times 400 \times 10^{6}} = 0.0405 \times 10^{-9}$$

and, $$r_{b'e} = \frac{h_{fe}}{g_m} = \frac{100}{100 \times 10^{-3}} = 1 \text{ k}\Omega$$

then, $$C = 0.04 \times 10^{-9} + 12^{-12} (1 + 100 \times 10^{-3} \times \frac{1}{2} \times 10^{3})$$

$$= 91 \text{ pF}$$

Now, from the circuit, voltage tranfer function,

$$A_{vs} = \frac{-g_m R_L G_{s'}}{G_{s'} + g_{b'e} + Sc}$$

which is in the form

$$A_{vs} = \frac{K_2}{S - S_1}$$

where $$S_1 = -\frac{G_{s'} + g_{b'e}}{C} = -1/_{RC}$$

Then $$f_H = \frac{|S_1|}{2\pi} = \frac{1}{2\pi Rc}$$

or, $$R = \frac{1}{2\pi f_H C} = \frac{1}{6.28 \times 5 \times 10^{6} \times 1 \times 10^{-12}}$$

$$= 350 \ \Omega$$

and $$R = R_{s'} \parallel r_{b'e}$$

or, $$0.350 \text{ K} = \frac{R_{s'} r_{b'e}}{R_{s'} + r_{b'e}}$$

or, $$0.35 = \frac{R_{s'}}{R_{s'} + 1 \text{ K}}$$

or, $$R_{s'} = 0.539 \text{ K}$$

As, $$R_{s'} = R_s + r_{bb'}$$

Then, $$R_s = R_{s'} - r_{bb'} = 539 - 100 = 439 \ \Omega$$

(b) For $S = jw$ and $w = 0$, the mid band gain,

$$A_{vs}\ (jw)/_{w=0} = 0 = \frac{-g_m R_L G_{s'}}{G_{s'} + g_{b'e}}$$

$$= \frac{-100 \times 10^{-3} \times \frac{1}{2} \times 10^{3} \times 1.86 \times 10^{-3}}{1.86 \times 10^{-3} + 10^{-3}} = -32.5 \textbf{ Ans}$$

Problem 8.11: The hybrid-II parameter of the transistor used in the circuit given

$g_m = 50$ m A/V
$r_{bb'} = 100\ \Omega$
$r_{b'e} = 1$ K
$r_{b'c} = 4$ M
$r_{ce} = 80$ K
$C_c = 3$ pF
$C_e = 100$ pF

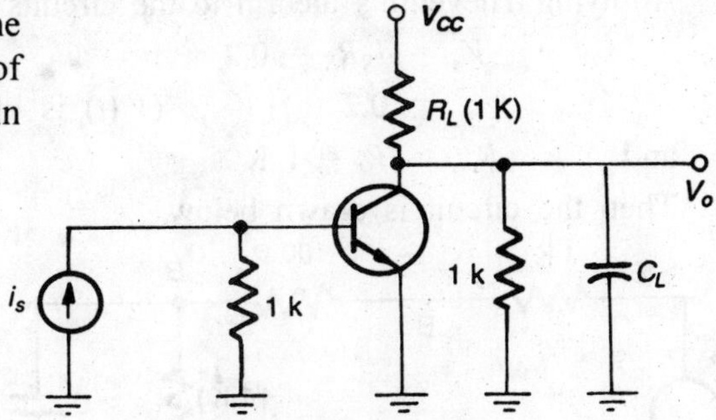

Fig. P. 8.11

The input to the amplifier is an abrupt current step 0.2 mA in magnitude.
Find the output voltage as a function of time (a) if $C_L = 0$. Neglect the output time constant. (b) If $C_L = 0.1\ \mu F$. Neglect the input time constant

Solution:

$g_m = 50$ m A/V $\quad r_{bb'} = 100\ \Omega$
$r_{b'e} = 1$ k $\quad r_{b'c} = 4$ M
$r_{ce} = 80$ k $\quad C_c = 3$ pF
$C_e = 100$ pF

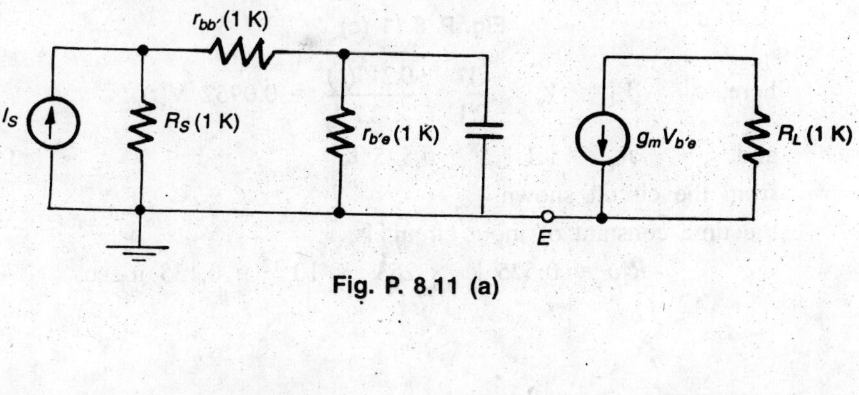

Fig. P. 8.11 (a)

(i) Using approximate analysis and using millers theorm, we have the circuit shown above.

Here $C = C_e + C_c\ (1 - \text{k})$

where $k = \dfrac{V_{ce}}{V_{b'e}} = -g_m R_L$

$\therefore$ $C = C_e + C_c\ (1 + g_m R_L)$

$= 100 + 3\ (1 + 50) = 253$ pF

Applying Thevenin's theorm to the circuits left to B, we have

$V_s = I_S R_S = 0.2 \times 1$

$= 0.2\ \mu(t)\ V$ ($V(t)$ is unit step function)

and, $R_{eg} = R_s = 1$ K

Then the circuit is drawn below,

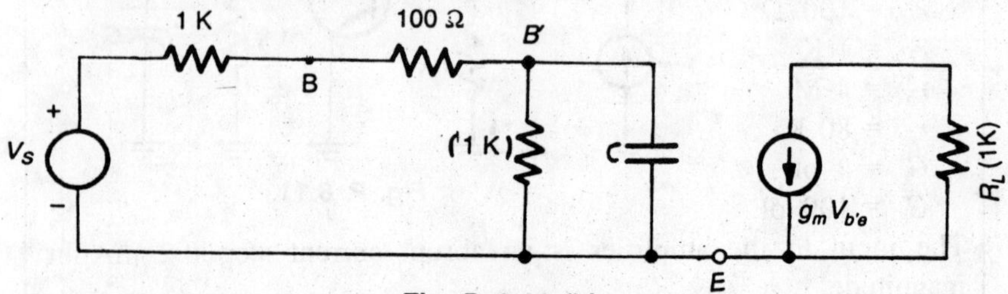

Fig. P. 8.11 (b)

Again, using Thevenin's theorm to the left of B′, we have circuit drawn below

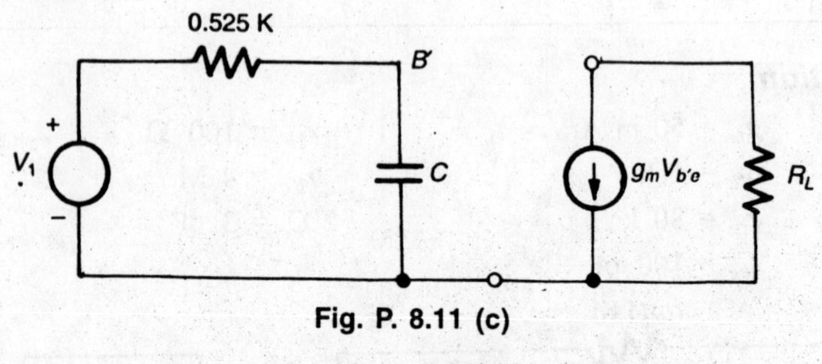

Fig. P. 8.11 (c)

here, $V_1 = V_s \times \dfrac{1}{2.1} = \dfrac{0.2V(t)}{2.1} = 0.0952\ \text{V}(t)$

and, $R_1 = 1.1 \parallel 1 = 0.525$ K

from the circuit shown,

the time constant of input circuit is,

$R_1 c = 0.525\ \text{k} \times 253 \times 10^{-12} = 0.133\ \mu\text{-sec.}$

Hence, $V_{b'e} = 0.952\ (1 - e^{-t/0.133})$

and, $V_o = -g_m R_L V_{b'e}$

$= -50 \times 0.0952 = (1 - e^{-t/0.133})$

$= 4.76\ (1 - e^{-t/0.133})$ where t is in seconds.

(b) In this case g_m, $V_{b'e}$ can be considered as the unit step because time constant of the output

$$R_L C_c = 10^3 \times 10^{-7} = 10^{-7} = 10^{-4} \gg R_1 C = 0.133 \times 10^{-6}$$

Hence $V_o = -50 \times 0.0952 \left(1 - e^{-t/10^{-4}}\right)$

$= -4.76 \left(1 - e^{-t/10^{-4}}\right)$ where t is given in seconds.

Problem 8.12: (a) For the emitter follower of Fig. 8.12 at high frequencies, obtain $K = V_e/V'_i$ and (with $g \equiv g_m + g_{b'e}$) verify that

$$K = \frac{gR_L}{1 + gR_L} \quad \frac{1 + jw(C_e / g)}{1 + jw\left(\dfrac{C_L + C_e}{1 + gR_L}\right)R_L}$$

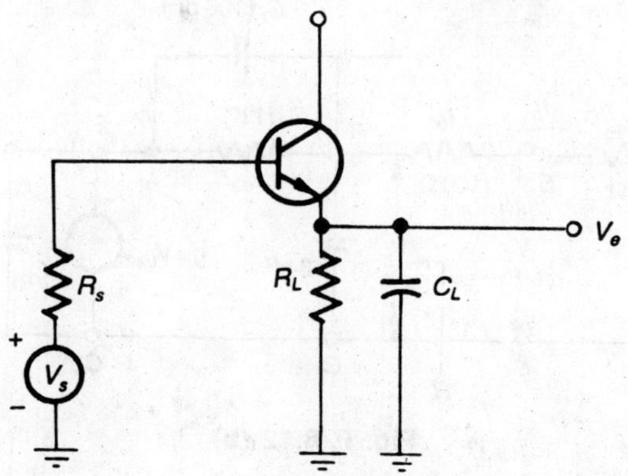

Fig. P. 8.12

(b) If $gR_L \gg 1$ and $C_L \gg C_e$, show that

$$k \approx \frac{1}{1 + Jf / f_H}$$

where

$$f_H = \frac{1}{2\pi} \frac{g}{C_L + C_e} = \frac{1}{2\pi} \frac{g_m + g_{b'e}}{C_L + C_e}$$

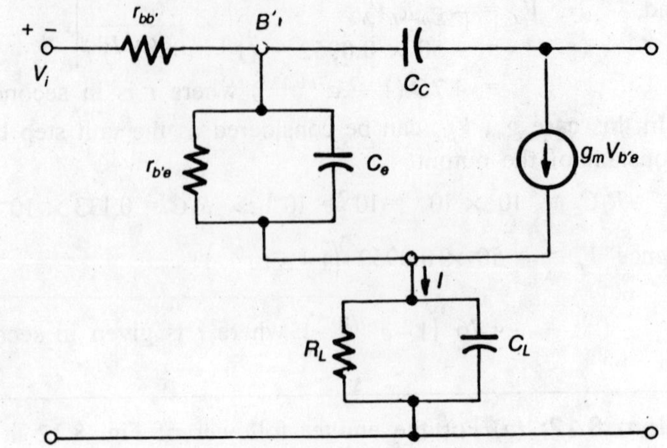

Fig. P. 8.12 (a)

Solution:

Drawing the equivalent circuit for emitter follower as shown below,

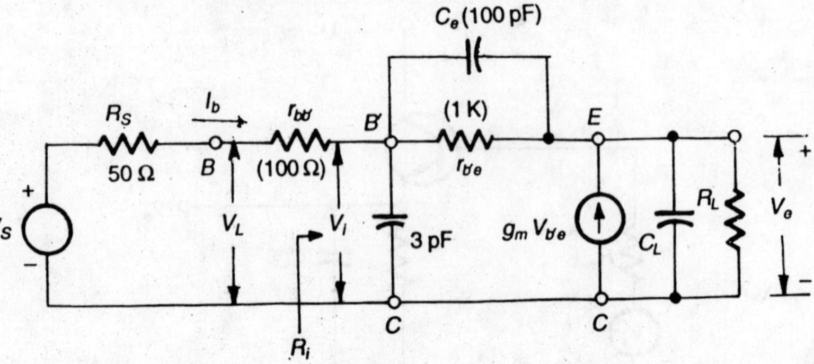

Fig. P. 8.12 (b)

From the circuit we have

$$V_e = IZ = \text{I.}\frac{1}{\dfrac{1}{R_L} + jwC_L} = \text{I.}\frac{R_L}{1 + jwC_L R_L} \quad \text{...(i)}$$

and,
$$\begin{aligned} \text{I} &= g_m V_{b'e} + V_{b'e}(g_{b'e} + jwc_e) \\ &= [g_m (g_{b'e} + jwc_e)(V_{i'} - V_e) \quad \text{...(ii)} \end{aligned}$$

Using equation (ii) in equation (i), we get,

$$V_e = [g_m (g_{b'e} + jwc_e)] [V_{i'} - V_e] \frac{R_L}{1 + jwC_L R_L}$$

or, $$V_e\left[1 + (g_m + g_{b'e}) + jwc_e)\frac{R_L}{1+jwC_L R_L}\right]$$

$$= V_{i'}\left[g_m \; (g_{b'e} + jwc_e)\right]\left(\frac{R_L}{1+jwC_L R_L}\right)$$

or, $$k = \frac{V_e}{V_{i'}} = \frac{(g_m + g_{b'e} + jwC_e)\frac{R_L}{1+jwC_L R_L}}{1+(g_m + g_{b'e} + jwC_e)\frac{R_L}{1+jwC_L R_L}}$$

$$= \frac{(g_m + g_{b'e} + jwC_e) R_L}{1+jwC_L R_L + (g_m + g_{b'e} + jwC_e) R_L}$$

$$k = \frac{g R_L\left(1 + jw\frac{C_e}{g}\right)}{1 + g R_L + jw R_L (C_L + C_e)}$$

$$= \frac{g R_L}{1+g R_L} \times \left(\frac{1+jw\frac{C_e}{g}}{1+\frac{jw R_L}{1+g R_L}(C_L + C_e)}\right)$$

where, $g = g_m + g_{b'e}$

(ii) Now, for, $gR_L \gg 1$, $C_L \gg C_e$

$$k = \frac{g + jwC_e}{g + jw(C_L + C_e)} \approx \frac{g}{g + jw(C_L + C_e)}$$

$$\approx \frac{1}{1 + \frac{jw(C_L + C_e)}{g}}$$

If, $$f_H = \frac{g}{2\pi(C_L + C_e)}$$

then, $$K = \frac{1}{1 + j\dfrac{f}{f_H}}$$

Problem 8.13: A JFET with the following parameters is used in single stage common source amplifier with a load resistance of 100 KΩ. Calculate the high frequency cut off (upper 3 dB cut-off frequency) of the amplifier.

$$g_m = 2 \text{ mA/V}$$
$$C_{gd} = 2 \text{ pF}$$
$$C_{ds} = 1.0 \text{ pF}$$
$$r_d = 100 \text{ K}\Omega$$
$$C_{gs} = 2.5 \text{ pF}$$

(GATE – 93)

Solution :

Small single equivalent **ciruit for Cs** amplifier drawn below.

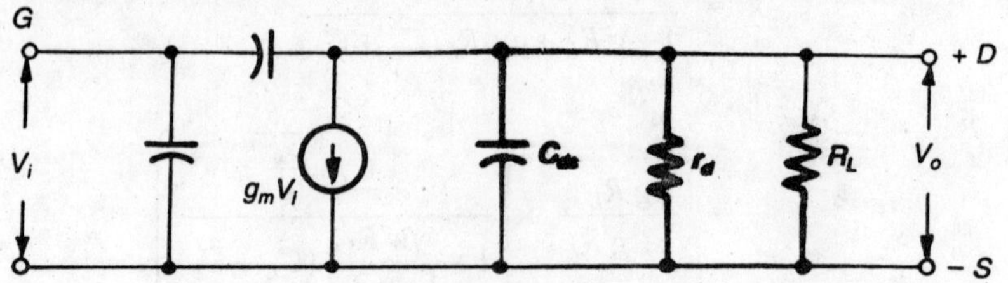

Fig. P. 8.13 (a)

Using Millers theorm, admittance γ_{gd} between gate and drain can be replaced by γ_{gd} (1 – K) between 'G' and 'S', and by $\gamma_{gd}\left(1-\dfrac{1}{K}\right)$ between 'D' and 'S', where $K = Av$. Then the circuit will be as show below.

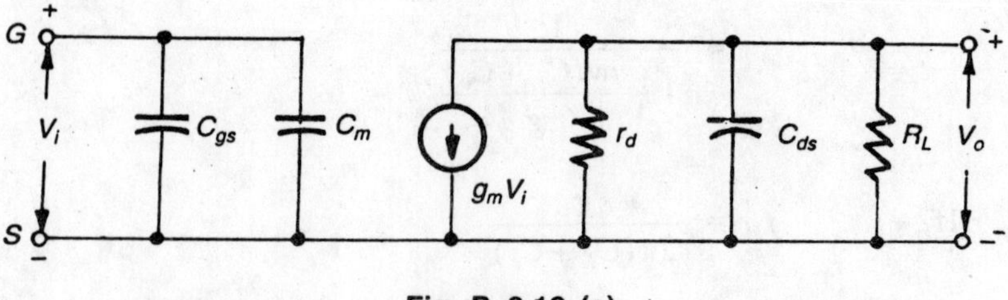

Fig. P. 8.13 (a)

Here, $C_M = C_{gd}(1 - K) = C_{gd}(1 - A_V)$

$$= C_{gd}\left(1+\frac{g_m r_d R_L}{r_d + R_L}\right)$$

$$= 2 \times 10^{-2}\left(1+\frac{200 \times 100\ \text{K}}{200\ \text{K}}\right)$$

$$= 202 \times 10^{-12}\ \text{F} = 202\ \text{pF}$$

Then input capacitance,

$$C_i = C_M + C_{gs} = (202 + 2.5)\ \text{pF} = 204.5\ \text{pF}$$

Now $C_M' = C_{gd}\left(1-\frac{1}{K}\right) = C_{gd}\left(1-\frac{1}{A_v}\right)$

$$= C_{gd}\left(1+\frac{r_d + R_L}{g_M r_d R_L}\right)$$

$$= 2 \times 10^{-12}\left(1+\frac{200\ \text{K}}{2 \times 100 \times 100\ \text{K}}\right) \approx 2\ \text{pF}$$

Then output capacitance, $C_o = C_{M'} + C_{ds}$

$$= 2 + 1 = 3\ \text{pF}$$

$$w_{H_1} = \frac{1}{R_{in}C_1} = \frac{1}{204.5 \times \infty} = 0\ [\because\ R_i = \infty \text{ for JEFT}]$$

and $w_{H_2} = \frac{1}{R_{L'}C_o} = \frac{1}{(r_d \parallel R_L)C_o}$

$$= \frac{1}{3 \times 10^{-12} \times 50 \times 10^3} = 6.67\ \text{MHz}$$

$\therefore$ High frequency cut off

$$= w_{H_1} + w_{H_2} = 6.67\ \text{MHz}.$$

Chapter 9

Multistage Amplifiers

Problem 9.1: (a) Show that the magnitude plot for a two-pole transfer function is equal to the sum of the magnitude plots of each pole considered separately.
(b) Repeat part a for the Bode phase plot.

Solution:

(a) The two pole transfer function is given by,

$$\frac{|A(if)|}{A_o} = \frac{1}{1+\left(\frac{if}{f_{p_1}}\right)\left(1+\frac{if}{f_{p_2}}\right)}$$

where f_{p1} and f_{p_2} are the two pole frequencies.

$$\frac{|A(if)|}{A_o} = \frac{1}{\left(1+\frac{f^2}{f_{p_1}^{\ 2}}\right)\left(1+\frac{f^2}{f_{p_2}^{\ 2}}\right)}$$

or,

$$20 \log \frac{A(if)}{A_o} = -10 \log \left[\left(1+\frac{f^2}{f_{p_1}^{\ 2}}\right)\left(1+\frac{f^2}{f_{p_2}^{\ 2}}\right)\right]$$

$$= -10 \log \left(1+\frac{f^2}{f_{p_1}^{\ 2}}\right) - 10 \log \left(1+\frac{f^2}{f_{p_2}^{\ 2}}\right)$$

Hence each of these ztwo terms represent the gain magnitude in dB due to one of the two poles. Thus, the overall gain magnitude in dB is the sum of gain magnitude in dB due to each pole.

Therefore the overall bode magnitude plot is the sum of the plots due to each pole.

(b) Now, $$\theta = -\text{arc tan}\left(\frac{f}{fp_1}\right) - \text{arc tan}\left(\frac{f}{fp_2}\right)$$

where $$\theta = \text{arc tan}\frac{A(if)}{A_o}$$

Each of these two curves represents the phase due to arc of the two poles.

Hence the overall Bode phase plot may be obtained by adding the phase due to each pole.

Problem 9.2: (a) Prove that the response of a two-stage (identical and noninteracting) low-pass amplifier to a unit step is

$$v_o = A_o^2 \; [1-(1+x)\,\varepsilon^{-x}]$$

Solution:

(a) The equivalent circuit for each of the stage is as shown

From this circuit,

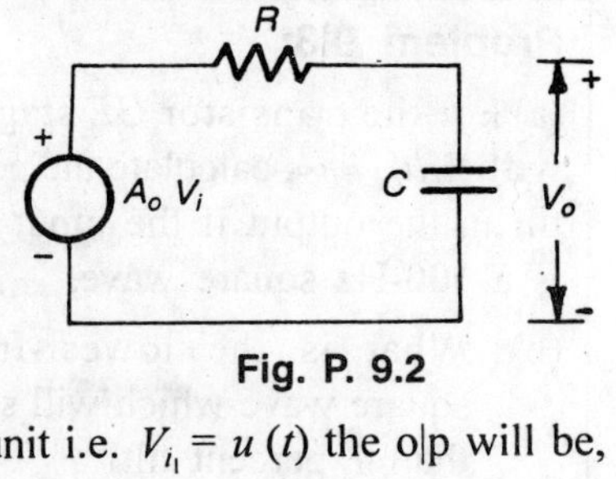

Fig. P. 9.2

$$\frac{dv_o}{dt} = \frac{v_o}{R_2C_L}$$

$$= \frac{A_o}{R_2C_L}V_i$$

If the input to the first stage is a unit i.e. $V_{i_1} = u(t)$ the o|p will be,

$$V_{o1} = A_o\left(1-e^{t/R_2C_2}\right)_{u(t)}$$

This will be the input to the second stage.

$$\therefore \frac{dv_{o2}}{dt} + \frac{V_{o2}}{R_2C_2} = \frac{\Delta_o^2}{R_2C_2}\left(1-e^{-t/R_2C_2}\right)$$

where general relation is

$$\therefore \qquad V_{o2} = A_o^2 = \left(1 - e^{\frac{t}{R_2C_2}}\right) + B\,t.\,(f)$$

B is found to be equal to $\dfrac{-A_o^2}{R_2C_2}$ by subsititution

$$V_{o2} = A_o^2\left[1 - \left(1 + \frac{t}{R_2C_2}e^{-t/R_2C_2}\right)\right]_{u(t)}$$

putting $\quad x = \dfrac{t}{R_2C_2}$

Then, $\quad V_o = A_o^2\,[1 - (1 + x)\,e^{-x}]$ for $t \geq 0$ proved.

(b) If it is assumed that $t \ll R_2C_2$

$$\therefore \qquad x \ll 1 \text{ and } e^{-x} \approx 1 - x + \frac{x^2}{2} \ldots$$

Then from output eq. of $V_{o,}$ we obtain,

$$V_o = A_o^2\left[1 - \left(1 - x + \frac{x^2}{2} - \ldots\right)\right] \equiv \left(A_o^2\,\frac{x^2}{2}\right)$$

Problem 9.3:

(a) For the transistor CE stage shown with $1/h_{oe} \approx \infty$, calculate the percentage tilt in the output if the input current I is a 100-Hz square wave.

(b) What is the lowest-frequency square wave which will suffer less than 1 percent tilt?

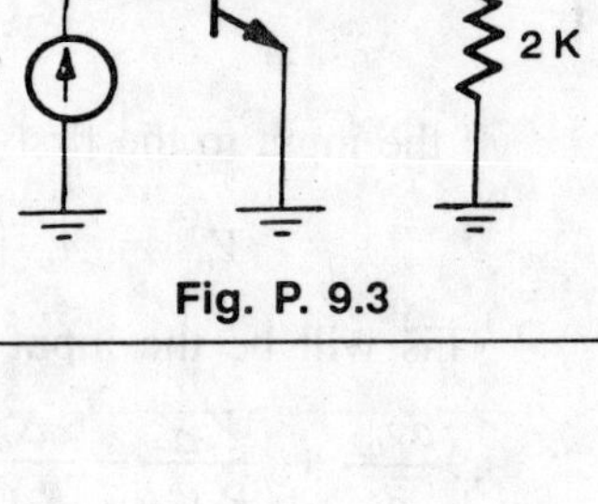

Fig. P. 9.3

Solution :

$$f_L = \frac{1}{2\pi\,(R_o' + R_i')\,c_b}$$

% tilt $\quad P = \dfrac{\pi \times 3.18}{100} \times 100 = 10\%$

(b) For the % tilt to be 1%

$$1 = \frac{100\,\pi\, f_L}{f} = p \text{ and } f = 100\,\pi\, f_L = 10^3 \text{ Hz.}$$

Problem 9.4: Consider a transfer with poles at 1 MHz and 2 MHz. Assume all other poles and zeros are larger that 2 MHz. Calculate the high 3-dB frequency.

Solution:

f_H^* can be calculated from the product

$$= \frac{1}{\sqrt{1+\left(\dfrac{f_H^*}{f_{H_1}}\right)^2}} \times \frac{1}{\sqrt{1+\left(\dfrac{f_H^*}{f_{H_2}}\right)}}$$

or, $$2 = \left(1+f_H^2\right)\left(1+\frac{f_H^2}{4}\right);\ f_H^4 + 5f_H^2 - 4 = 0$$

$$f_H^2 = 0.702$$

or $\quad f_H = 0.83$ MHz

Also, from equation

$$\frac{1}{f_H} = 1.1\sqrt{\frac{1}{f_1^2}+\frac{1}{f_1^2}+\ldots} = 1.1\left(1+\frac{1}{4}\right)^{\frac{1}{2}} = 1.23$$

$\therefore \quad f_H = 6.814$ MHz

$\therefore \quad \text{Error} = \dfrac{0.016}{0.83} = 1.9\ \%$

Problem 9.5: Three identical cascaded stages have an overall upper 3-dB frequency of 20 kHz and a lower 3-dB frequency of 20 Hz. What are f_L and f_H of each stage? Assume noninteracting stages.

Solution:

The high 3 dB frequency for the identical cascaded stages in,

$$f_H^* = f_H\left(2^{1/V} - 1\right)\frac{1}{2} \quad \text{...(i)}$$

where, $V = 3$

and, f_H = upper 3 dB frequency of the individual stage

Now, low 3 dB frequency for cascaded amplifier

$$f_H^* = \frac{f_L}{\sqrt{2^{1/V} - 1}} \quad \text{...(ii)}$$

where, f_L = lower 3-dB frequency of the individual stage

Given, f_H^* = 20 KHz and

f_L^* = 20 Hz

then, from eq (i),

we have $20 \times 10^3 = f_H\left(2^{1/3} - 1\right)^{\frac{1}{2}}$

or, $$f_H = \frac{20 \times 10^3}{\left(2^{\frac{1}{3}} - 1\right)^{\frac{1}{2}}} = 39.23 \text{ KHz}$$

from equation (ii)

$$f_L = f_H^* \cdot \sqrt{2^{\frac{1}{3}} - 1} = 20 \times 0.509 = 10.2 \text{ Hz } \textbf{Ans}$$

Problem 9.6: The parameters of the transistors in the circuit shown are $h_{fe} = 50$, $h_{ie} = 1.1$ K, $h_{re} = h_{oe} = 0$. Find (a) the midband gain, (b) the value of C_b necessary to give a lower 3-dB frequency of 20 Hz. Assume that C_z represents a short-circuit at this frequency. (c) Find the value of C_b necessary to ensure less than 10 percent tilt for a 100-Hz square-wave input.

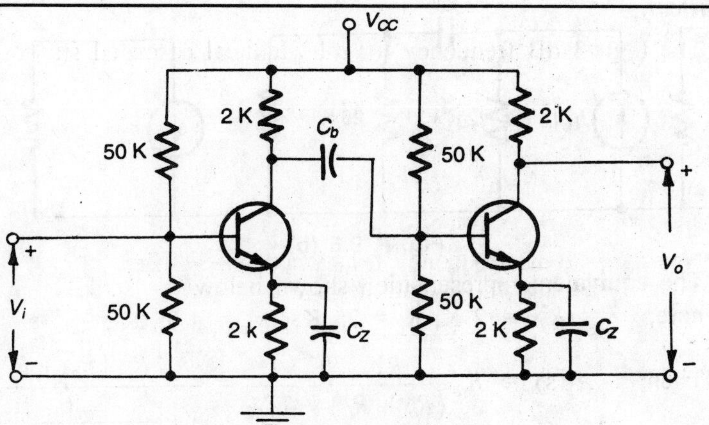

Fig. P. 9.6

Solution:

From small-signal operation C_b is short circuited .

For the second stage of amplifier,

$$R_{L_2} = 2 \text{ K}$$

$$R_{i2} = h_{ie} = 1.1 \text{ K}$$

$$A_{I_2} = -50$$

and $$A_{v_2} = \frac{A_{I_2} \times R_{L_2}}{R_{i_2}} = \frac{-50 \times 2}{1.1} = -91$$

For the first stage,

$$R_{L_1} = 2 \parallel 50 \parallel 50 \parallel R_{i_2} = \left(\frac{1}{1.45}\right) = 0.69 \text{ k}\Omega$$

and, $$R_{i_1} = h_{ic} = 1.1 \text{ k}\Omega$$

and, $$A_{I_1} = -50$$

Then, $$A_{V_1} = \frac{A_{I_1} \times R_{L_1}}{R_{i_1}} = \frac{-50 \times (0.69)}{1.1}$$

$$= -31.4$$

Hence, $$A_v = A_{V_1} \times A_{v_2}$$

$$= (-31.4) \times (-91) = 2857.4$$

(b) Drawing the equivalent circuit, of the given amplifier,

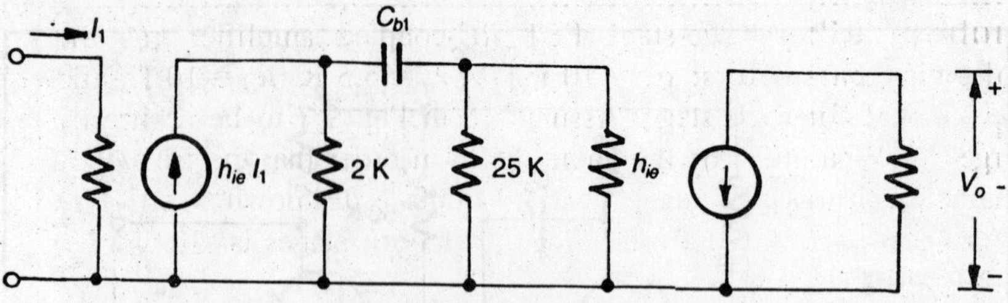

Fig. P. 9.6 (b)

The equivalent representation shown below

here, $R_o' = 2$ K; $R_i = 25$ K $\| h_{ie} = 1.05$ K

Then, $$A_1(s) = h_{fe} \frac{R_o'}{(R_o' + R_i')} + \frac{1}{SC_b} = \frac{h_{fe} R_o' (R_o' + R_i')}{1 + \frac{1}{SC_b (R_o' + R_i')}}$$

Then, the lower 3-dB frequency is given by

$$f_L = \frac{1}{2\pi C_b (R_o' + R_i')};$$

$$C_b = \frac{1}{2\pi \times 20 \times (3.05) \times 10^3} = 2.65 \ \mu F$$

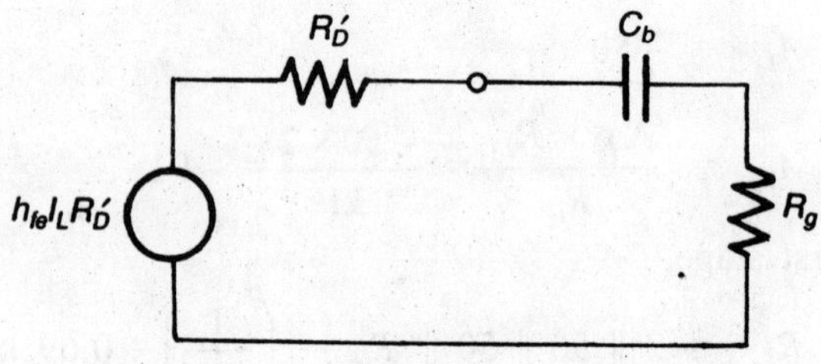

Fig. P. 9.6 (c)

(c) Percentage lift,

$$p = \frac{\pi f_L}{f} \times 100$$

or, $$10 = \frac{\pi \times 1}{2\pi C_b \times (R_o' + R_i') \times f} \times 100$$

or, $$C_b = \frac{10}{100 \times 3.05 \times 10^3} = 16.4 \ \mu\text{F}$$

Problem 9.7: A two-stage FET *RC*-coupled amplifier has the following parameters: $g_m = 10$ mA/V, $r_d = 5.5$ K, $R_d = 10$ K, and $R_g = 0.5$ M for each stage. Assume C_s in Fig. 9.7 to be arbitrarily large. (a) What must be the value of Cb in order that the frequency characteristic of each stage be flat within 1 dB down to 10 Hz? (b) Repeat part *a* if the overall gain of both stages is to be down 1 dB at 10 Hz.

Solution:

Two stage FET RC coupled amplifier is shown below.

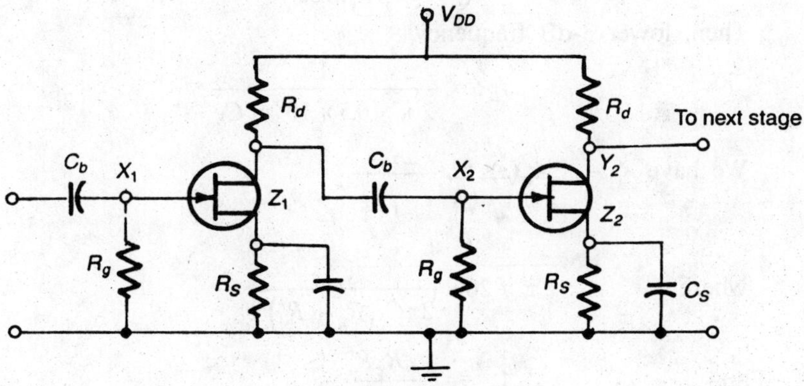

Fig. P. 9.7

Drawing the equivalent one

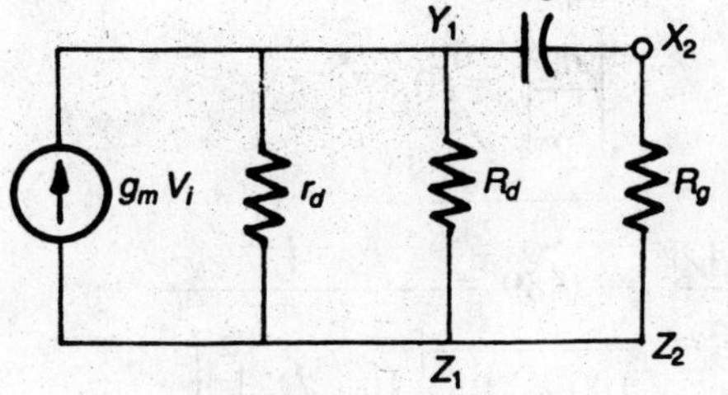

Fig. P. 9.7 (a)

Taking Thevenin's equivalent to the left of terminal $r_1 z_1$, we have circuit as shown below,

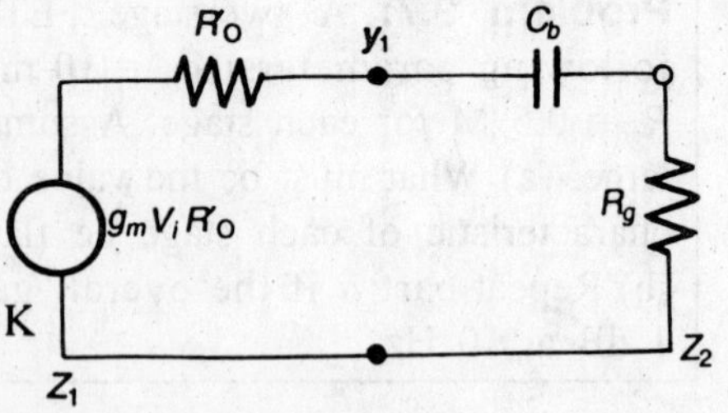

Fig. P. 9.7 (b)

Here,

$$R'_o = r_d \parallel R_d$$
$$= 5.5 \text{ K} \parallel 10 \text{ K} = 3.55 \text{ K}$$

From the circuit,

$$\text{Voltage gain, } A_V(s) = \frac{g_m R_g R'_o}{\left(R_g + R'_o + \dfrac{1}{SC_b}\right)}$$

Then, lower 3-dB frequency,

$$f_L = \frac{1}{2\,\pi \times 0.5 \times 10^6 \times C_b}$$

We have
$$A_v(s) = \frac{A_o}{1/j\dfrac{f_L}{f}}$$

where,
$$f_L = \frac{1}{2\pi C_b\left(R_g + R'_0\right)}$$

and
$$A_o = \frac{g_m R_g R'_o}{R_g + R'_o}$$

or,
$$\left|\frac{A_V}{A_o}\right| = \frac{1}{\left[1+\left(\dfrac{f_L}{f}\right)^2\right]^{\frac{1}{2}}}$$

given,
$$20 \log \left|\frac{A_V}{A_o}\right| = -1 \text{ dB}$$

or,
$$\left|\frac{A_V}{A_o}\right| = 0.89$$

Therefore,

$$\left|\frac{A_V}{A_o}\right| = 0.89 = \frac{1}{\left[1+\left(\dfrac{f_L}{f}\right)^2\right]^{\frac{1}{2}}}$$

$$\text{or, } 1+\left(\frac{f_L}{f}\right)^2 = 1.26$$

$$\text{or, } \quad f_L = \sqrt{0.26\times f^2} = \sqrt{0.26\times 100} = 5.09 \text{ Hz}$$

We have

$$C_b = \frac{1}{2\,\pi\times 0.5\times 10^6\times f_L} = 0.0621\ \mu\text{F} = 62.1\ \mu\text{F}$$

(ii) For two stages,

$$20\log\left|\frac{A_V}{A_o}\right|^2 = -20\log\left[1+\left(\frac{f_L}{f}\right)^2\right] = -1\text{ dB}$$

$$\text{or, } 1+\left(\frac{f_L}{f}\right)^2 = 1.125$$

$$\Rightarrow \quad \frac{f_L}{f} = 0.352$$

$$f_L = 0.352\times 10 = 3.52\text{ Hz}$$

$$\text{then, } \quad C_b = \frac{1}{2\pi\times 0.5\times 10^6\times 3.52} = 90\ \mu\text{F}$$

Problem 9.8: (a) Show that the relative voltage gain of an amplifier with an emitter resistor R_e bypassed by a capacitor C_z may be expressed in the form

$$\frac{A_V}{A_o} = \frac{1+jwR_eC_z}{B+jwR_eC_z}$$

where $B = 1 + R'/R$, $R' = R_e(1+h_{fe})$, and $R = R_s + h_{ie}$.

(b) Prove that the lower 3-dB frequency is

$$f_L = \frac{\sqrt{B^2-2}}{2\,\pi\,R_e\,C_z}$$

what is the physical meaning of the condition $B<\sqrt{2}$?

(c) If $B^2 \gg 2$, Show that $f_L \approx f_p$,

Solution :

(a) Here from eq of

$$\frac{A_V}{A_o} = \left(\frac{1}{1+\frac{R'}{R}}\right) \cdot \left(\frac{1+\frac{jf}{f_o}}{1+\frac{jf}{f_p}}\right)$$

when $$f_o = \frac{1}{2\pi C_Z R_e}, f_p = \left(\frac{1+\frac{R'}{R}}{2\pi C_Z R_e}\right)$$

$$\therefore \quad \frac{A_V}{A_o} = \frac{1}{1+\frac{R'}{R}} \times \frac{1+jwC_Z R_e}{1+jw\frac{C_Z R_e}{R'/R+1}}$$

$$= \frac{1+jwC_Z R_e}{\left(1+\frac{R'}{R}\right)+jwC_Z R_e} = \frac{1+jwC_Z R_e}{B+jw\, C_Z R_e}$$

where $$B = \left(1+\frac{R'}{R}\right)$$

(b) $$\left|\frac{A_V}{A_o}\right|^2 = \left(\frac{1+w_L{}^2 C_Z{}^2 R_e{}^2}{B^2+w_L{}^2 C_Z{}^2 R_e{}^2}\right) = \frac{1}{2}$$

$$2 + 2W_L{}^2 C_Z{}^2 R_e{}^2 = B^2 + w_L{}^2 C_Z{}^2 R_e{}^2$$

solving this eq.

$$w_L = \left(\frac{\sqrt{B^2-2}}{C_Z R_e}\right) \text{ hence } f_L = \left(\frac{\sqrt{B^2-2}}{2\pi C_Z R_e}\right)$$

when $B < \sqrt{2}$ (since we know that $B > 0$)
We conclude that $B^2 < 2$

or, $\sqrt{B^2-2} = j\sqrt{(B^2-2)}$ but that means no kal value of w exists,

i.e. circuit has no lower 3-dB frequency

(c) for $B^2 \gg 2$; $f_L \equiv \dfrac{B}{2\pi C_Z R_e}$

$$= \frac{1+\dfrac{R'}{R}}{2\pi C_Z R_e} = f_p$$

Problem 9.9: Show that the low-frequency voltage gain of the FET stage shown with $r_d \gg R_L + R_s$ is given by

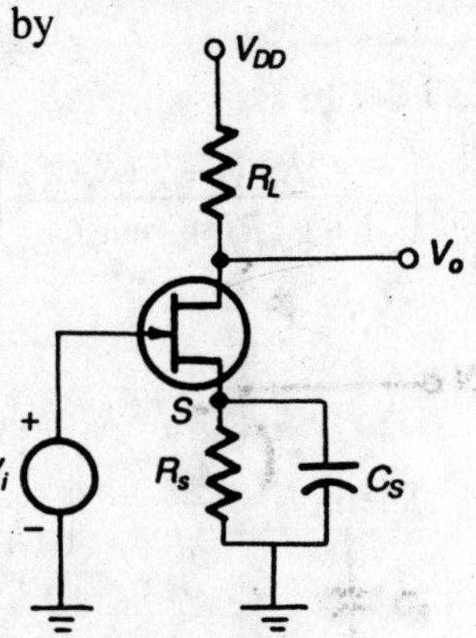

(a) $$\frac{A_V}{A_o} = \frac{1}{1+g_m R_s}\,\frac{1+\dfrac{jf}{f_o}}{1+\dfrac{jf}{f_p}}$$

where, $A_o = -g_m R_L$

$$f_o = \frac{1}{2\pi C_s R_s}$$

$$f_p = \frac{1+g_m R_S}{2\pi C_s R_s}$$

Fig. P. 9.9

(b) if $g_m R_s \gg 1$ $g_m = 5$ mA/v, find C_s so that a 50-Hz square-wave input will suffer no more that 10 percent tilt.

Solution:

The equivalent ckt. of the given stage is indicated.

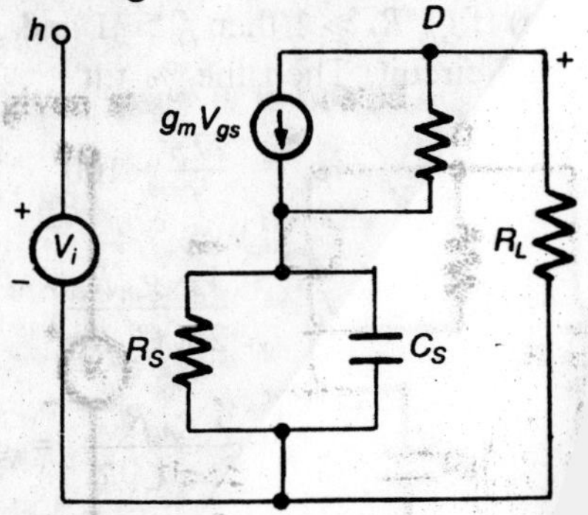

since $r_d \gg R_L + R_s$,

Hence can be neglected

and $z_k = R_s \parallel \dfrac{1}{jwc_s}$

Then, $V_{gs} = -g_m V_{gs} Z_s + V_s$

or, $V_{gs} = \left(\dfrac{V_s}{1+g_m Z_s}\right)$

But $V_o = -g_m V_{gs} R_L$

Fig. P. 9.9 (a)

$$= \frac{-g_m R_L}{1+g_m Z_S} - V_S$$

Hence $$A_V = \frac{V_o}{V_S}$$

$$= \frac{-g_m R_L}{1+\dfrac{g_m R_L}{1+jwR_sC_s}}$$

$$\left(\frac{-g_m R_L(1+jwR_sC_s)}{1+g_m R_L + jwR_sC_s}\right) = \frac{-g_m R_L}{(1+g_m R_L)} \frac{(1+jwR_sC_s)}{\left(1+j\dfrac{wR_sC_s}{1+g_m R_s}\right)}$$

Then, $$\frac{A_V}{A_o} = \left(\frac{1}{1+g_m R_L}\right)\cdot\left(\frac{1+\dfrac{jf}{f_o}}{1+\dfrac{jf}{f_p}}\right);$$

where $$f_o = \left(\frac{1}{2\pi C_s R_s}\right)$$

and $$f_p = \left(\frac{1+g_m R_L}{2\pi R_s C_s}\right)$$

(b) If $g_m R_s \gg 1$ then $f_p \gg 1$ and circuit will behave as a single time constant circuit. Then the % tilt

$$p = \frac{\pi f_p}{f} \times 100$$

where, $$f_p = \frac{1+g_m R_s}{2\pi C_S R_S}$$

$$= \frac{g_m R_s}{2\pi C_s R_s} = g_m/2\pi C_s$$

[W]e want, $$p = 10 = \frac{\pi f_p}{f} \times 100$$

$$= \frac{\pi f_p}{50} \times 100$$

or, $$f_p = \frac{5}{\pi}\left(\max^{m}\right)$$

Hence $$C_s = \frac{g_m}{2\pi f_p} = \frac{g_m}{10} = 0.5 \times 10^{-9}\text{F } \textbf{Ans}$$

Problem 9.10: (a) Find the noise bandwidth B_n for an amplifier for which $A_{V_o} = 1$, $f_L = 0$ Hz. and

$$|A_V(f)| = \frac{1}{\sqrt{1+\left(\frac{f}{f_H}\right)^2}}$$

(b) Compute B_n if $f_H = 10$ kHz.

Solution:

(a) The noise bandwidth equation,

$$B = \frac{1}{\left|A_{V_0}\right|^2} \int_0^{\infty} |A_V(f)|^2\, df$$

$$= \int_0^{\infty} \frac{df}{1+\left(\frac{f}{f_H}\right)^2}$$

or, $$= f_H^2 \int_0^{\infty} \frac{df}{f_H^2 + f^2}$$

$$= f_H^2\left(\frac{1}{f_H}\tan^{-1}\left(\frac{f}{f_H}\right)\right)_0^{\infty}$$

$$= \frac{\pi}{2} f_H$$

(b) if $f_H = 10$ kHz, $B = 5\pi$ kHz.

Problem 9.11:

(a) Find the mean-square value V_o^2 of the output noise voltage for the circuit shown. The circle represents a generator supplying Johnson noise to the RC combination.

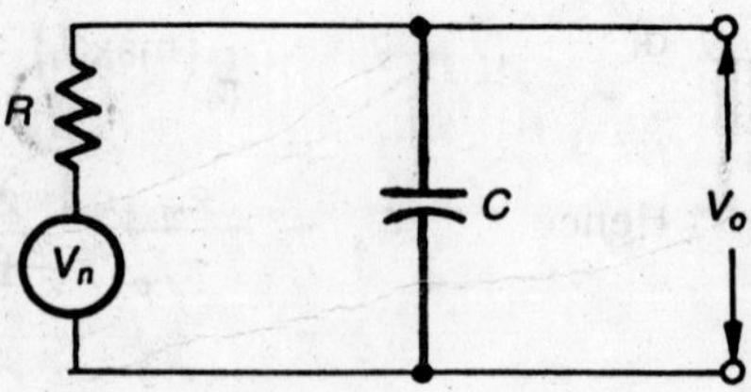

Fig. P. 9.11

$$\frac{1}{2}CV_o^2 = \frac{1}{2}\overline{K}T$$

This result is known as the equipartition theorem.

Solution:

The output noise voltage of the given circuit is gives by the expansion.

$$V_o^2 = 4\overline{K}\,TRB = \frac{4\,\overline{K}TR}{|A_{V_0}|^2}\int_o^{\infty}|A_V(f)|^2\,df$$

$$|A_{V_o}| = 1 \text{ and } A_v(f) = \frac{1}{jwRc+1}$$

$$\therefore \quad V_o^2 = 4\,\overline{K}\,TR\int_0^{\infty}\frac{df}{1+(2\,\pi\,f\,Rc)^2}$$

$$= \frac{4\,\overline{K}TR}{2\,\pi\,Rc}\left(\tan^{-1}\infty - \tan^{-1}0\right) = \left(\frac{\overline{K}T}{C}\right)$$

(b) $$\frac{1}{2}CV_o^2 = \frac{1}{2}\left(\frac{\overline{K}T}{C}\right)\times C = \frac{1}{2}\overline{K}T.$$ **Proved**

Chapter 10

Feedback Amplifiers

Problem 10.1: For the circuit, with

R_c = 4 K
R_L = 4 K
R_b = 20 K
R_s = 1 K

and the transistor parameters given in the following Table A, find

(a) The current gain $I_L / I_s = A_I$
(b) The voltage gain V_o / V_s, where $V_s \equiv I_s R_s$
(c) The transconductance $I_L / V_s = G_M$
(d) The transresistance $V_o / I_s = R_M$
(e) The input resistance seen by the source
(f) The output resistance seen by the load.

Make reasonable approximations. Neglect all capacitive effects.

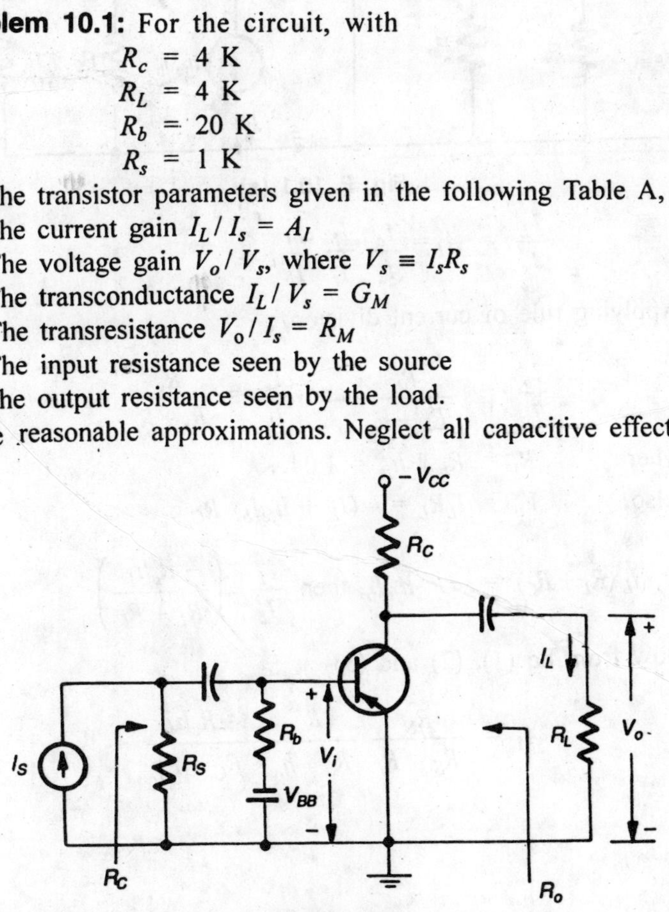

Fig. P. 10.1

Table-A

Typical h-parameter values for a transistor (at I_E = 1.3 mA)

Parameter	*CE*	*CC*	*CB*
$h_{11} = h_i$	1,100 Ω	1,100 Ω	21.6 Ω
$h_{12} = h_r$	2.5×10^{-4}	~ 1	2.9×10^{-4}
$h_{21} = h_f$	50	– 51	– 0.98
$h_{22} = h_o$	24 μA/V	25 μA/V	0.49 μA/V
I/*ho*	40 K	40 K	2.04 M

Solution:

The equivalent circuit is drawn with shorted dc source and the capacitor.

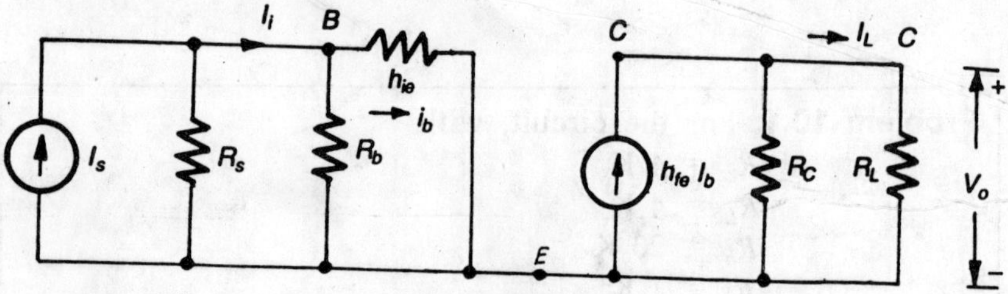

Fig. P. 10.1 (a)

(a) $$\frac{I_L}{I_s} = A_I = \frac{I_c}{I_s} \cdot \frac{I_b}{I_i} \cdot \frac{I_L}{I_b} \quad \text{...(1)}$$

Applying rule of current division,

$$\frac{I_b}{I_i} = \frac{R_b}{R_b + h_{ie}} \text{ and } \frac{I_i}{I_b} = \frac{R_s}{R_s + R_i} \quad \text{...(2)}$$

where $R_i = R_b \parallel h_{ie} = 1.04$ K

Also, $V_o = I_L R_L = -(I_L + h_{fe} I_b)\, R_C$

or, $I_L (R_L + R_C) = -R_C h_{fe} I_b$, then $\frac{I_L}{I_b} = \left(\frac{-R_c h_{fe}}{R_L + R_C}\right)$ ...(3)

Now from eq (1), (2) and (3)

$$A_I = \frac{R_S}{R_S + R_i} \cdot \frac{R_b}{R_b + h_{ie}} \cdot \frac{-R_c h_{fe}}{R_C + R_L}$$

$$= \frac{1}{2.04} \times \frac{20}{21.1} \times \frac{-4}{8} \times 50 = -11.65$$

(b) $$\frac{V_o}{V_s} = \text{voltage gain, } A_V = \frac{I_L R_L}{I_S R_S} = \frac{R_L}{R_S} \times A_I$$

$$= -4 \times 11.65 = -46.5$$

(c) Transconductance,

$$\frac{I_L}{V_S} = \frac{V_o}{R_L V_S} = \frac{1}{R_L} A_v = \frac{-46.5}{4} = -11.65 \text{ mA/V}$$

(d) Transresistance,

$$R_M = \frac{V_o}{I_S} = \frac{I_L}{I_S} \times R_L = -11.65 \times 4 = -46.5 \text{ k}\Omega$$

(e) Input resistance,

$$R_i \cong R_b \parallel h_{ie} = \frac{20 \times 1.1}{21.1} = 1.04 \text{ k}\Omega$$

(f) Output resistance,

$$R_o \cong \frac{1}{h_{oe}} \parallel R_c = \frac{40 \times 4}{44} \cong 3.64 \text{ k}\Omega$$

Problem 10.2: Repeat Prob. 10.1 for the circuit shown, with $g_m = 5$ mA/V and $r_d = 100$ K.
Note that $V_s \equiv I_s R_s$

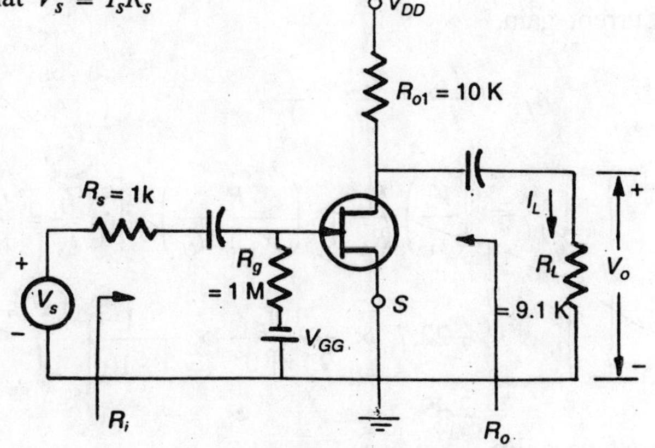

Fig. P. 10.2

Solution:

The equivalent ckt is shown as below,

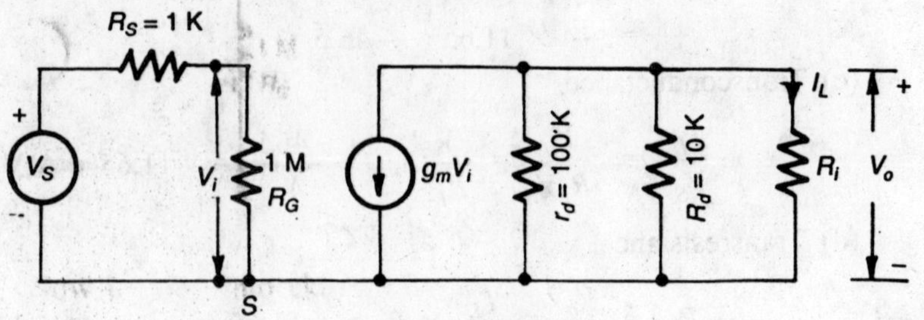

Fig. P. 10.2 (a)

Now looking into ckt, we have,

$$V_o = -g_m V_i \,(r_d \parallel R_d \parallel R_L);$$

or, $$\frac{V_o}{V_i} = -g_m \times (r_d \parallel R_d \parallel R_L)$$

$$= -5 \times 10^{-3} \times \frac{9.1}{2} \times 10^3 = -22.7$$

Also, $$V_i = \left(\frac{R_G}{R_G + R_S}\right) V_s$$

(a) Current gain,

$$A_I = \frac{I_L}{I_S} = \frac{I_L}{I_i} \cdot \frac{I_i}{I_S}$$

$$= \left(\frac{V_o}{V_i}\right) \frac{R_G}{R_L} \times \left(\frac{R_S}{R_S + R_G}\right) \quad \left[\because \frac{I_i}{I_S} = \frac{R_S}{R_S + R_G}\right]$$

$$= -22.7 \times \frac{10^6}{9.1 \times 10^3} \times \frac{1}{1 + 10^3}$$

$$\cong 2.49$$

(b) Voltage gain,

$$A_v = \frac{V_o}{V_S} = \frac{I_L R_L}{I_i (R_S + R_G)}$$

$$= \left(\frac{I_L}{I_i}\right)\left(\frac{R_L}{R_S + R_G}\right) = -2500 \times \frac{9.1}{10000+1} = 22.7$$

(c) Transconductance,

$$\frac{I_L}{V_S} = G_m = \frac{V_o}{R_L} \cdot \frac{1}{V_S}$$

$$= \left(\frac{V_o}{V_S}\right) \times \frac{1}{R_L} \cdot \frac{-22.7}{9.1 \times 10^3} = -2.5 \text{ mA/V}$$

(d) Transresistance,

$$R_M = \frac{V_o}{I_S} = R_s \cdot \frac{V_o}{I_S R_S}$$

$$= R_S \cdot \left(\frac{V_o}{V_S}\right) = 10^3 (-22.7) = -22.7 \text{ k}\Omega$$

(e) $$R_i = \frac{V_i}{R_i} = R_G = 1 \text{ M}$$

(f) Output resistance,

$$R_o = r_d \parallel R_d = \frac{100 \times 10}{110} = 9.1 \text{ k}\Omega$$

Problem 10.3:

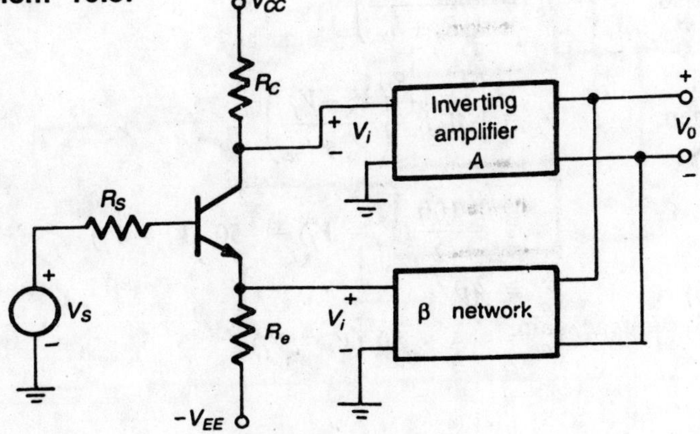

Fig. P. 10.3

(a) For the circuit shown, find the *ac* voltage V_i as a function of V_s and V_f. Assume that the inverting -amplifier input resistance is infinite, that $A = A_v = -1{,}000$, $\beta = V_f / V_o = \frac{1}{100}$, $R_S = R_C = 1$ K, $h_{ie} = 1$ K, $h_{re} = h_{oe} = 0$, and $h_{fe} = 100$. (b) Find $A_{vf} = V_o / V_s = AV_i / V_s$.

Solution:

From the transistor ckt given in the problem, one can draw the equivalent ckt as follows:

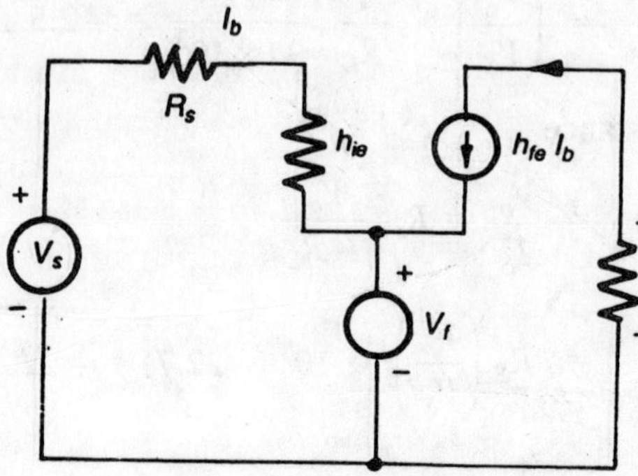

Fig. P. 10.3 (a)

It is given that. $\frac{V_f}{V_o} = \beta$ i.e. $V_f = \beta V_o$

and $V_o = A_{V_i}$

(a) $V_i = -h_{fe} I_b R_c$

Also, $I_b = \left(\frac{V_S - V_f}{R_s + h_{ie}}\right)$

then $V_i = -h_{fe} R_c \left(\frac{V_s - V_f}{R_s + h_{ie}}\right)$

$$= -\frac{100}{2}(V_s - V_f) = -50\ (V_s - V_f)$$

(b) $\because$ $V_o = AV_i$

$V_o = -A \times 50\ (V_s - V_f)$

$$= 5 \times 10^4 (V_s - V_f)$$

$$(\because A = -1000)$$

$$= 5 \times 10^4 (V_S - \beta V_0)$$

$$= 5 \times 10^4 (V_s - 0.01\, V_o)$$

or, $$V_o(500 + 1) = 5 \times 10^4\, V_s$$

Therefore $$A_{V_f} = \frac{V_o}{V_S} = \frac{50000}{501} \cong 100$$

Problem 10.4 : An amplifier consists of there identical stages connected in cascade. The output voltage is sampled and returned to the input in series opposing. If it is specified that the relative change dA_f/A_f in the closed-loop voltage gain A_f must not exceed ψ_f, Show that the minimum value of the open-loop gain A of the amplifer is given by

$$A = 3A_f \frac{|\psi_1|}{|\psi_f|}$$

where $\psi_1 \equiv dA_1/A_1$, is the relative change in the voltage gain of each stage of the amplifier.

Solution:

The closed loop gain,

$$A_f = \left(\frac{A}{1+\beta A}\right) \qquad \text{...(1)}$$

Now, differentiating (1) with respect to A,

$$\frac{dA_f}{dA} = \left(\frac{A_f}{A}\right)\cdot\left(\frac{1}{1+\beta A}\right)$$

$$\left|\frac{d_{Af}}{A_f}\right| = \left|\frac{d_A}{A}\right| \times \left(\frac{1}{1+\beta A}\right) \qquad \text{...(2)}$$

Also, given $\psi_f = \left|\frac{dA_f}{A_f}\right|$, then $|\psi_f| = \left|\frac{dA}{A}\right|\cdot\left(\frac{1}{1+\beta A}\right)$ ...(3)

Now, open loop gain, A = A_1^3; or $dA = 3A_1^2\, dA_1$

Then, $$\left(\frac{dA}{A}\right) = \left(\frac{3dA_1}{A_1}\right) \qquad \ldots(4)$$

Therefore, from equation (3) and (4),

$$|\psi_f| = 3\left|\frac{dA_1}{A_1}\right| \cdot \frac{1}{|1+\beta A|}$$

$$= 3|\psi_1| \cdot \frac{1}{|1+\beta A|} \qquad \left(\because \psi_1 = \frac{dA_1}{A_1}\right)$$

$$\therefore \quad |\psi_f| = 3|\psi_1|\left(\frac{A_f}{A}\right)$$

$$\therefore \quad A = 3A_f \times \frac{|\psi_1|}{|\psi_f|} \qquad \textbf{Proved}$$

Problem 10.5: An amplifier with open-loop voltage gain A_v = 1,000 ± 100 is available. It is necessary to have an amplifier whose voltage gain varies by no more than ± 0.1 percent.
(a) Find the reverse transmission factor β of the feedback network used.
(b) Find the gain with feedback.

Solution:

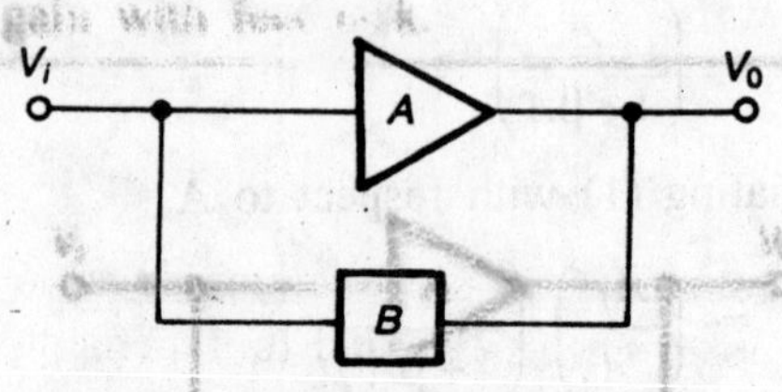

Fig. P. 10.5

(a) Closed loop gain, $A_f = \left(\frac{A}{1+\beta A}\right)$

Now when differentiating

$$\left|\frac{dA_f}{A_f}\right| = \left|\frac{dA}{A}\right|\cdot\left(\frac{1}{1+\beta A}\right) \quad \ldots(1)$$

Hence $\left|\dfrac{dA_f}{A_f}\right| = 0.001$

and $\left|\dfrac{dA}{A}\right| = \dfrac{100}{1000} = 0.1$

Then from eq (1),

$$0.001 = \left(\frac{1}{\beta A + 1}\right) \times 0.1$$

and $(1 + \beta A) = 100$ $\quad (\therefore \beta A = 99)$

$$\therefore \quad \beta = \frac{99}{1000} = 0.099$$

(b) Now the gain with feed back, $A_f = \left(\dfrac{A}{1+\beta A}\right)$

$$= \frac{1000}{1+1000\times 0.099} = 10$$

Problem 10.6: An amplifier without feedback gives a fundamental output of 36 V with 7 percent second-harmonic distortion **when the** input is 0.028V

(a) If 1.2 percent of the output is feed back into the input in a negative voltage series feedback circuit, what is the output voltage?

(b) If the fundamental output is maintained at 36 V but the second-harmonic distortion is reduced to 1 percent, what is the input voltage?

Solution:

(a) $$A = \frac{36}{0.026} = 1285$$

and $\beta = \frac{V_f}{V_o} = \frac{1.2}{100} = 0.012$

Also, $A_f = \left(\frac{A}{1+\beta_A}\right) = \frac{1285}{1+1285\times 0.012} \cong 78.2$

and $V_o = A_f V_s = 78.2 \times 0.0289 = 2.19$ V

(b) If the ouput is maintained constant at 36V then the distortion generated by the device is unchanged. The reduction of the total distortion is caused by feedback is given by,

$$B_{2f} = \left(\frac{B_2}{1+\beta A}\right) \text{ or, } (1 + \beta A) = D = \frac{B_2}{B_{2f}} = 7$$

then $\beta A = 6$, Now from equation, $A_f = \frac{A}{1+\beta A} = \frac{1285}{7}$

$$\therefore \quad V_s = \left(\frac{V_0}{A_f}\right) = \frac{36\times 7}{1285} = 0.196 \text{ V}$$

Problem 10.7: An amplifier with an open-loop voltage gain of 1,000 delivers 10 W of output power at 10 percent second-harmonic distortion when the input signal is 10 mV. If 40-dB negative voltage-series feedback is applied and the output power is to remain at 10 W, determine (a) the required input signal, (b) the percent harmonic distortion.

Solution:

(a) since, $A_f = \left(\frac{A}{1+\beta A}\right)$, or $20 \log \left|\frac{A_f}{A}\right| = 20 \log \left|\frac{1}{1+\beta A}\right|$

or, $-40 = -20 \log |1+\beta A|$, or $|1 + \beta A| = 100$

or, $A_f = \frac{1000}{100} = 10$

when the amplifier delivers low, its output voltate, $V_o = AV_S$

$= 1000 \times 10 \times 10^{-3} = 10$ V

If the ouput power is remain at 10w, then the output voltage also remain must at 10V.

Hence the required input signal, when feedback is applied

$$V_s = \frac{V_o}{A_f} = \frac{10}{10} = 1 \text{ V}$$

(b) The distortion of the amplifier with feedback will be reduced by the feedback factor i.e. by $|1+ \beta A| = 100$

Hence, $$D_f = \frac{10}{100}\,\% = 0.1\%$$

Problem 10.8: Prove that for voltage-series feedback, with $R_s = 0$, $A_{If} = A_I \cdot A_v = A_I R_L / R_i$

Solution:

Now we have,

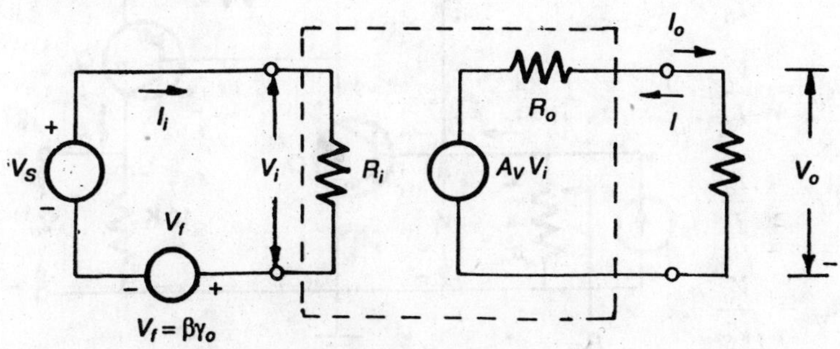

Fig. P. 10.8

$$A_{If} = \frac{I_o}{I_i}$$

and, $$A_I = \left.\frac{I_o}{I_i}\right|_{\beta=0}$$

But, $$I_o = \left(\frac{A_V V_i}{R_o + R_L}\right)$$

and $$V_i = I_i \times R_i$$

Thus, $\dfrac{I_o}{I_i} = \left(\dfrac{R_i}{R_0 + R_L}\right) A_v$ and it is independent of β

Hence, $\dfrac{I_o}{I_i} = \left.\dfrac{I_o}{I_i}\right|_{\beta=0}$

or, $A_{if} = A_I$

Problem 10.9: In the two-stage feedback amplifier shown, the transistors are identical and have the following parameters:

$h_{fe} = 50$
$h_{ie} = 2\text{ K}$
$h_{re} = 0$
and $h_{oe} = 0$.
Calculate :

(a) $A_{If} = \dfrac{I_o}{I_S}$

(b) $R_{if} = \dfrac{V_i}{I_s}$

(c) $A'_{If} = \dfrac{I_o}{I'_i}$

(d) $A_{vf} = \dfrac{V_o}{V_S}$

where $V_s = I_S R_S$

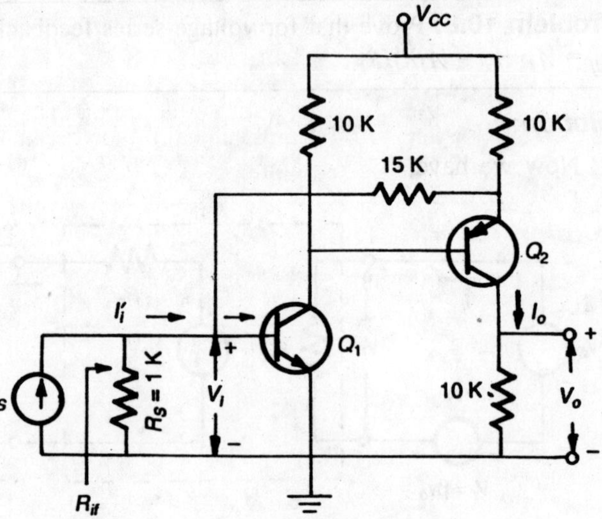

Fig. P. 10.9

Solution:

Since the mixing network is of shunt mixing type.

So, $I_f = \dfrac{V_i - V_{e_2}}{15\text{ K}} \approx \dfrac{-V_{e_2}}{15\text{ K}} = \dfrac{10(-I_f + I_o)}{15}$

or, $I_f = \dfrac{10}{25} I_o$

Thus, $\beta = \dfrac{10}{25}$

Since the feedback current is proportional to the output current , so. circuit end up to be of current shunt feedback amplifier.

Now, to obtain basic amplifier without feedback, opening the output loop ($I_o = 0$) to find the input circuit and by setting $V_i = 0$ find the output circuit so, the equivalent circuit without feedback will be as shown below.

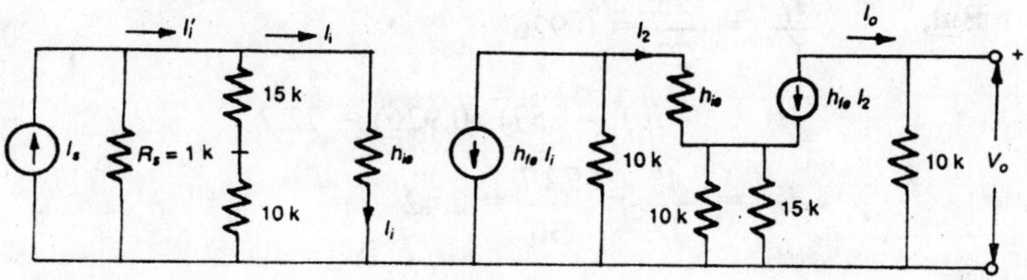

Fig. P. 10.9 (a)

For the amplifier without feedback,

$$A_I = \frac{I_o}{I_S} = \frac{I_o}{I_2} \times \frac{I_2}{I_i} \times \frac{I_i}{I_S} \qquad \text{...(i)}$$

Now, from the circuit,

$$\frac{I_o}{I_2} = -h_{fe}$$

and $$\frac{I_2}{I_i} = -h_{fe} \frac{10}{10 + R_{i_2}} \qquad \text{...(ii)}$$

where, $R_{i2} = h_{ie} + (1 + h_{fe})\ (10 \parallel 15) = 2 + 51\left(\frac{150}{25}\right) = 308\ \text{K}$

$$\therefore \quad \frac{I_2}{I_i} = -50 \times \left(\frac{10}{10 + 308}\right) = -1.57 \qquad \text{...(iii)}$$

and, $$\frac{I_i}{I_S} = \frac{(1 \| 25)}{h_{ie} + (1 \| 25)} = \frac{0.962}{2.962} = 0.324 \qquad \text{...(iv)}$$

$$A_I = (0.324)\ (-1.57)\ (-50) = 25.4$$

Then, desensitivity, $D = 1 + \beta A, = 1 + 25.4 \times \frac{10}{25} = 11.2$

(i) $$A_{if} = \frac{A_I}{D} = \frac{25.4}{11.2} = 2.27$$

(ii) $$R_i = (1 \parallel 25 \parallel h_{ie})\ \text{K} = (0.962 \parallel 2) = 0.65\ \text{k}\Omega$$

Then, $R_{if} = \dfrac{R_i}{D} = \dfrac{0.65K}{11.2} = 58\ \Omega$

(iii) From the circuit, we have

$$A'_I = \left(\frac{I_o}{I_2}\right)\left(\frac{I_2}{I_1}\right)\left(\frac{I_i}{I_{i'}}\right) = (-50)\,(-1.57) \times \frac{I_i}{I_{i'}}$$

But, $\dfrac{I_i}{I_{i'}} = \dfrac{25}{27} = 0.926$

$$A'_I = (-50)\,(-1.57)\,(0.926) = 72.7$$

$$A'_{If} = \frac{A'_I}{D} = \frac{72.7}{30} = 2.42$$

(iv) $A_{vf} = \dfrac{V_o}{V_S} = \dfrac{I_o R_{C2}}{I_S R_S} = A_{if} \cdot \dfrac{R_{C2}}{R_S}$

From part (i) $A_{if} = 2.27$

$\therefore$ $A_{vf} = 2.27 \times \dfrac{10K}{1K} = 22.7$ **Ans**

Problem 10.10:

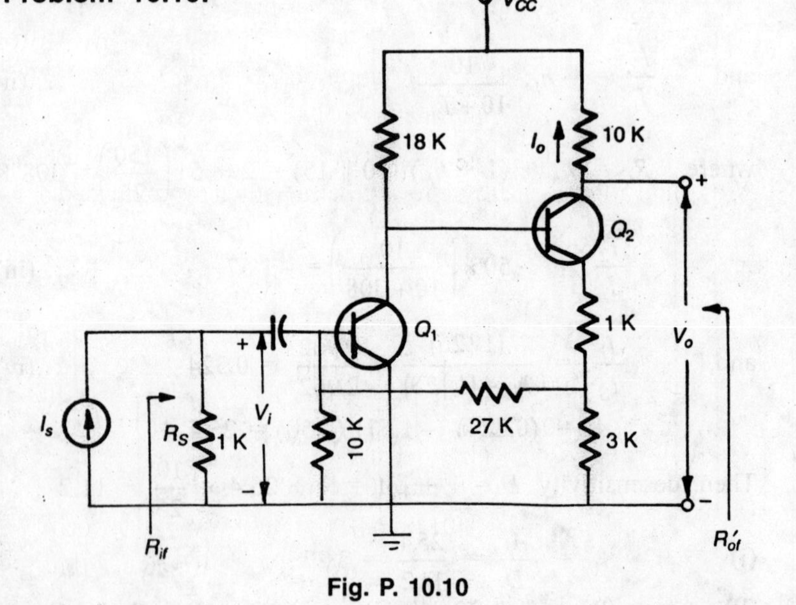

Fig. P. 10.10

For the circuit shown (and with the *h*-parameter values given in Prob.10.9
find

(a) $A_{if} \equiv \dfrac{I_o}{I_S}$ (b) R_{if}

(c) $A_{vf} \equiv \dfrac{V_o}{V_S}$ where $I_S \equiv \dfrac{V_2}{R_S}$

(d) $A'_{vf} \equiv \dfrac{V_o}{V_i}$ (e) R'_{of}

Solution:

(a) In the mixture of shunt type and if $= \dfrac{V_i - V}{27\text{ K}}$, where V is the voltage drop across the 3 K resistor.

since $\qquad V \gg V_i$

Then $\qquad I_f = \dfrac{-V_o}{27} = \dfrac{3}{27}[I_o - I_f]$

or, $\qquad I_f = \dfrac{3}{30} I_o$

or, $\qquad \text{If} = \beta I_o$

Hence $\qquad \beta = \dfrac{1}{10}$ and this is a current shunt feed back.

Now, to obtain the basic amplifier without feed back we open the output loop $I_o = 0$ and we thus have at the input a resistor (27 K + 3 K) = 30 K parallel with R_s.

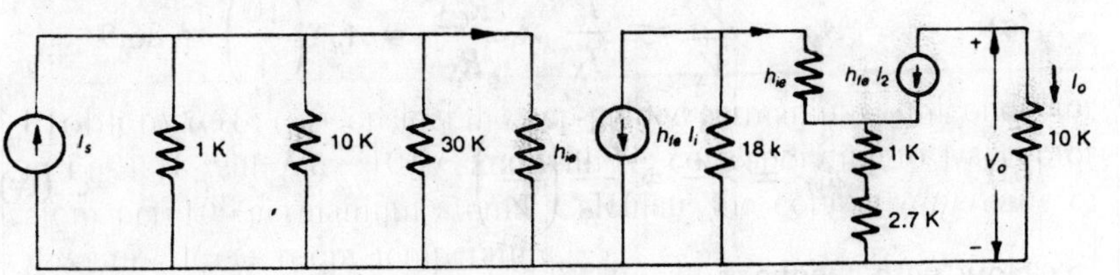

Fig. P. 10.10 (a)

Then we let $V_i = 0$ and we have at the output loop a resistor 27 $\|$ 3K = 2.7 K.

Hence the equivalent circuit is as shown below.

Since the circuit is a current shunt feedback amplifier, the current gain will be stabilized. We now find

$$A_I = \frac{I_o}{I_2}\left(\frac{I_2}{I_i}\right)\left(\frac{I_i}{I_s}\right) \quad \text{...(i)}$$

From the circuit,

$$h_{fe}I_2 = -I_o$$

or,

$$\frac{I_2}{I_o} = -h_{fe} \quad \text{...(ii)}$$

and,

$$\frac{I_2}{I_1} = -h_{fe}\frac{18}{18+R_{i2}}$$

where,

$$R_{i2} = h_{ie}(1+h_{fe})(1+2.7)$$
$$= 0.2 + 51 \times 3.7 = 190\ \text{k}\Omega \quad \text{...(iii)}$$

then,

$$\frac{I_2}{I_1} = -50\cdot\left(\frac{18}{18+190}\right) = -4.32$$

$$\frac{I_i}{I_s} = \frac{1\|10\|30}{h_{ie}+(1\|10\|30)} = \frac{0.884}{2+0.884} = 0.308$$

hence,

$$A_I = (0.308)\times(-4.32)\times(-50) = 66.6 \quad \text{(Using eq.(i))}$$

and,

$$A_{if} = \frac{A_I}{D} = \frac{66.6}{1+\beta A_1} = \frac{66.6}{1+\frac{1}{10}\times 66.6} = 8.69$$

(b)

$$R_i = 1\|10\|30\|h_{ie} = 0.384\|2 = 0.615\ \text{k}\Omega$$

$$R_{if} = \frac{R_i}{D} = \frac{0.615}{7.66} = 80.2\ \Omega$$

(c)

$$A_{vf} = \frac{V_o}{V_s} = \frac{I_o}{I_S}\times\frac{R_{C2}}{R_S} = A_{if}\cdot\left(\frac{10}{1}\right) = 86.9$$

(d)

$$A'_{vf} = \frac{V_o}{V_i} = \frac{V_o}{V_S}\left(\frac{V_S}{V_i}\right) \quad \text{...(iv)}$$

Now with feedback

$$I_s R_{if} = V_i$$

and, $$I_s = \frac{V_s}{R_s}$$

hence, $$\frac{V_s}{V_i} = \frac{R_s}{R_{if}} = \frac{1}{0.08} = 12.5$$

putting value of $\frac{V_S}{V_i}$ in eq (iv), we get,

$$A'_{vf} = 12.5 \times 86.9 = 1070$$

(e) $$R'_{of} = R'_o \left(\frac{1+\beta A_i}{1+\beta A_i}\right)$$

since $Ai = \underset{R_L \to 0}{Lt}$ AI and AI is independent of RL, then

$$A_i = A_I$$

and. $$R'_{of} = R'_o = 10 \text{ k}\Omega$$

Problem 10.11: For the transistor feedback-amplifier stage shown, h_{fe} = 100, h_{ie} = 1 k, while h_{re} and h_{oe} are negligible. Determine with R_e = 0

(a) $R_{Mf} = \frac{V_o}{I_s}$

where $I_S = \frac{V_s}{R_s}$

(b) $A_{vf} = \frac{V_o}{V_s}$

(c) R_{if} (d) R'_{of}

(e) Repeat the four preceding calculations if R_e = 1 K

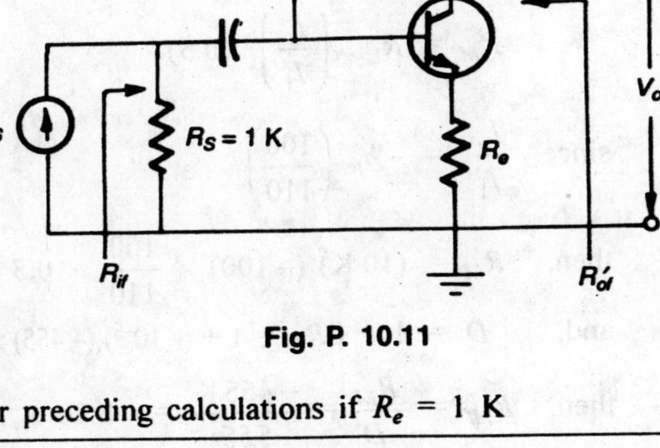

Fig. P. 10.11

Solution:

(a) The given transistor feedback-amplifier is a voltage-shunt configuration, so there is Norton's equivalent of the source.

since, $$I_f = \frac{V_i - V_o}{100\text{ K}} \approx \frac{-V_o}{100\text{ K}} \approx \beta V_o$$

then, $$\beta = -10^{-5}$$

Now, the amplifier without feedback drawn by setting $V_o = 0$ to find input circuit and by setting $V_i = 0$ to find the output circuit.

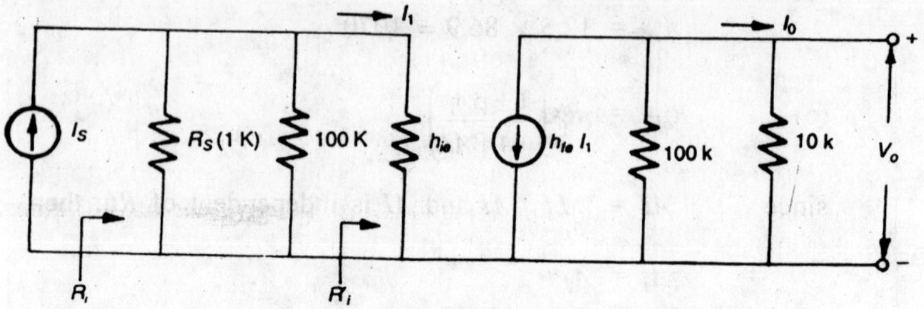

Fig. P. 10.11 (a)

From the circuit,

$$R_M = \frac{V_o}{V_s} = \frac{R_L I_o}{V_S} = R_L \cdot \frac{I_o}{I_1} \times \frac{I_1}{I_s} \quad \text{...(i)}$$

By the current division,

$$\frac{I_1}{I_S} = \frac{(R_s \| 100\text{ K})}{h_{ie} + (R_s \| 100\text{ K})} \approx \frac{1}{2} = 0.5$$

Then from eq (i)

$$R_M = R_L \cdot \left(\frac{I_o}{I_1}\right) \cdot (0.5)$$

since, $$\frac{I_o}{I_1} = -h_{fe}\left(\frac{100}{110}\right)$$

then, $$R_M = (10\text{ K})(-100) \times \frac{100}{110} \times 0.5 \times 10^3 = -455\text{ K}$$

and, $$D = 1 + \beta R_M = 1 + (-10^{-5})(-455) \times 10^3 = 1 + 4.55 = 5.55$$

then, $$R_{Mf} = \frac{R_M}{D} = \frac{-455\text{ K}}{5.55} = -8.20 \times 10^4\ \Omega = -82\text{ k}\Omega$$

(b) $$A_{vf} = \frac{V_0}{V_s} = \frac{V_0}{I_s R_s} = \frac{R_{Mf}}{R_s} = \frac{-82 \times 10^3}{1 \times 10^3} = -82$$

(c) $$R_{if} = R_s \parallel h_{ie} \parallel 100\text{ K}$$

$$1\text{ K} \parallel 1\text{ K} \parallel 100\text{ K} = 0.5\text{ K}$$

then, $$R_{if} = \frac{R_i}{D} = \frac{0.5K}{5.55} = 90.1\text{ K}$$

(d) $$R_{of} = \frac{R_o}{1+\beta R_m} \text{ but}$$

$$R_m = \underset{R_L \to \infty}{Lt} \; Rm$$

then, $R_o = 100\text{ K}$ $[\because R_L = \infty]$

and, $V_o = -h_{fe}\, I_1 \times 100\text{ K}$

thus, $$R_m = \frac{V_0}{I_S} = -h_{fe} \times 100 \times \frac{I_1}{I_S} = -100 \times 10^5 \times 0.5$$

$$= -5 \times 10^6\ \Omega$$

then, $$R_{of} = \frac{R_o}{1+\beta R_m} = \frac{100}{1+\left(-10^{-5}\right)\times\left(-5\times 10^6\right)}$$

$$= \frac{100K}{51} \approx 1.96\text{ K}$$

thus, $R'_{of} = R_{of} \parallel 10K = 1.96\text{ K} \parallel 10\text{ K} \approx 1.64\text{ K}$

(e) when $R_e = 1$ K, the equivalent circuit shown, without feed back,

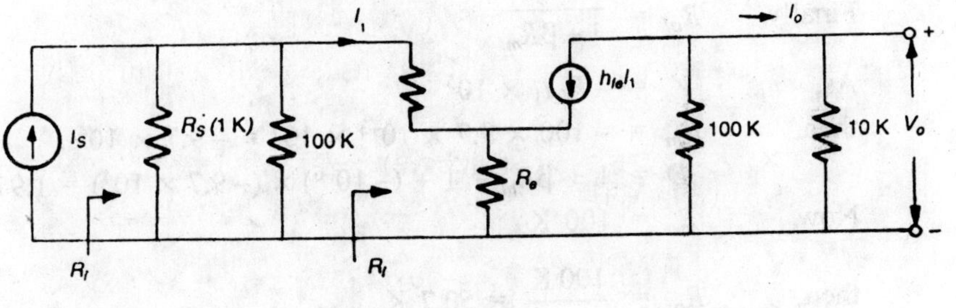

Fig. P. 10.11 (b)

From the circuit,

$$\frac{I_o}{I_1} = -h_{fe}\cdot\left(\frac{100}{110}\right) = -91$$

and, $$\frac{I_1}{I_S} = \frac{R_s \| 100}{R_{i'} + (R_s \| R_{100})} \quad \ldots(i)$$

here, $$R_i' = h_{ie} + (h_{fe} + 1)\, R_e = 1.1 + 101 \times 1 = 102.1 \text{ K}$$

$$\frac{I_1}{I_s} = \frac{1\text{ K}}{103\text{ K}} \approx 9.7 \times 10^{-3} \text{ K}$$

then, $$R_M = \frac{V_o}{I_s} = \frac{I_o R_L}{I_S} = R_L \cdot \frac{I_o}{I_1} \cdot \frac{I_1}{I_s}$$

$$= (10 \times 10^3) \times (-91) \times (9.7 \times 10^{-3})$$

$$= -8.20 \times 10^3\ \Omega$$

$$R_{Mf} = R_M / D$$

$$= -\frac{8.82 \times 103}{1.088} = -8.10 \qquad [\because D = 1 + \beta R_M]$$

$$= 1 + (-10^{-5})(-8.82 \times 10^3) = 1.088$$

$$A_{vf} = \frac{R_{Mf}}{R_S} = \frac{-8.10\text{ K}}{1\text{ K}} = -8.10$$

Now, $$R_i = 1\text{ K} \| 100\text{K} \| R_i' = 1\text{K} \| 100 \| 102.1\text{ K} \times 980$$

then $$R_{if} = \frac{R_i}{D} = \frac{980}{1.088} = 910\ \Omega$$

Finally, $$R_{of} = \frac{R_o}{1 + \beta R_m}$$

As, $$V_o = -h_{fe} I_1 \times 10^5$$

then, $$R_m = -100 \times 9.7 \times 10^{-3} \times 10^5 = -9.7 \times 10^4$$

$$D = 1 + \beta R_m = 1 + (-10^{-6}) \times (-9.7 \times 10^4) = 1.97$$

Now, $$R_o = 100\text{ K}$$

then, $$R_{of} = \frac{100\text{ K}}{1.97} = 50.7\text{ K}$$

thus, $\quad R'_{of} = R_{of} \parallel 10\text{ K} = 8.35\text{ K}.$

Problem 10.12:
For the circuit shown, prove that

$$A_{Vf} = \frac{V_o}{V_s} = -\frac{R'}{R}\frac{1}{1+\frac{R'}{R_m}\left(\frac{R_i+R'}{R'}+\frac{R'}{R}\right)}$$

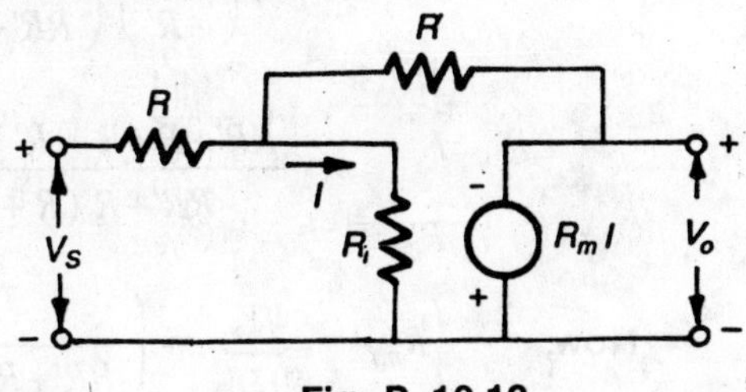

Fig. P. 10.12

Solution:

This is a voltage-shunt amplifier. Then to find input circuit set $V_o = 0$ and $V_i = 0$ to find the output circuit. Then the equivalent circuit is as shown below, using Norton's eqivalent circuit for the source V_S

Also, $\quad I_f = \frac{V_1 - V_o}{R'} \approx -\frac{V_o}{R'} = \beta V_o$

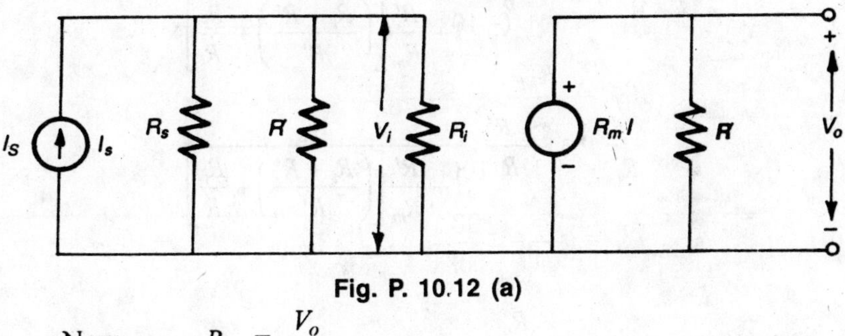

Fig. P. 10.12 (a)

Now, $\quad R_m = \frac{V_o}{I_s}$

but, $\quad V_o = -R_M I$

where, $\quad I = I_s\left(\frac{R \parallel R'}{R \parallel R' + R_i}\right) = I_s \frac{RR'}{R_i(R+R')+RR'}$ ...(i)

∴ Using expression for I from equation (i),

$$R_m = \frac{V_o}{I_s} = \left(\frac{-R_m RR'}{RR' + R_i(R+R')}\right) \quad ...(ii)$$

Then, resensitivity, $D = 1 + \beta R_m$

$$= 1 + \left(\frac{1}{-R'}\right)\left(\frac{-R_m RR'}{RR' + R_i(R+R')}\right)$$

$$= \frac{RR' + R_i(R+R^1) + R_m R}{RR' + R_i(R+R^1)} \quad \text{...(iii)}$$

Now, $R_{mf} = \frac{R_m}{D} = \left(\frac{-R_m RR'}{RR' + R_i(R+R') + R_m R}\right)$

Now, using eq. (iii), we have,

$$A_{Vf} = \frac{V_o}{V_s} = \frac{V_o}{I_s R} = \frac{R_{mf}}{R} = \left(\frac{-R_m R'}{RR' + R_i(R+R') + R_m R}\right)$$

factoring out $\frac{R'}{R}$, we get.

$$A_{Vf} = \frac{-R'}{R} \cdot \frac{1}{1 + \frac{R'}{R_m}\left[\left(\frac{R_i + R'}{R'}\right) + \frac{R_i}{R}\right]}$$

$$= \frac{-R'}{R} \cdot \frac{1}{1 + \frac{R'}{R_m}\left[\left(\frac{R_i + R'}{R'}\right) + \frac{R_i}{R}\right]}$$

Chapter 11

Stability and Oscillators

Problem 11.1: Verify the equation $-\beta = \dfrac{V_f'}{V_o} = \left(\dfrac{1}{1-5\alpha^2-(6\alpha-\alpha^3)}\right)$ in figure as below, assuming that this network does not load the amplifier. Prove that the phase shift of $\dfrac{V_f'}{V_o}$ is 180° for $\alpha^2 = 6$ and that at this frequency $\beta = +\dfrac{1}{29}$.

Solution: Assuming $x = \dfrac{1}{wC}$

Now the phase shift network is drawn :

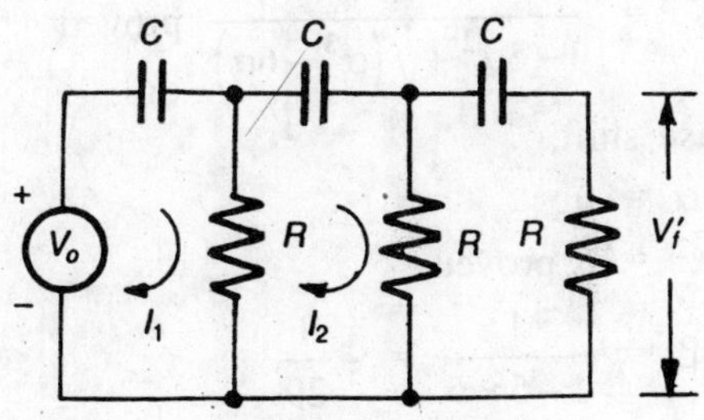

Fig. P. 11.1

Applying the mesh analysis :

$$V_o = I_1 (R - jx) - I_2 R + I_3 (0)$$
$$0 = -I_1 R + I_2 (2R - jx) - I_3 R$$
$$0 = 0 \times I_1 - I_2 R + I_3 (2R - jx)$$

Now I_3 can be found,

$$I_3 = \frac{\Delta_3}{\Delta}$$

Where $\Delta_3 = \begin{vmatrix} R - jx & -R & V_o \\ -R & 2R - jx & 0 \\ 0 & -R & 0 \end{vmatrix} = R^3 [1 - 5\alpha^2 + j(\alpha^3 - 6\alpha)]$

and $\Delta = \begin{vmatrix} R - jx & -R & 0 \\ -R & 2R - jx & -R \\ 0 & -R & 2R - jx \end{vmatrix}$

Assuming again,

$$\alpha = \frac{x}{R} = \frac{1}{wRC}$$

$$\therefore \quad I_3 = \frac{\Delta_3}{\Delta} = \frac{R^2 V_o}{R^3\left[1 - 5\alpha^2 + j\left(\alpha^2 - 6\alpha\right)\right]}$$

Hence, $-\beta = \dfrac{V'_f}{V_o} = \dfrac{I_3 R}{V_o}$

$$= \frac{1}{1 - 5\alpha^2 + j\left(\alpha^3 - 6\alpha\right)} \text{ proved}$$

For 180° phase shift,

$$\alpha^3 - 6\alpha = 0$$

or, $\alpha^2 = 6$ proved

and $-\beta = \dfrac{1}{1 - 5 \times 6} = -\dfrac{1}{29}$

$$\therefore \quad \beta = \frac{1}{29}$$

Problem 11.2: (a) For the network of Prob. 11.1, show that the input impedance is given by

$$Z_i = R\,\frac{1-5\alpha^2 - j\left(6\alpha - \alpha^3\right)}{3-\alpha^2 - j4\alpha}$$

(b) Show that the input impedance at the frequency of the oscillator,

$$\alpha = \sqrt{6}\text{, is } (0.83 - j2.70)R.$$

Note : That if the frequency is varied by varying *C*, the input impedance remains constant. However, if the frequency is varied by varying *R*, the impedance is varied in proportion to *R*.

Solution: From equations derived in the last problem,

$$I_1 = \frac{\Delta_1}{\Delta};$$

where
$$\Delta_1 = \begin{vmatrix} V_o & -R & 0 \\ 0 & 2R-jx & -R \\ 0 & -R & 2R-jx \end{vmatrix}$$

$\therefore$
$$\Delta_1 = V_o\,[(2R - jx)^2 - R^2]$$
$$= V_oR^2\,[3 - \alpha^2 - j4\alpha]$$

where
$$\alpha = \frac{x}{R} = \left(\frac{1}{wCR}\right)$$

$\therefore$
$$Z_i = \frac{V_o}{I_i} = \frac{V_o\Delta}{\Delta_1}$$

$$= \left[\frac{R\left[1-5\alpha^2 + j\left(\alpha^3 - 6\alpha\right)\right]}{3-\alpha^2 - j4\alpha}\right] \quad \text{Proved}$$

(b) For α^2 6 i.e., $\alpha = \sqrt{6}$

$$Z_i = R\left[\frac{1-5\times 6 + j\left(6\sqrt{6} - 6\sqrt{6}\right)}{3-6-j4\sqrt{6}}\right]$$

$$= R\left(\frac{-29}{-3-j4\sqrt{6}}\right)$$

$$= R\,(0.83 - j2.7) \text{ Proved}$$

Problem 11.3: Design a phase-shift oscillator to operate at a frequency of 5 kHz. Use a MOSFET with $\mu = 55$ and $r_d = 5.5$ K. The phase-shift network is not to load down the amplifier.

(a) Find the minimum value of the drain-circuit resistance R_d for which the circuit will oscillate.

(b) Find the product RC.

(c) Choose a reasonable value for R, and find C.

Solution: To design a phase shift using MOSFET,

$$\text{Gain,} \qquad |A| = \frac{(\mu \times R_d)}{r_d + R_d} = 29 \ (\text{As } \beta = \frac{1}{29})$$

$$\therefore \qquad R_d = \frac{29 r_d}{\mu - 29}$$

Here $\mu = 55$ and $r_d = 5.5$ K

$$\text{Then} \qquad R_d = \left(\frac{29 \times 5.5}{55 - 29}\right) = 6.12 \text{ K}$$

(b)

$$\alpha = \frac{x}{R} =$$

$$\frac{1}{2\pi f RC} = \sqrt{6}$$

$$\therefore \qquad RC = \frac{1}{2\pi \times f \times \sqrt{6}}$$

$$= \frac{1}{2\pi \times 5 \times 10^3 \times \sqrt{6}} = 12.9 \ \mu\text{sec.}$$

(c) The input impedance of the phase shift network must be high compared to the output impedance of the amplifier, in order to not to load down the amplifier.

If R is chosen too large, however C will be impractically small, e.g., if $R = 1$ M, $C = 12.9$ pF, which is of the same order of magnitude as stray winding capacitance.

If we take this FET, the amplifier output impedance is equal to

$$\frac{r_d R_d}{r_d + R_d} = \frac{5.5 \times 6.12}{5.5 + 6.12} \cong 2.9 \text{ K}$$

If we choose R to be, say 10 times this value or 30 K then the load of the phase shift network will be negligible.

Then $$C = \frac{1.29 \times 10^{-6}}{3 \times 10^{4}} = 430 \text{ pF}$$

Problem 11.4: (a) A two-stage FET oscillator uses the phase-shifting network shown. Prove that

$$\frac{V_i}{V_f'} = \frac{1}{3 + j(\omega RC - 1/\omega RC)}$$

(b) Show that the frequency of oscillation is $f = 1/2\pi RC$ and that the gain must exceed 3.

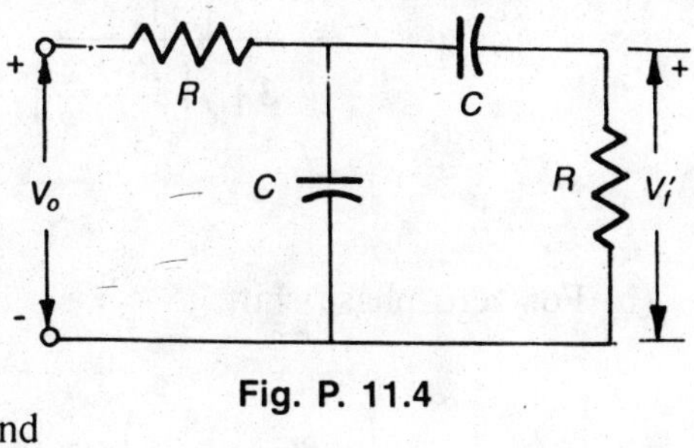

Fig. P. 11.4

Solution:

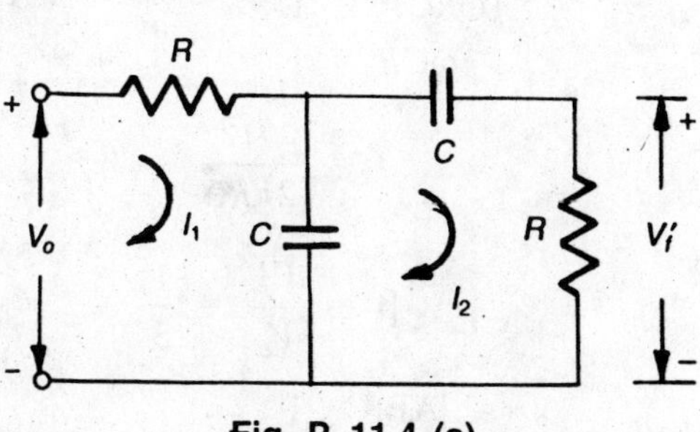

Fig. P. 11.4 (a)

Assuming $x = 1/\omega C$ then applying mesh rule,

$$V_o = I_1 (\text{R} - jx) - I_2 (-jx) \qquad \ldots(1)$$

$$0 = I_1(-jx) + I_2(R - 2jx) \quad \ldots(2)$$

From (2), we have,

$$I_1 = \left(\frac{R - j_2x}{-jx}\right) = (2 + j\ R/x)\ I_2 \quad \ldots(3)$$

Therefore, $$V_o = I_2\left[(R - jx)\left(2 + \frac{jR}{x}\right) + jx\right]$$

$$= I_2\left[3R + j\left(\frac{R^2}{x} - x\right)\right]$$

Now, $$\frac{V_f'}{V_o} = \frac{I_2 R}{V_o} = \frac{R}{3R + j\left(\frac{R^2}{x} - x\right)}$$

$$= \frac{1}{3 + j\left(\frac{R}{x} - \frac{x}{R}\right)} = \frac{1}{3 + j\left(\omega RC - \frac{1}{\omega CR}\right)} \quad \text{Proved}$$

(b) For zero phase shift,

$$\left(\omega RC - \frac{1}{\omega CR}\right) = 0$$

or, $$(\omega RC)^2 = 1$$

$$\omega CR = 1$$

and $$f = \frac{1}{2\pi RC}$$

Since, $$\beta = \frac{V_f'}{V_o} = \frac{1}{3}$$

and $$|AB| > 1$$

then $$A > 3 \ \textbf{proved}$$

Problem 11.5:
(a) Find V_i/V'_f for the network shown.
(b) Sketch the circuit of a phase-shift FET oscillator, using this feedback network.
(c) Find the expression for the frequency of oscillation, assuming that the network does not load down the amplifier.
(d) Find the minimum gain required for oscillation.

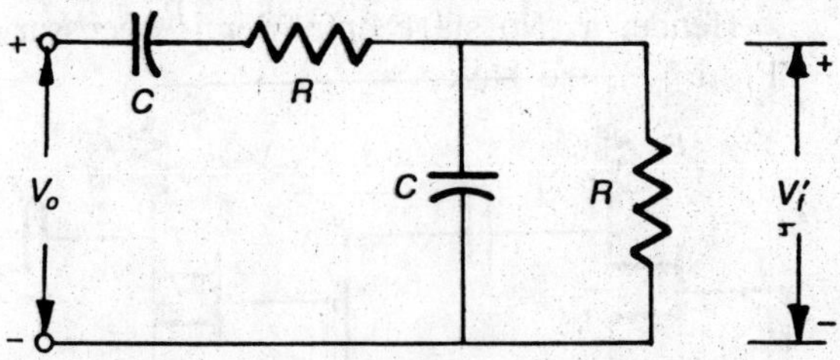

Fig. P. 11.5

Solution: (a) By using voltage division rule, we set,

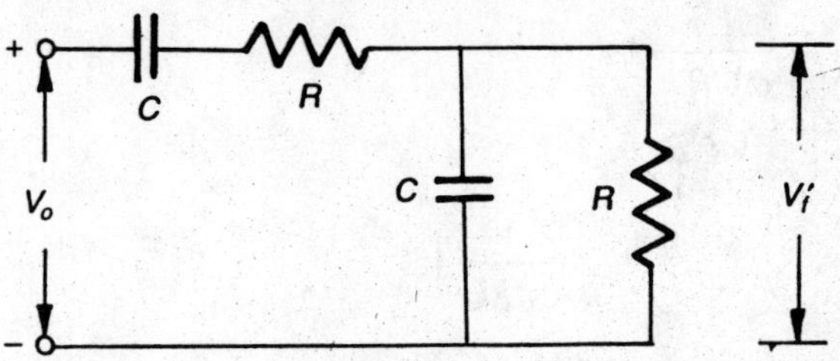

Fig. P. 11.5 (a)

$$\frac{V'_f}{V_o} = \frac{R(-jx)}{R - jx}$$

$$= \frac{1}{(R - jx) + \dfrac{R(-jx)}{(R - jx)}} = \frac{-j \times R}{R^2 - 2jRx - x^2 - jxR}$$

$$= \frac{1}{3 + j\left(\dfrac{R}{x} - \dfrac{x}{R}\right)} = \frac{1}{3 + j\left(\omega CR - \dfrac{1}{\omega CR}\right)}$$

(b) At the frequency at which β real,

$$\beta = \frac{1}{3}$$

Since $A\beta > 1$ and then $A > 3$, must be (+)ve.

Hence, a two stage amplifier is necessary

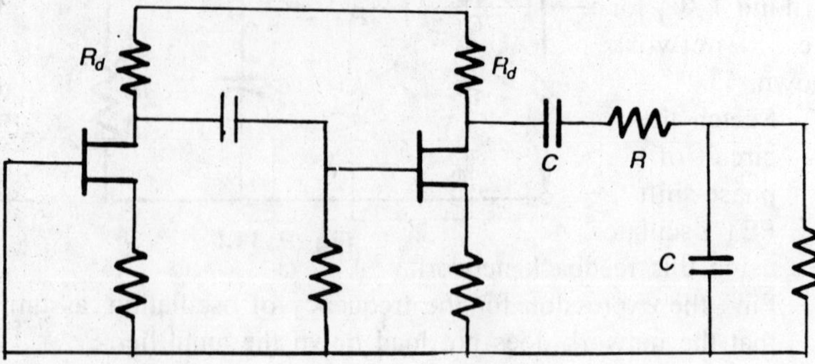

Fig. P. 11.5 (b)

(c) For zero phase shift,

$$\omega CR - \frac{1}{\omega CR} = 0$$

or, $\quad wCR = 1$

and $\quad f = \left(\frac{1}{2\pi RC}\right)$

(d) The minimum value of gain required for oscillation is $A = 3$.

Problem 11.6: Consider the two-section RC network shown. Find the V_i/V'_f function, and verify that it is not possible to obtain 180° phase shift with a finite attenuation.

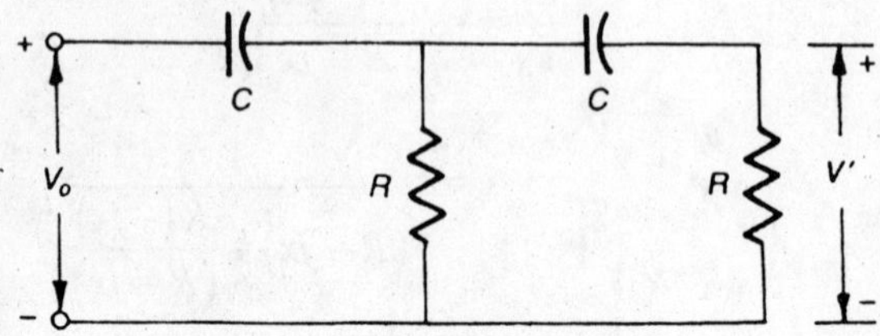

Fig. P. 11.6

Solution: Assuming here also $x = \left(\frac{1}{\omega C}\right)$

Applying Mesh rule, we have,

$$V_o = I_1(R - jx) - I_2R \quad \text{...(i)}$$

and $\quad 0 = -I_1R + I_2(2R - jx) \quad \text{...(ii)}$

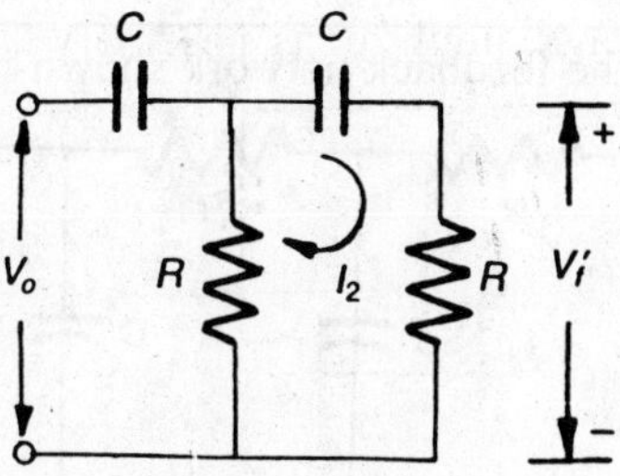

Fig. P. 11.6 (a)

Solving (i) & (ii), we get,

$$V_o = \frac{I_2}{R}\,[R^2 - x^2 - j3Rx]$$

$$\therefore \quad \frac{V_f'}{V_o} = \frac{I_2 R}{V_o} = \frac{1}{1-\left(\frac{x}{R}\right)^2 - j\frac{3x}{R}}$$

$$= \frac{1}{1-\alpha^2 - j3\alpha} \quad \left(\text{Where } \alpha = \left(\frac{1}{\omega RC}\right) = \frac{x}{R}\right)$$

In order to have phase shift of 180°,

$$\tan 180^\circ = 0 = \frac{3\alpha}{1-\alpha^2}$$

Hence either $\alpha = 0$

or $\alpha = \infty$.

If $\alpha = 0$,

Then $\dfrac{V_f'}{V_0} = 1$

and phase shift is zero.

Note that $\alpha = 0$ requires either $R = \infty$ or $C = \infty$, which means that the output is connected directly to the input.

On the other hand if $\alpha = \infty$ then $\dfrac{V_f'}{V_o} = 0$ and this would require an amplifier with infinite gain. Note that $\alpha = \infty$ means either $C = 0$ or $R = 0$ (output short circuited), so the attenuation is infinite.

Problem 11.7: For the feedback network shown find (a) the transfer function, (b) the input impedance. (c) If this network is used in a phase-shift oscillator, find the frequency of oscillation and the minimum amplifier voltage gain. Assume that the network does not load down the amplifier.

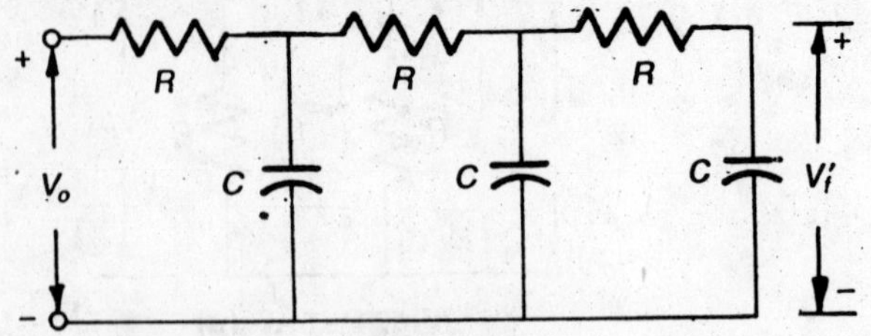

Fig. P. 11.7

Solution:

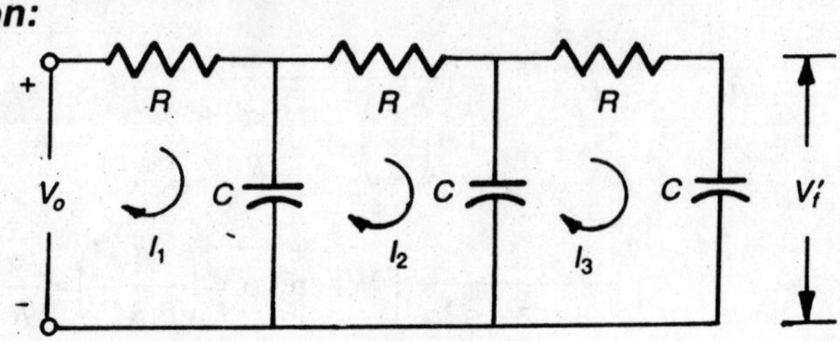

Fig. P. 11.7 (a)

Assume $\gamma = \omega RC$

and $x = \dfrac{1}{\omega C}$

Applying mesh rule,

$$V_o = (R - jx)\, I_1 - I_2\,(-jx) \qquad \text{...(i)}$$

$$0 = -I_1(-jx) + (R - 2jx)\, I_2 \qquad \text{...(ii)}$$

and $$0 = -I_2\,(-jx) + (R - 2jx)\, I_3 \qquad \text{...(iii)}$$

Solving, we get,

$$\frac{V_f'}{V_o} = \frac{1}{1 - 5\gamma^2 + j\left(6\gamma - \gamma^3\right)}$$

Where $\gamma = \omega RC$

(b) Now, $$Z_i = -\frac{1}{\omega c}\left[\frac{\left(1-5\gamma^2\right)+j\left(6\gamma-\gamma^2\right)}{3-\gamma^2+j4\gamma}\right]$$

(c) For β be real,

$$6\gamma - \gamma^2 = 0$$

$$\therefore \quad \gamma^2 = 6 = \omega^2 R^2 C^2$$

then, $$f = \left(\frac{\sqrt{6}}{2\pi RC}\right)$$

$$-\beta = \frac{V_f'}{V_o} = \frac{1}{1 - 5 \times 6} = -\frac{1}{29}$$

or, $$|AB| > 1$$

$$\therefore \quad |A| > 29$$

Problem 11.8:
For the FET oscillator shown,
find :
(a) V'_f/V_o,
(b) the frequency of oscillations.

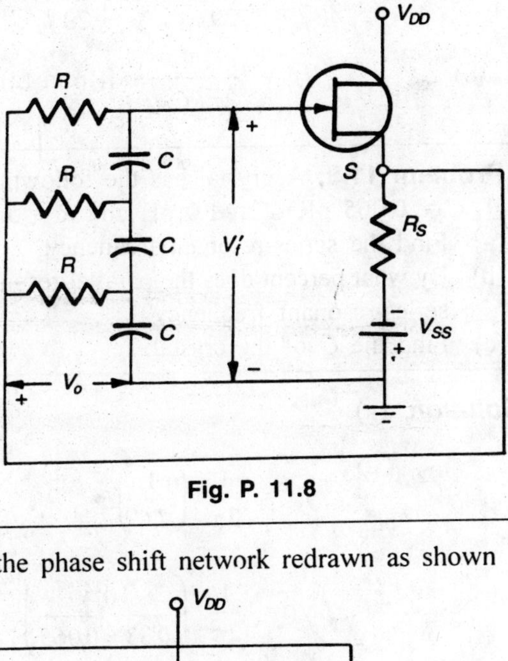

Fig. P. 11.8

Solution: (a) Consider the phase shift network redrawn as shown

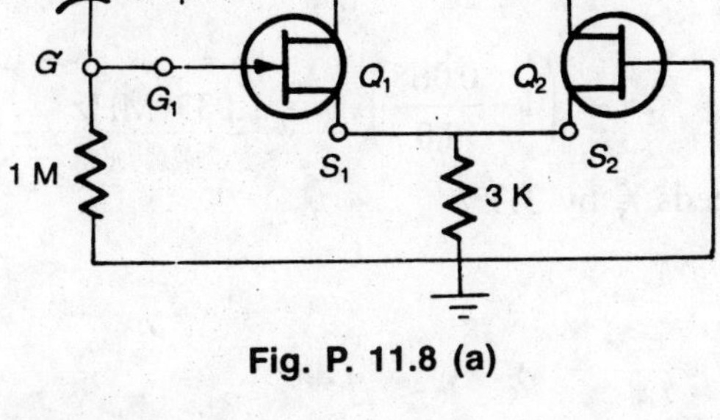

Fig. P. 11.8 (a)

$$\therefore \qquad \alpha^2 = 6 = \left(\frac{1}{2\pi fRC}\right)^2$$

then the phase shift of $\frac{V_o'}{V_o}$ is 180°

$$\text{or,} \qquad V'_o = -\frac{1}{2a}V_o$$

Applying KVL around the outside loop, we have,

$$V'_f = -V_o' + V_o = \frac{1}{29}V_o + V_o$$

$$\text{or,} \qquad \frac{V_f'}{V_o} = \frac{1}{29} + 1 = \left(\frac{30}{29}\right) = -\beta$$

(b) $\therefore$ $$f = \frac{1}{2\pi RC\sqrt{6}} \quad \text{[(from (a) obtained]}$$

Problem 11.9: A crystal has the following parameters : $L = 0.33$ H, $C = 0.065$ pF, $C' = 1.0$ pF, and $R = 5.5$ K.

(a) Find the series-resonant frequency.

(b) By what percent does the parallel-resonant frequency exceed the series-resonant frequency?

(c) Find the Q of the crystal.

Solution: (a)

$$f_s = \frac{1}{2\pi}\left(\frac{1}{LC}\right)^{1/2}$$

$$= \frac{1}{2\pi}\left(\frac{10^{12}}{0.33 \times 0.065}\right)^{1/2} = 1.09 \text{ MHz}$$

(b) $$\frac{f_p}{f_s} = \left(1 + \frac{C}{C'}\right)^{\frac{1}{2}}$$

$$= \left(1 + \frac{0.065}{1.0}\right)^{1/2} = 1.033 \text{ MHz}$$

$\therefore$ f_p exceeds f_s by 3.3%

(c) $$Q = \frac{\omega_s L}{R} = \frac{1}{R}\sqrt{\frac{L}{C}}$$

$$= \frac{1}{5.5 \times 10^3}\left(\frac{0.33}{0.065 \times 10^{-12}}\right)^{1/2} = 410$$

Problem 11.10:
(a) At what frequency will the circuit shown oscillate, if at all?
(b) Find the minimum value of R needed to sustain oscillations. The FETs are identical with
$g_m = 1.6$ mA/V
and $r_d = 44$ K.

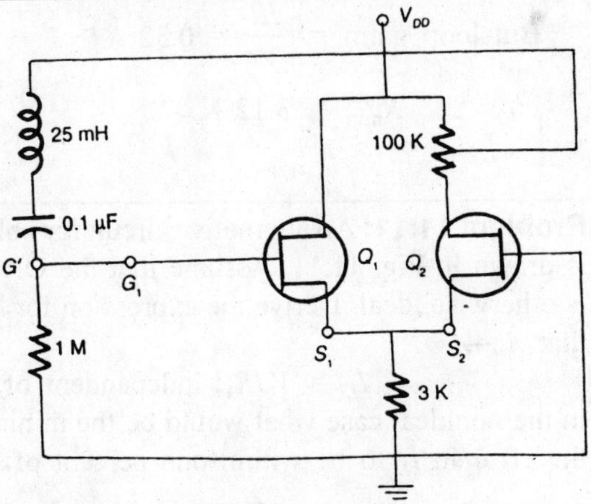

Fig. P. 11.10

Solution: (a) If a signal causes the potential at G_1 to increase, then the source will follow this increase, since the gate of Q_2 is grounded, then the voltage increase at S_2 will cause D_2 to rise. Part of this increase in voltage in feedback to G_1. Hence there will be phase shift around the loop provided that none is introduced by the reactive elements.

For the series resonance,

$$f = \frac{1}{2\pi\sqrt{LC}}$$

$$= \frac{1}{2\pi\sqrt{25 \times 10^{-3} \times 10^{-7}}} = 3183 \text{ Hz}$$

(b) We may neglect the loading of 1 M on *RC* 100 K

If V is applied to G_1, then the equivalent ckt. seen looking into each source is shown.

The parallel combination of 3 K and $\dfrac{r_d + R_d}{\mu + 1} = 2.03$ K is $\dfrac{3 \times 2.03}{5.03}$ K = 1.21 K,

$$\text{Then} \quad I = \frac{70\,V}{71}\left(\frac{1.21\text{ K}}{0.62\text{ K} + 1.21\text{ K}}\right) \times \left(\frac{1}{2.03}\right)$$

$$= 0.32 \text{ V}$$

$$\text{But loop gain} = \frac{IR}{V} = 0.32\,R > 1$$

$$\therefore \quad R_{\min} = 3.12 \text{ K}$$

Problem 11.11: An alternative circuit for voltage to current converter is drawn in Fig. 11.11. Assume that the OPAMP has finite gain but is otherwise ideal. Derive the expression for load current i_L and show that $A \to \infty$

$$I_L = V_s/R_1; \text{ independent of } Z_L$$

in the nonideal case what would be the minimum value of gain A for the error in i_L to be within one percent of the above expression.

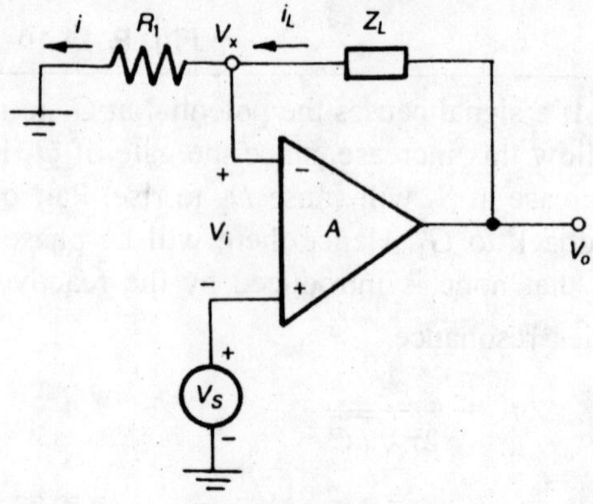

Fig. P. 11.11

Solution:

A = finite gain; OPAMP is otherwise ideal, i.e. its input current and output resistance are zero.

$$\therefore \quad i = i_L \qquad \text{...(i)}$$

$$\text{or} \quad \frac{v_x}{R_1} = \frac{v_o - v_x}{Z_L} \qquad \text{...(ii)}$$

Also $$v_o = -A\,(v_x - v_s) \qquad \text{...(iii)}$$

Substituting for v_o from eq. (iii) in eq. (ii)

$$\frac{v_x}{R_1} = \frac{-A\left(v_x - v_s\right) - v_x}{Z_L'}$$

We know, discharging time

$$t_2 = (R_{B1} + R_1)\, C\, \ln\left(\frac{V_P}{V_V}\right)$$

We have, cycle time

$$T = t_1 + t_2$$

Also, oscillation frequency

$$f_{OSC} = 1/T$$

Approximations

$$t_1 \gg t_2,\ V_D \text{ ignored}$$

Then $$f_{OSC} = \frac{1}{R_T C_T \ln\left[1/(1-\eta)\right]}$$

Chapter 12

Integrated Circuits as Analog System Building Blocks

Problem 12.1: For the given circuit, show that the output voltage is

$$-v_o = \frac{R_2}{R_1} v + \left(R_2 C + \frac{L}{R_1}\right)\frac{dv}{dt} + LC\frac{d^2v}{dt^2}$$

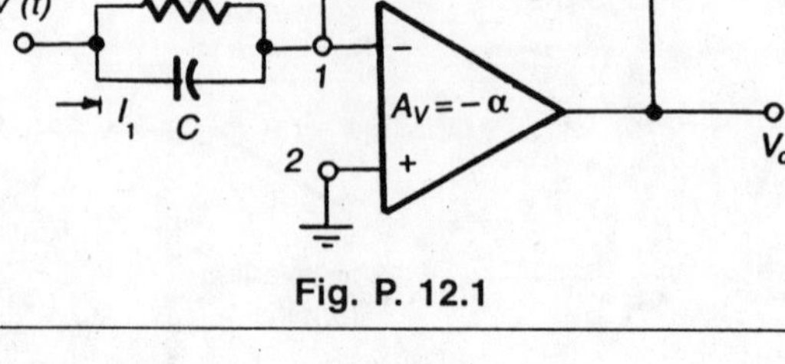

Fig. P. 12.1

Solution:

The phase-shift network is redrawn below,

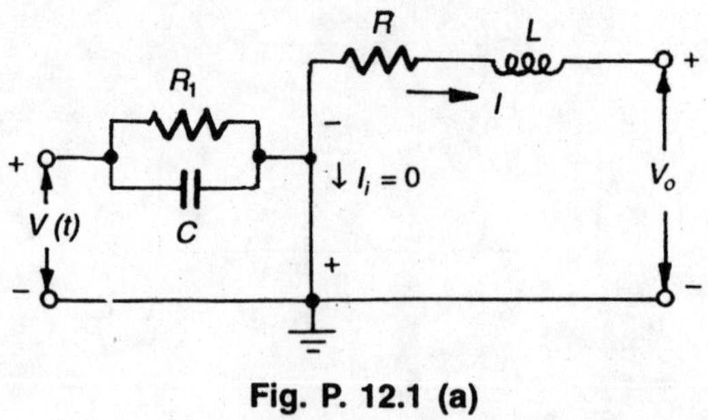

Fig. P. 12.1 (a)

since both the non inverting terminals are virtually grounded

then, $$i = \frac{v}{R_1} + C\frac{dv}{dt} \text{ (using KCL)} \qquad \ldots\text{(i)}$$

and, $$v_o = -R_2 I - \frac{Ldi}{dt} \text{ (using KVL)} \qquad \ldots\text{(ii)}$$

using eq.(i) in eq. (ii), we have

$$v_o = -R_2\left(\frac{v}{R_1} + C\frac{dv}{dt}\right) - L\left(\frac{dv}{dt}\cdot\frac{1}{R_1} + C\frac{d^2v}{dt^2}\right)$$

or, $$-v_o = \frac{R_2}{R_1}\cdot v + \left(R_2C + \frac{L}{R_1}\right)\frac{dv}{dt} + LC\frac{d^2v}{dt^2}$$

Problem 12.2: Consider the operational amplifier circuit of Fig. 12.2 with Z consisting of resistor R in parallel with a capacitor C, and Z' consisting of a resistor R'. The input is a sweep voltage $v = \alpha t$. Show that the output voltage v_o is a sweep voltage that starts with an initial step. Thus prove that

$$v_0 = -\alpha R'C - \alpha\frac{R'}{R}t$$

Assume infinite open-loop gain.

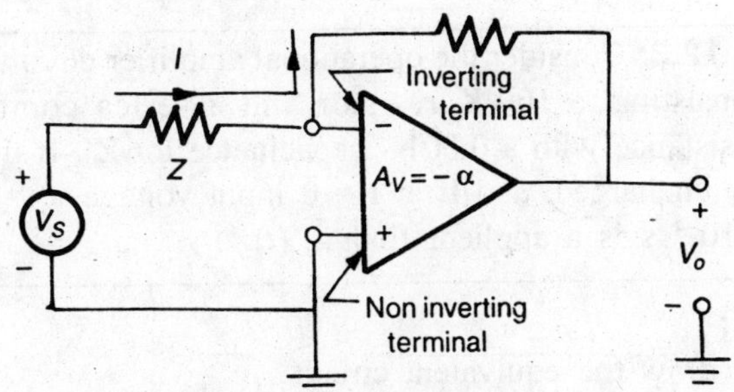

Fig. P. 12.2 (a)

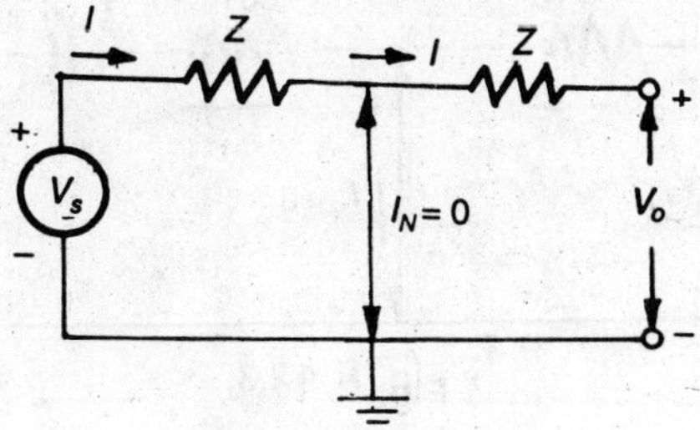

Fig. P. 12.2 (b)

Solution:

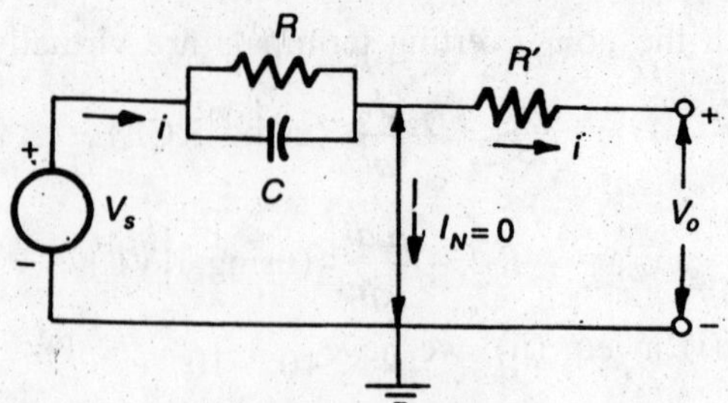

Fig. P. 12.2 (a)

Given, $V_s = \alpha t$

since equivalent circuit has a virtual ground, v_s will appear across 'R' and 'C' in parallel, then

$$i = C\frac{dv_s}{dt} + \frac{v_s}{R} \qquad \ldots\text{(i)}$$

if $v_s = \alpha t$, then,

$$i = \frac{\alpha t}{R} = C\alpha \qquad \ldots\text{(ii)}$$

hence, $v_o = -iR' = -C\alpha R' - \alpha\dfrac{R'}{R}t$ from eq (ii)

Problem 12.3: Consider the operational amplifier circuit of Fig.12.2 with Z consisting to 100-K resistor and a series combination of a 50-K resistance with a 0.001 - capacitance for Z'. If the capacitor is initially uncharged, and if at $t = 0$ input voltage $v_s = 10\varepsilon^{-t/T}$ with $T = 5 \times 10^{-4}$ s is a applied, find v_0 (t).

Solution :

Let us draw the equivalent circuit

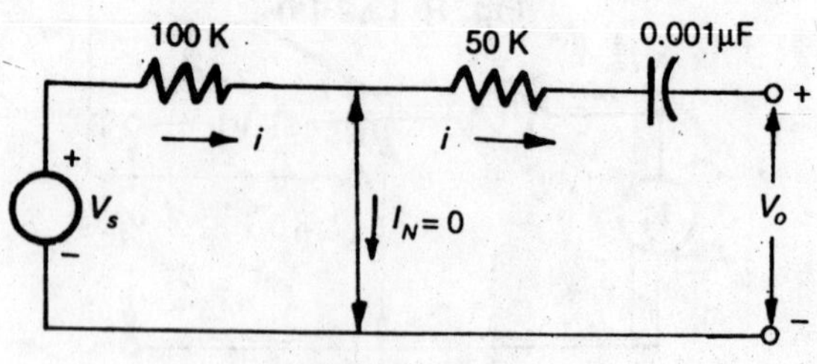

Fig. P. 12.3

Given, $v_s = 10e^{-t/T} = 10e^{-2\times10^3 t}$

then, $i = \dfrac{v_S}{100\text{ K}} = 10^{-4}e^{-2\times10^3 t}$

Now, $v_o(t) = -50\times10^3\, i - \dfrac{1}{10^{-9}}\displaystyle\int_0^t i\,dt$

$= -50\times10^3\times10^{-4}\ 10^{-4}e^{-2\times10^3 t}$

$\dfrac{-1}{10^{-5}}\displaystyle\int_0^t e^{-2\times10^3 t}dt = 45e^{-2\times10^{-3} t} - 50$

Problem 12.4: For the figure shown below show that I_L is equal to $\dfrac{-V_S}{R_2}$ if $\dfrac{R_3}{R_2'} = \dfrac{R'}{R_1}$

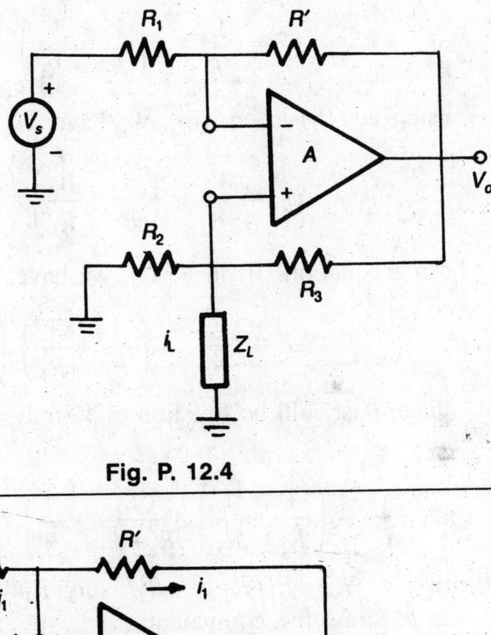

Fig. P. 12.4

Solution:

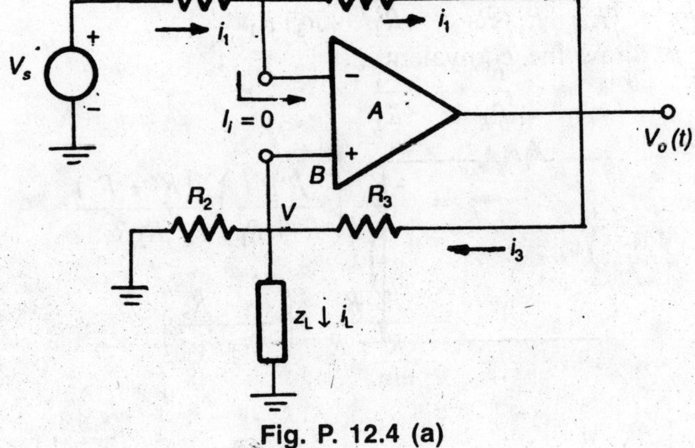

Fig. P. 12.4 (a)

From the above circuit, we have,

$$V_A = V_B = V$$

and, $$i_1 = \frac{V_S - V}{R_1} = \frac{V - V_o}{R'} \quad \ldots(i)$$

or, $$\frac{V_S}{R_1} + \frac{V_o}{R'} = V\left(\frac{1}{R'} + \frac{1}{R_1}\right) \quad \ldots(ii)$$

and, $$i_3 = \frac{V_o - V}{R_3} = i_L + \frac{V}{R_2}$$

or, $$i_L = \frac{V_o}{R_3} - \left(\frac{V}{R_2} + \frac{V}{R_3}\right) \quad \ldots(iii)$$

using eq. (ii) in eq. (iii), we have,

$$i_L = \frac{V_o}{R_3} - \left(\frac{1}{R_2} + \frac{1}{R_3}\right)\left(\frac{V_S}{R_1} + \frac{V_o}{R'}\right)\frac{R'_1}{R' + R_1}$$

using equation (ii) in eq (i), we have,

$$i_L = \frac{V_o}{R_3} - \left(\frac{1}{R_2} + \frac{1}{R_3}\right) \cdot \left(\frac{V_S}{R_1} + \frac{V_0}{R'}\right)\frac{R' R_1}{R' + R_1}$$

since, i_L should be function of V_s only, the co-efficient of V_o must be zero.

i.e. $$\frac{1}{R_3} - \left(\frac{1}{R_2} + \frac{1}{R_3}\right) \cdot \frac{R' R_L}{R' + R_1} \times \frac{1}{R'} = 0$$

or, $$(R' + R_1) R_2 = (R_2 + R_3) R_1$$

or, $$\frac{R'}{R_L} = \frac{R_3}{R_2} \quad \ldots(iv)$$

Then, $$i_L = \frac{-V_S}{R_1}\left(\frac{R' R_1}{R' + R_1}\right)\frac{(R_2 + R_3)}{R_2 R_3}$$

$$= \frac{V_S R'}{R' + R_1}\left(\frac{R_2 + R_3}{R_2 R_3}\right)$$

$$= \frac{-V_S}{R_2} \cdot \left(\frac{R'R_2 + R'R_3}{R_3 R' + R_1 R_3} \right)$$

$$= \frac{-V_S}{R_2} \left(\frac{R_3(R_1 + R')}{R_3(R_1 + R')} \right) = \frac{-V_S}{R_2}$$

Problem 12.5: The differential input operational amplifier shown consists of a base amplifier of infinite gain. Show that $V_O = (R_2/R_1)(V_2 - V_1)$

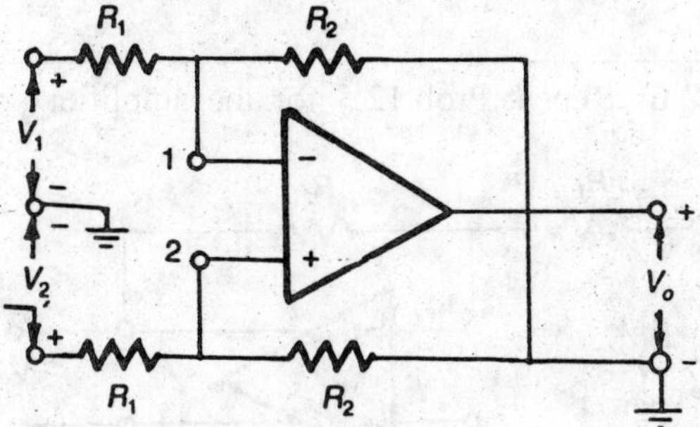

Fig. P. 12.5

Solution:

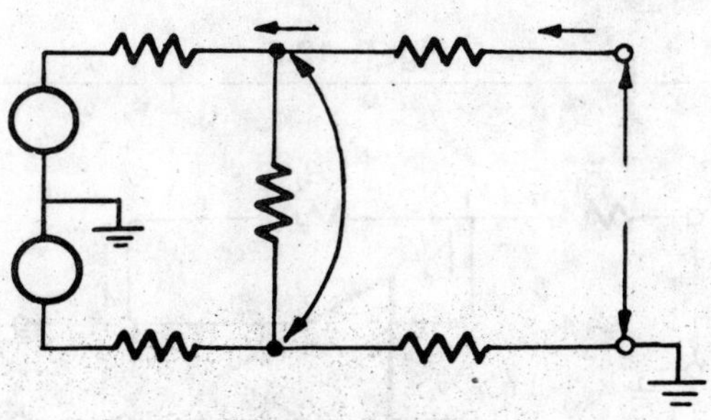

Fig. P. 12.5 (a)

With V_o finite,

$$V_i = 0, \text{ for } A = -\infty$$

then $$I_i = 0$$

Thus, $$I = \frac{V_o - V_1}{R_1 + R_2}$$

and $$I' = \frac{V_2}{R_1 + R_2}$$

here, $V_i = V_A - V_B$

or, $V_i = V_1 + IR_1 - V_2 + I'R_2$

or, $V_i = V_1 - V_2 + \frac{R_1}{R_1 + R_2}(V_o - V_1) + R_2 \frac{V_2}{R_1 + R_2} = 0$

or, $(V_1 - V_2)\left(1 - \frac{R_1}{R_1 + R_2}\right) = \frac{-R_1}{R_1 + R_2} \cdot V_o$

$$V_o = \frac{R_2}{R_1} \cdot (V_2 - V_1)$$

Problem 12.6: Repeat Prob.12.5 for the amplifier shown.

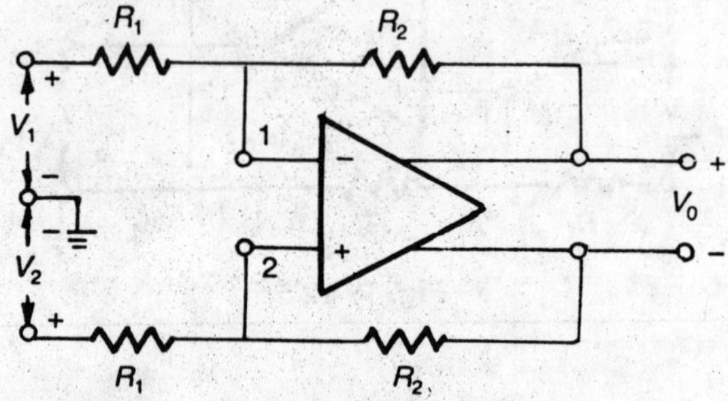

Fig. P. 12.6

Solution:

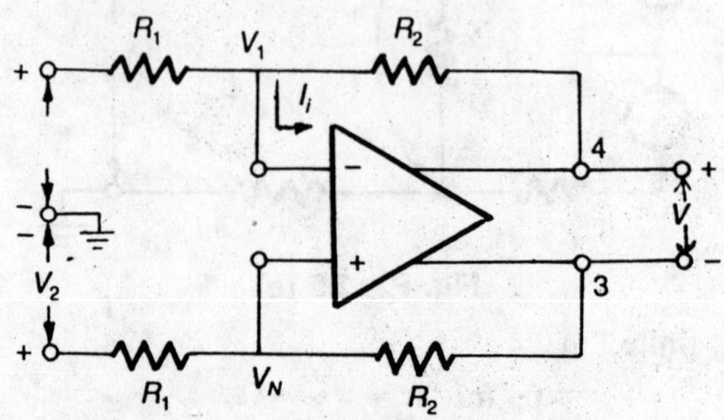

Fig. P. 12.6 (a)

Let the voltage at the inverting and non-inverting terminals are V_1 and V_N respectively. Then,

$$V_1 = (V_4 - V_1)\left(\frac{R_1}{R_1 + R_2}\right) + V_1 \quad \text{...(i)}$$

and, $$V_N = (V_3 - V_2)\left(\frac{R_1}{R_1 + R_2}\right) + V_2 \qquad \text{...(ii)}$$

since, $I_i = 0$, $V_I = V_N$

Then equating equation (i) and (ii), we have

$$(V_4 - V_1)\,R_1 + V_1\,(R_1 + R_2) = (V_3 - V_2)\,R_1 + V_2\,(R_1 + R_2)$$

or, $$V_4 - V_3 = \frac{R_2}{R_1}\,(V_2 - V_1)$$

or, $$V_o = \frac{R_2}{R_1}(V_2 - V_1) \text{ since, } V_o = V_4 - V_3$$

Problem 12.7: For the base differential-input amplifier shown, assume infinite input resistance, zero output resistance, and finite differential gain $A_V = V_o/(V_1 - V_2)$.

(a) Obtain an expression for the gain $A_{vf} = V_o/V_S$.

(b) Show that $\lim A_{vf} = n + 1$, $A_v \to \infty$.

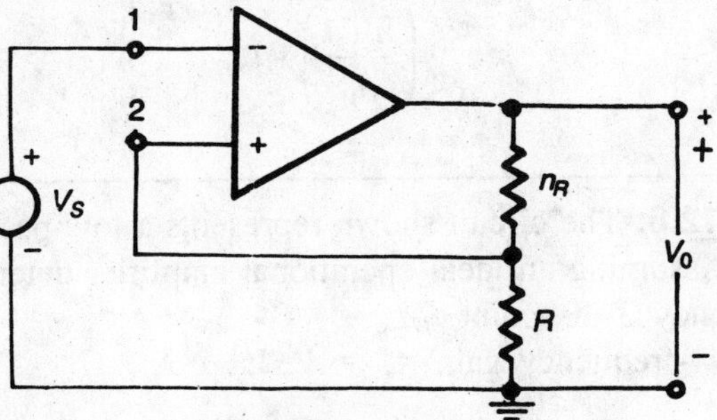

Fig. P. 12.7

Solution:

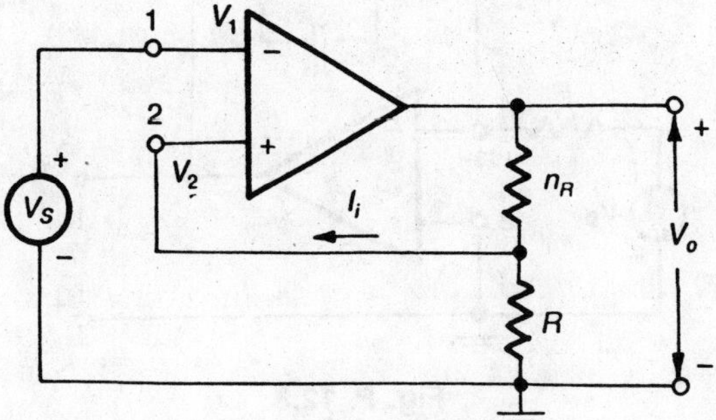

Fig. P. 12.7 (a)

(i) From the circuit by voltage division,

$$V_2 = \frac{V_o . R}{R + nR} = \frac{V_o}{n+1} \quad \text{since } I_i = 0 \text{ as } R_i = \infty$$

and $\quad V_1 = V_S$

Given, $\quad V_o = A_V (V_1 - V_2)$

or, $$V_o = A_V \left(Vs - \frac{V_o}{n+1} \right)$$

or, $$A_{vf} = \frac{V_o}{V_S} = \frac{A_V}{1 + \frac{A_V}{n+1}}$$

(ii) when $\quad A_V \to \infty$.

$$A_{vf} = \underset{A_v \to \infty}{Lt} \frac{A_V}{1 + \left(\frac{A_V}{n+1} \right)}$$

$$= \underset{A_v \to \infty}{Lt} \frac{n+1}{\left(\frac{n+1}{A_V} \right) + 1} = n+1$$

Problem 12.8: The circuit shown represents a low-pass dc-coupled amplifier. Assuming an ideal operational amplifier determine (a) the high-frequency 3-dB point f_H;
(b) the low-frequency gain $A_V = V_o/V_S$.

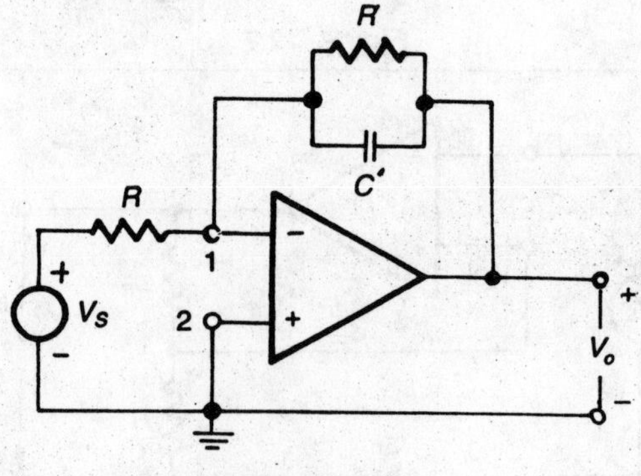

Fig. P. 12.8

Solution:

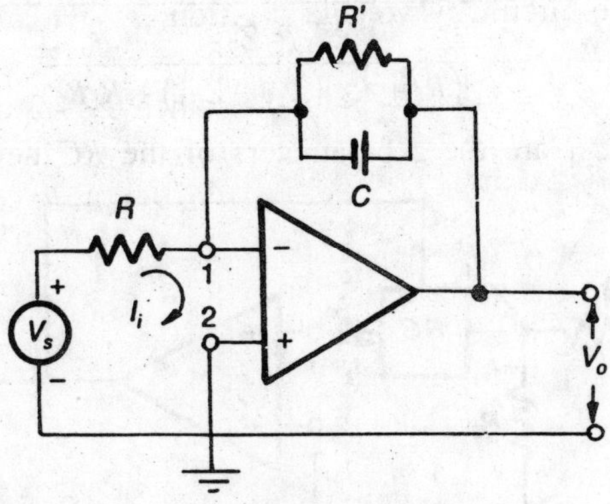

Fig. P. 12.8 (a)

Given circuit is redrawn in following manner,

(i) Since, OPAMP is ideal, it having infinite input resistance, then $I_i \approx 0$ and $V_- = V_+ = 0$ (virtually grounded)

Now KCL at node (i) gives,

$$\frac{0-V_s}{R}+\frac{0-V_o}{R'/(R'C_S'+1)} = 0$$

$$\text{or,}\quad \frac{V_o}{V_S} = \left(\frac{-R'}{R}\right)\left(\frac{1}{1+jwC'R'}\right)$$

$$\text{This gives, } w_H = \frac{1}{R'C'}$$

$$\text{or,}\quad f_H = \frac{1}{2\pi R'C'}$$

(ii) Now, 10w frequency gain,

$$A_v = \left.\frac{V_o}{V_S}\right|_{W\to 0} = -\frac{R'}{R}$$

Problem 12.9: Define the z parameters of a two-port network by the following relations:

$$V_1 = Z_{11}I_1 + Z_{12}I_2$$
$$V_2 = Z_{21}I_1 + Z_{22}I_2$$

For the circuit shown prove that the voltage gain $A_v = V_o/V_S$ is given by

$$\frac{V_o}{V_S} = -\frac{R_2 Z_{21}}{(R_1 + R_2)(Z_{11} - Z_{21}) + R_1 R_2}$$

where z_{11} and z_{21} are the z parameters of the RC network.

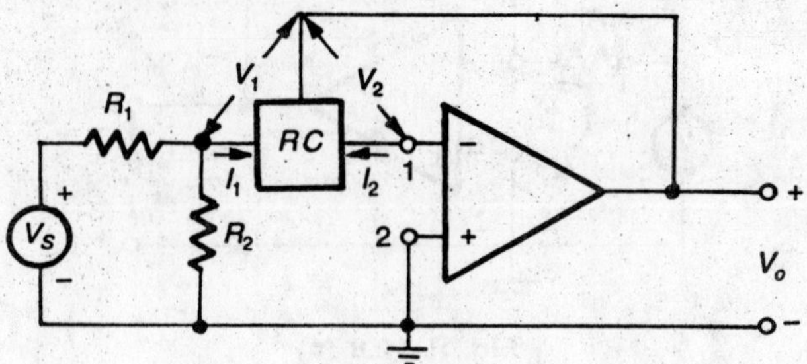

Fig. P. 12.9

Solution:

As it is easily seen, OPAMP is ideal,

$$I_2 = 0$$

then, $V_1 = Z_{11} I_1$ and $V_2 = z_{21} I_1$

As the circuit has virtual ground,

$$V_o = -V_2 = -Z_{21} I_1$$

The current in the resistor $R_2 = \dfrac{V_1 + V_o}{R_2}$

Now, using KVL for the loop $V_s - R_1 - R_2$, we have

$$V_s = I_1 R_1 + \left(\frac{R_1}{R_2}\right)(V_1 + V_o) + V_1 + V_o \qquad \text{...(i)}$$

$$\text{or,} \quad = I_1 R_1 + \left(\frac{R_1}{R_2}\right)(V_o + Z_{11} I_1) + Z_{11} I_1 + V_o \qquad [\because V_1 = Z_{11} I_1]$$

$$\text{or,} \quad = I_1\left[R_1 + Z_{11}\left(\frac{R_1}{R_2}\right) + Z_{11}\right] + V_o\left(\frac{R_1}{R_2} + 1\right)$$

$$\text{or,} \quad = \frac{-V_o}{Z_{21}}\left[R_1 + Z_{11}\left(\frac{R_1}{R_2} + 1\right) + V_o\left(\frac{R_1}{R_2} + 1\right)\right]$$

$$\text{or,} \quad = \frac{-V_o}{R_2 Z_{21}}\left[R_1 R_2 + (Z_{11} - Z_{12})(R_1 + R_2)\right]$$

or, $$\frac{V_o}{V_S} = \frac{-R_2 Z_{21}}{(R_1 + R_2)(Z_{11} - Z_{21}) + R_1 R_2}$$

Problem 12.10: Obtain the transfer function for the following delay equalizer shown below :

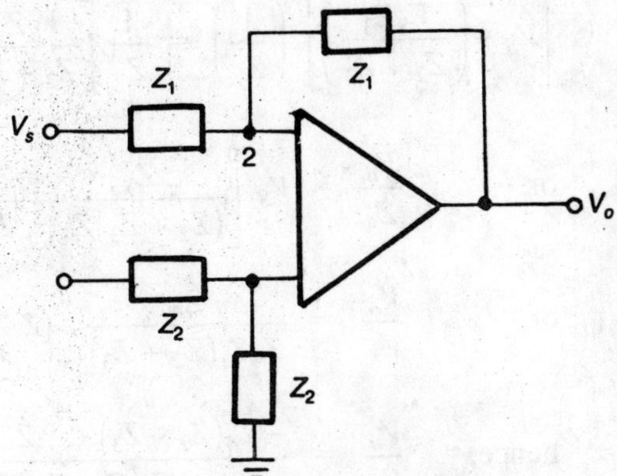

Fig. P. 12.10

Solution:

Given, circuit is reskeched in following manner will give,

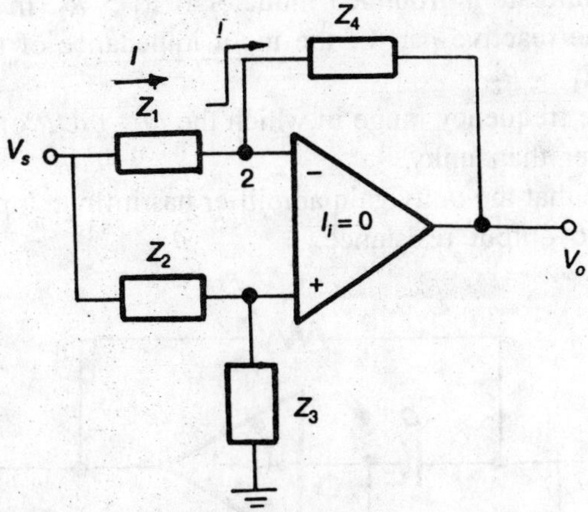

Fig. P. 12.10 (a)

Since OPAMP is ideal therefore.

$$V_1 = V_2 = V \qquad (\because I_i = 0)$$

Now, $$I = \frac{V_S - V}{Z_1} = \frac{V - V_o}{Z_4} \qquad \text{...(i)}$$

and, $$V = \left(\frac{V_S}{Z_2 + Z_3}\right) . Z_3 \qquad \text{...(ii)}$$

putting the expression for V in eq. (i), we have,

$$\left[V_S - \left(\frac{V_S Z_3}{Z_2 + Z_3}\right)\right] \cdot \left(\frac{1}{Z_1}\right) = \frac{1}{Z_4}\left[\frac{V_S Z_3}{Z_2 + Z_3} - V_o\right]$$

or, $$\frac{V_o}{Z_4} = V_S\left[\frac{Z_3}{(Z_2 + Z_3)Z_4} - \frac{Z_2}{Z_1}\left(\frac{1}{Z_2 + Z_3}\right)\right]$$

or, $$\frac{V_o}{V_S} = \frac{Z_4}{Z_4 Z_1 (Z_2 + Z_3)} . \; [Z_1 Z_3 - Z_2 Z_4]$$

hence, $$\frac{V_o}{V_S} = \frac{-Z_4(Z_2 + Z_3) + Z_3 Z_4 + Z_3 Z_1}{Z_1(Z_2 + Z_3)}$$

Problem 12.11: (a) Show that the circuit of the accompanying figure can simulate a grounded inductor if $R_1 > R_2$. In other words, show that the reactive part of the input impedance of this circuit is positive if $R_1 > R_2$.

(b) Find the frequency range in which the $Q = \omega L/R$ of the inductor is greater than unity.

Assume that the unity gain amplifier has infinite input resistance and zero output resistance.

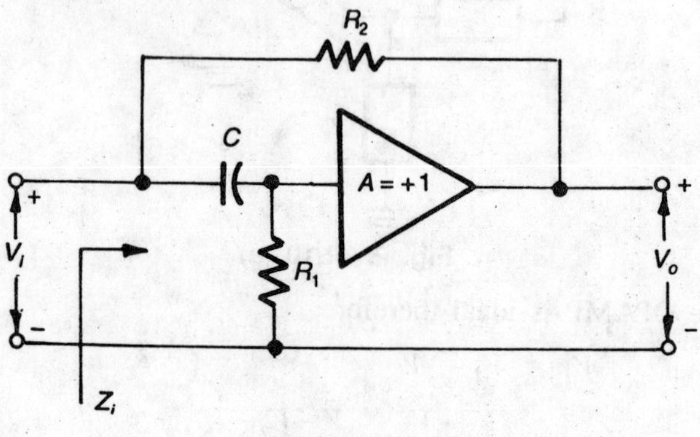

Fig. P. 12.11

Solution:

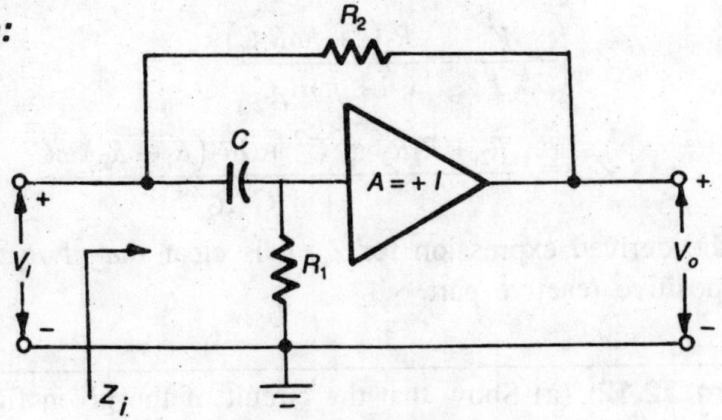

Fig. P. 12.11 (a)

Since OPAMP is ideal, so having infinite input resistance, then $I_i = 0$

$$I_2 = \frac{V_i}{R_1 + \frac{1}{j\omega C}} \qquad \text{...(i)}$$

and, $$I_1 = \frac{V_i - V_o}{R_2} \qquad \text{...(ii)}$$

At node (i), $I = I_1 + I_2$ ...(iii)

and voltage at node (2), is $V = V_o$, since $A = +1$

Then, $$V = \frac{V_i}{R_1 + \frac{1}{j\omega C}} \times R_1$$

or, $$V_o = \frac{V_i}{R_1 + \frac{1}{j\omega C}} \times R_1 \qquad \text{...(iv)}$$

putting the expression for V_0 in equation (ii), we have,

$$I_1 = \frac{\left[V_i - \frac{V_i R_1}{R_1 + \frac{1}{j\omega C}}\right]}{R_2} = \frac{V_i}{[1 + j\omega C R_1] R_2} \qquad \text{...(v)}$$

Now, putting the expression for I_1 and I_2 in equation (iii), we get,

$$I = \frac{V_i(j\omega C)}{R_1 j\omega C + 1} + \frac{V_i}{[j\omega C R_1 + 1] R_2}$$

$$Z_{in} = \frac{V_i}{I} = \frac{R_2(1+j\omega R_1 C)}{1+j\omega R_2 C}$$

$$Z_{in} = \frac{R_2 + R_1 R_2^{\,2}\omega^2 C^2 + JR_2(R_1 - R_2)\,\omega C}{1+\omega^2 C^2 R_2^{\,2}}$$

By the derived expression for Z_{in} it is clear that if $R_1 > R_2$ then Z_{in} have positive reactive part.

Problem 12.12: (a) Show that the circuit of the given figure can simulate a grounded inductor if $A > 1$. In other words, show that the reactive part of Z_i is positive.

(b) Show that the real of Z_i becomes zero ($Q = \infty$) at the frequency

$$\omega = \frac{1}{R_2 C}\sqrt{\frac{R_1 + R_2}{R_1(A-1)}}$$

Assume that the input resistance of the amplifier of gain A is infinite.

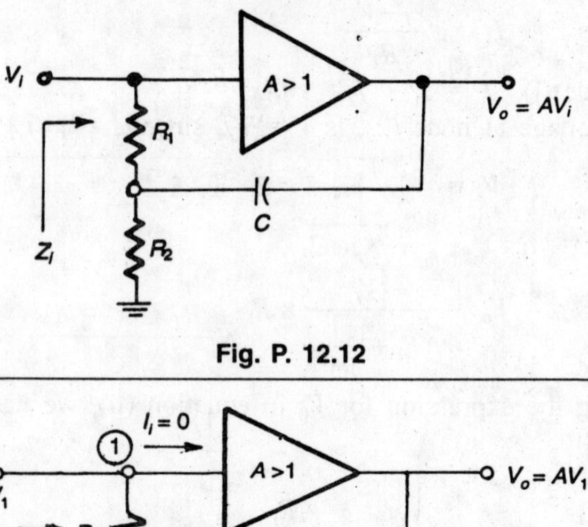

Fig. P. 12.12

Solution:

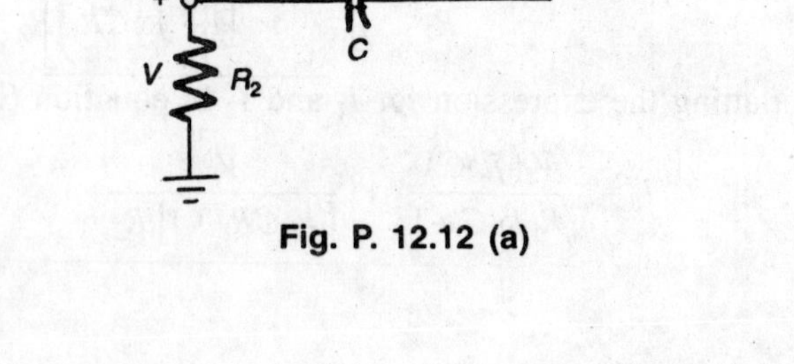

Fig. P. 12.12 (a)

Since OPAMP is ideal then. $I_i = 0$

and, $$I_1 = \frac{V_i - V}{R_1} \quad \text{...(i)}$$

Also, $$I_1 = \frac{V}{R_2} + (V - V_o)\, j\omega C$$

or, $$I_1 = \frac{V}{R_2} + (V - AV_i)\, j\omega C$$

or, $$\frac{I_1 j\omega C V_i A}{j\omega C + \dfrac{1}{R_2}} = V \quad \text{...(ii)}$$

using equation (i) in equation (ii), we have

$$I_1 R_1 = V_i - \frac{I_1 R_2}{1 + j\omega C R_2} - jAV_i \frac{\omega C R_2}{1 + j\omega C R_2}$$

or, $$I_1 \left[\frac{R_1 + R_2 + j\omega C R_1 R_2}{1 + j\omega C R_2}\right] = V_i \left[\frac{1 + j(1 - A)\omega R_2 C}{1 + j\omega C R_2}\right]$$

or, $$Z_i = \frac{R_1 R_2 + j\omega R_1 R_2 C}{1 + j(1 - A)\,\omega R_2 C}$$

$$= \frac{R_1 + R_2 + \omega^2 R_1 R_2{}^2 C^2 (1 - A) + J\omega C\left[R_1 R_2 - (1 - A) R_2 (R_1 + R_2)\right]}{1 + (1 - A)^2 \omega^2 R_2{}^2 C^2}$$

Here, z_i is positive if $A > 1$
equating $R_e\,[z_i] = 0$

or, $$\frac{R_1 + R_2 + \omega^2 R_1 R_2{}^2 C^2 (1 - A)}{1 + (1 - A)^2 \omega^2 R_2{}^2 C^2}$$

or, $$R_1 + R_2 = \omega^2 R_1 R_2^2 (A - 1)\, C^2$$

or, $$w = \frac{1}{R_2 C} \sqrt{\frac{R_1 + R_2}{R_1 (A - 1)}}$$

Problem 12.13: (a) Verify $T = \dfrac{R' C \ln\left[1 + \left(\dfrac{V_1}{V_2}\right)\right]}{1 - \beta}$ for the pulse width of a monostable multivibrator.
(b) If $V_z \gg V_1$ and $\beta = 1/2$. what is T?

Solution:

When triggering pulse introduced, the capacitor C starts charging from V_1 to the final value $-V_z$, then,

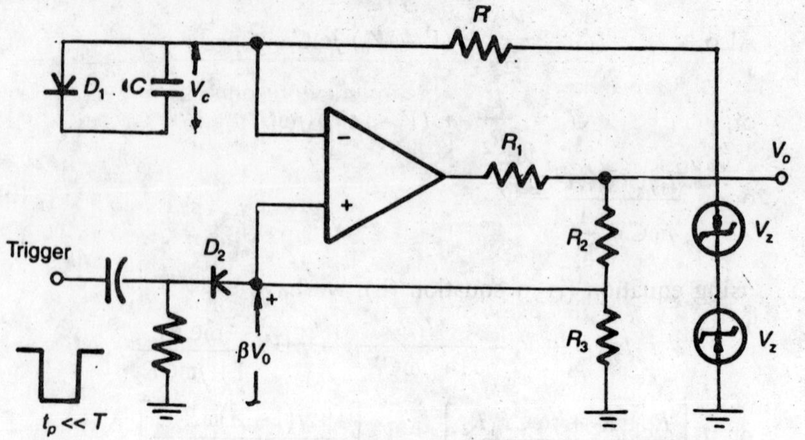

Fig. P. 12.13 (a)

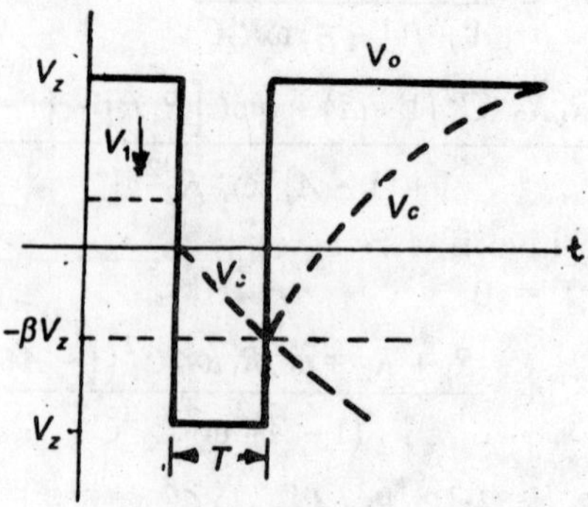

Fig. P. 12.13 (b)

$$V_c = -V_z (V_1 + V_Z) e^{-t/\tau}$$

where, $\tau = R'C$

At, $t = T, \; V_c = -\beta V_z$

Then, $V_e = -\beta - V_Z = -V_z + (V_1+V_Z) e^{-T/R'C}$

or, $\dfrac{V_Z(1-\beta)}{V_1+V_Z} = e^{-T/R'C}$ or, $e^{-T/R'C} = \dfrac{1+(V_1/V_Z)}{1-\beta}$

or, $$T = R'C \ln \frac{\left[1+\left(\frac{V_1}{V_Z}\right)\right]}{1-\beta}$$

(b) when $V_z \gg V1$ and $\beta = 0.5$

then, $$T = R'C \ln\left(\frac{1}{0.5}\right) = 0.695\, R'C$$

Problem 12.14: The circuit shown represents a low-pass d.c coupled amplifier. Assuming an ideal OPAMP. Find

(i) high frequency 3 dB point 'f_H'.

(ii) The low frequency gain $A_V = \dfrac{V_o}{V_S}$

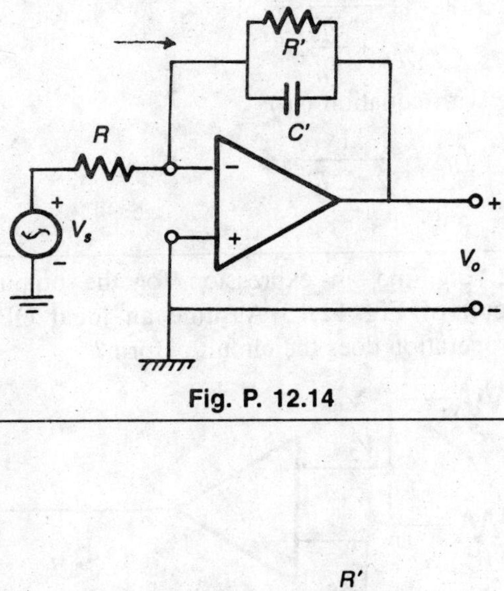

Fig. P. 12.14

Solution:

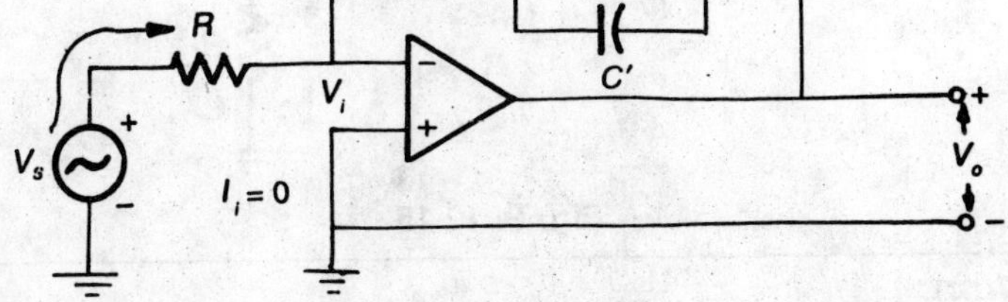

Fig. P. 12.14 (a)

From the circuit, it can be infered that, both inverting and non-inverting terminals are virtually grounded.

i.e $$V_- = V_+ \approx 0$$

then, $$I_1 = \frac{V_S}{R} \quad \text{...(i)}$$

and, $$V_o = -I_1 . \frac{R' \frac{1}{SC'}}{R' + \frac{1}{SC'}} = -I_1 \frac{R'}{R'C'S+1} \quad \text{...(ii)}$$

hence, $$\frac{V_o}{V_S} = \frac{-R'/R}{R'C'S+1} \quad \text{...(iii)}$$

(i) Using the transfer function for low-pass filter.

$$|A_H(jf)| = \frac{A_V}{1+\frac{jf}{f_H}}$$

comparing with equation (iii)

$$f_H = \frac{1}{2\pi C'R'}$$

Problem 12.15: Find an expression for the output 'V_0'of the amplifier circuit of Fig. 12.15 Assume an ideal OPAMP. What mathematical operation does the circuit peform?

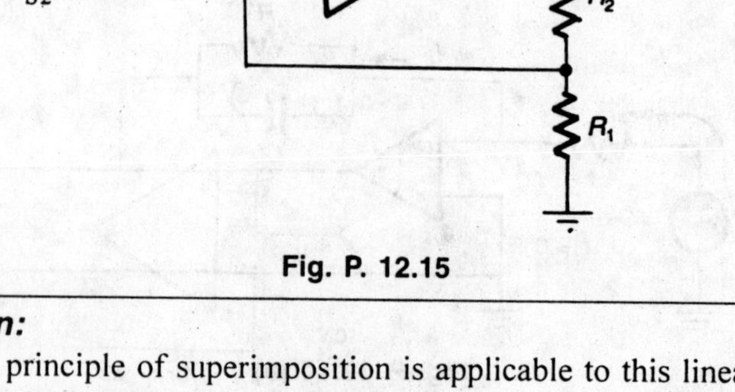

Fig. P. 12.15

Solution:

The principle of superimposition is applicable to this linear circuit with $V_{S2} = 0$ (shorted), the voltage appearing at the noninventive terminal is found by voltage division to be

$$V_2 = \frac{R}{R+R} V_{S1} = \frac{V_{S1}}{2} \qquad \ldots\text{(i)}$$

Let, V_{o1} be the value of V_0 with $V_{s2} = 0$

$$\therefore V_{o1} \left(1+\frac{R_2}{R_1}\right) V_2 = \left(1+\frac{R_2}{R_1}\right) \frac{V_{s1}}{2}$$

Similarly, with $V_S = 0$

By superposition, the total output is then

$$V_o = V_{o1} + V_{o2} = \frac{1}{2}\left(1+\frac{R_2}{R_1}\right)(V_{S1} + V_{S2})$$

The circuit is noninverting adder.

Problem 12.16: The output of a flow meter is voltage proportional to flow rate with proportionately constant $K = 25$ mV/(cm^3/s)with an effective output resistance of 2.5 KΩ.
Design a circuit to trip a relay after 250 cm^{3}of fluid has passed across the metering point. The relay is to be designed to trip when excited with 10 V.

Solution:

Voltage output of meter, $v_1 = Kq$; q = flow rate in cm^3/s
output resistance, R_1 = 2.5 KΩ

An OPAMP integrator will give the total flow at any time and an OPAMP computer with a buffer will feed the relay to trip it when the buffer output voltage is +10V as given in the schematic diagram of Fig. 12.16

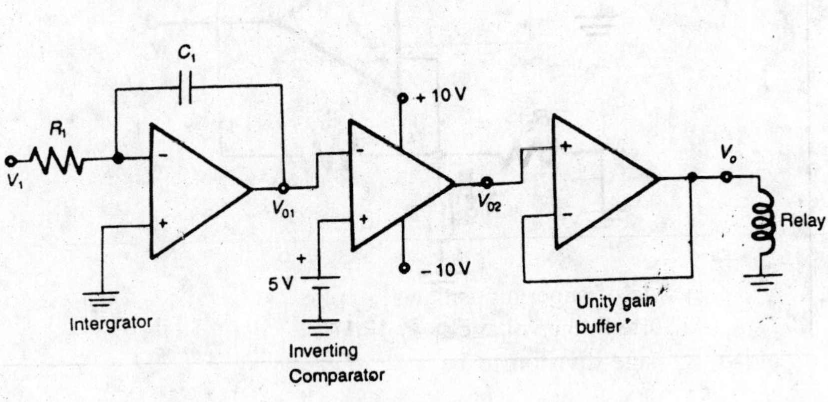

Fig. P. 12.16

$$V_{o1} = -\frac{1}{R_1C_1}\int_0^t v_1 dt; \; v_1 = Kq = 25 \text{ mV/(cm}^3\text{/s)}$$

$$= -\frac{K}{R_1C_1}\int_0^t q dt$$

$$V_{o1} = 5 \text{ V (say) at } Q = 250 \text{ cm}^3$$

$$R_1 = 2.5 \text{ K}\Omega$$

Substituting values,

$$5 = \frac{25 \times 250}{2.5C_1}$$

or, $C_1 = 50\ \mu\text{F}$

So, $v_{o1} = -5$ V when $Q = 250$ cm^3

This voltage is fed to an inverting comparator and +10V input of the comparator is fed to the relay through a unity gain buffer to match the relay impedance.

Problem 12.17: For the OPAMP circuit of Fig. 12.17 provided $R_3/R_2 = R'/R_1$. OPAMP is ideal.

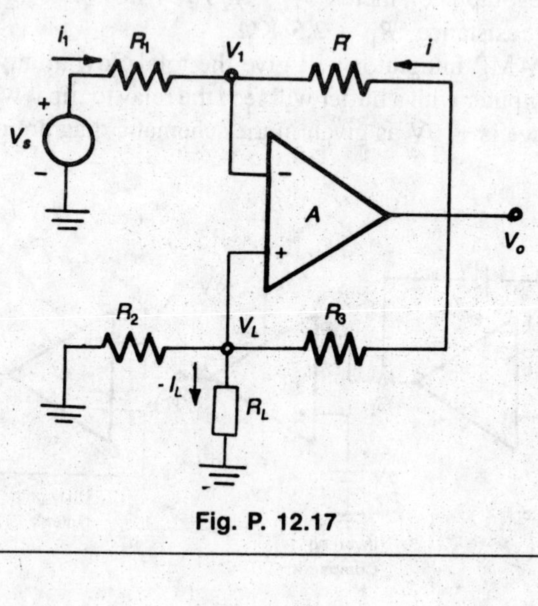

Fig. P. 12.17

Solution:

Since the OPAMP is ideal

Also $V_1 = V_L$ ...(i)

$$\frac{V_S - V_1}{R_1} = -\frac{V_o - V_1}{R'} \quad \text{...(ii)}$$

$$\frac{V_o - V_L}{R_3} = i_L + \frac{V_L}{R_2}; \; V_L = V_1 \quad \text{...(iii)}$$

Therefore

$$\frac{V_o - V_1}{R'} = \left(\frac{R_3}{R'}\right)\left(i_L + \frac{V_1}{R_2}\right) \quad \text{...(iv)}$$

Substituting Eq. (iv) in Eq. (ii) and reorganizing

$$\frac{V_S}{R_1} = -\left(\frac{R_3}{R'}\right)i_L + \frac{V_1}{R'}\left(\frac{R'}{R_1} - \frac{R_3}{R_2}\right)$$

$$\frac{R'}{R_1} = \frac{R_3}{R_2},$$

We get $$\frac{V_s}{R_1} = -\left(\frac{R_2}{R'}\right)i_L$$

$$i_L = -\frac{V_S}{R_2}$$

Problem 12.18: In the OPAMP circuit of Fig. 12.18. V_1 and V_2 represent variables measured at two points in a complex system—like the human body. These measurements include the common-mode signal (voltage) and differential-mode signal. The common-mode component is large mostly a dc component caused by drifts (changes in operating temperature etc.) and also sinusoidal "hum" attributed to inductive pickup from power lines. Our interest is mainly in the small differential voltage ($V_n - V_p$), which should get amplified, while the common mode voltage is to be eliminated as far as possible.

An OPAMP circuit which has this capability is shown in Fig 12.18. It is desired to establish the condition that must be met for this purpose.

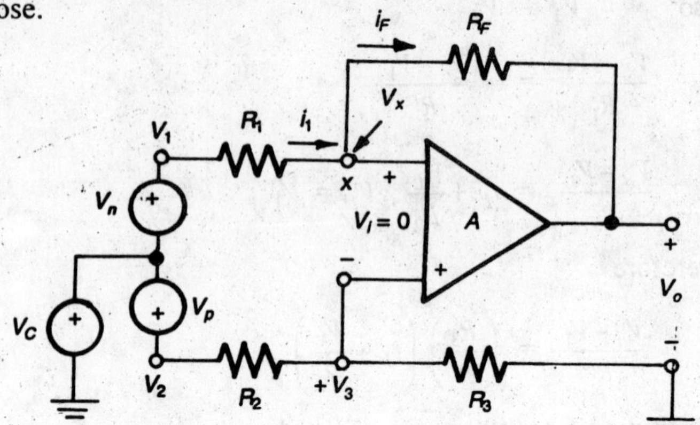

Fig. P. 12.18

Solution:

In an ideal OPAMP

$$V_i = 0 \text{ and so}$$

$$V_3 = V_x$$

or,
$$V_3 = (V_C + V_p) \cdot \left(\frac{R_3}{R_2 + R_3}\right) = V_x \quad \text{...(i)}$$

Also as $\quad i_i = 0$ then $i_1 = i_F$

or,
$$\frac{(V_C + V_n) - V_x}{R_1} = \frac{V_X - V_o}{R_F} \quad \text{...(ii)}$$

Substituting for V_x from Eq. (i) in Eq. (ii) and simplifying we have

$$V_o = V_C \left(\frac{R_3}{R_2 + R_3} - \frac{R_F}{R_1} + \frac{R_F}{R_1} \cdot \frac{R_3}{R_2 + R_3}\right)$$

$$+ V_p \left(\frac{R_3}{R_2 + R_3} + \frac{R_F}{R_1} \cdot \frac{R_3}{R_2 + R_3}\right) - V_n \left(\frac{R_F}{R_1}\right) \quad \text{...(iii)}$$

For V_c term to be reduced to zero

$$\frac{R_3}{R_2 + R_3} - \frac{R_F}{R_1} + \frac{R_F}{R_1} \cdot \frac{R_3}{R_2 + R_3} = 0$$

which yields a simple result

$$\frac{R_F}{R_1} = \frac{R_3}{R_2} \qquad \text{...(iv)}$$

Using this condition in Eq. (iii), we get

$$V_o = -\frac{R_F}{R_1}(V_n - V_p) \qquad \text{...(v)}$$

Problem 12.19: Consider the OPAMP frequency sensitive network of Fig. 12.19. Derive the expression for its gain and sketch its dB–log ω plot. What conclusion do you draw from it about the nature of its frequency response? The values of external circuit elements are as under:

C = 0.1 μF,
R = 100 kΩ;
C_1 = 1 μF,
R_1 = 10 kΩ

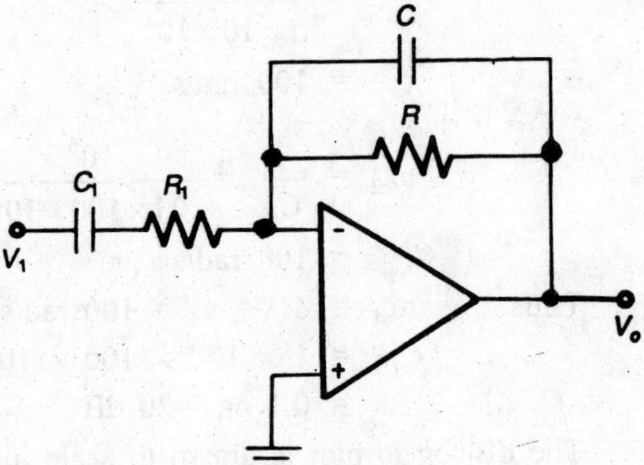

Fig. P. 12.19

Solution:

$$\overline{Z}_F(j\omega) = \frac{1}{j\omega C} \parallel R = \frac{R}{1 + jCR\omega}$$

$$\overline{Z}_1(j\omega) = \frac{1}{j\omega C_1} + R_1 = \frac{1 + jCR_1\omega}{jC_1\omega}$$

$$\text{Gain,} \quad \overline{G}(j\omega) = \frac{V_o(j\omega)}{V_1(j\omega)}$$

$$= -\frac{\overline{Z}_F(j\omega)}{Z_1(j\omega)}$$

$$= \frac{jC_1R\omega}{(1+jC_1R_1\omega)(1+jCR\omega)}$$

Critical frequencies are:

$$\omega_{c1} = \frac{1}{C_1R_1}$$

$$= \frac{10^6}{1\times 10\times 10^3}$$

$$= 100 \text{ rad/s}$$

$$\omega_{c2} = \frac{1}{CR} = \frac{10^6}{0.1\times 100\times 10^3}$$

$$= 100 \text{ rad/s}$$

Thus $\omega_{c1} = \omega_{c2} = \omega_c = 100$ rad/s

$$C_iR = 1\times 10^{-6}\times 100\times 10^3$$

$$= 0.1 \text{ or, } -20 \text{ dB}$$

The dB-log ω plot is drawn to scale in Fig. 12.19 (a). It is seen from this plot that the circuit suppresses both low and high frequencies but passes frequencies around ω_c =100 rad/s. So it is a band-pass network but not a highly selective one.

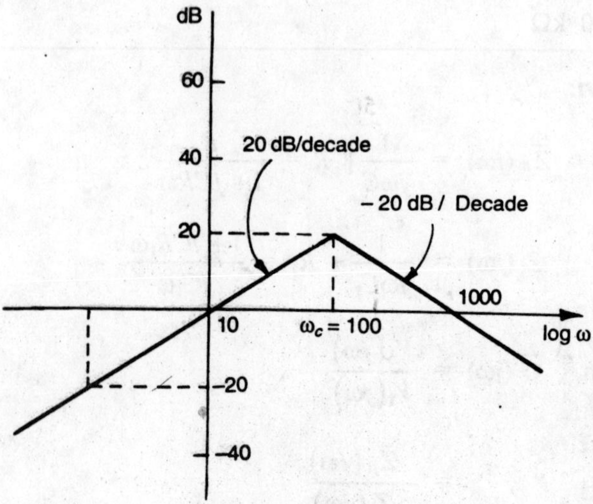

Fig. P. 12.19 (a)

Problem 12.20: For the OPAMP circuit shown in Fig. 12.20. find the values of R_1 and R_2 for the output to be

$$V_o = -5V_a + 3\ V_b$$

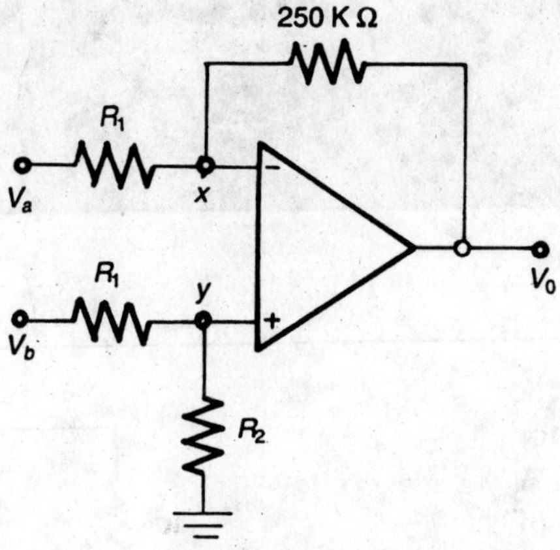

Fig. P. 12.20

Solution:

We shall use superposition.

(a) when v_b is short circuited, i.e. $V_b = 0$

$$\frac{V_o}{V_{oa}} = -\frac{250}{R_1} = -5$$

or, $R_1 = 50\ \text{K}\Omega$

(b) When V_a is short circuited, i.e. $V_a = 0$

$$V_x = \left(\frac{50}{50+250}\right) V_{ob} = (1/6)\ V_{ob}$$

$$V_y = \left(\frac{R_2}{50+R_2}\right) V_b$$

But $V_x = V_y$

or, $\left(\frac{1}{6}\right) v_{ob} = \left(\frac{R_2}{50+R_2}\right) V_b$

But $V_{ob} = 3\ v_b$

Therefore $(1/6) \times 3 = \dfrac{R_2}{50+R_2}$

or, $R_2 = 50\ \text{K}\Omega$

These values of R_1 and R_2 will yield the output as specified.

Chapter 13

Power Circuit and Systems

Problem 13.1: A power transistor operating class A in the circuit of Fig. 13.1 is to deliver a maximum of 5 W to a 4-Ω load (R_L = 4 Ω). The quiescent point is adjusted for symmetrical clipping, and the collector supply voltage is V_{CC} = 20 V. Assume ideal characteristics, as in Fig. 13.1 (a), with V_{min} = 0.

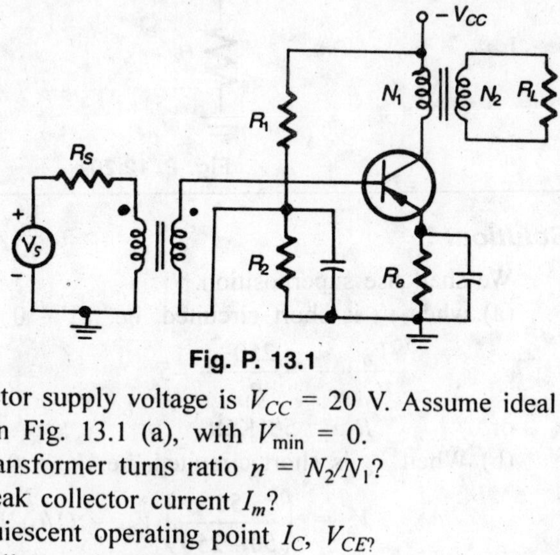

Fig. P. 13.1

(a) What is the transformer turns ratio $n = N_2/N_1$?

(b) What is the peak collector current I_m?

(c) What is the quiescent operating point I_C, V_{CE}?

(d) What is the collector-circuit efficiency?

Solution: Redrawing the given circuit,

Here, $\dfrac{N_1}{N_2} = a$

From the Fig. 13.1 note that,

$$R_{ac} = a^2 R_L = R'_L$$

Also, since the resistance of the windings of an ideal transformer is negligible, then,

$R_{dc} = R_e$

Now, the load line drawn as shown below.

Since, the given transformer is ideal, hence loss less, then the power delivered to the primary is identically the power delivered to R_L

Now we have,

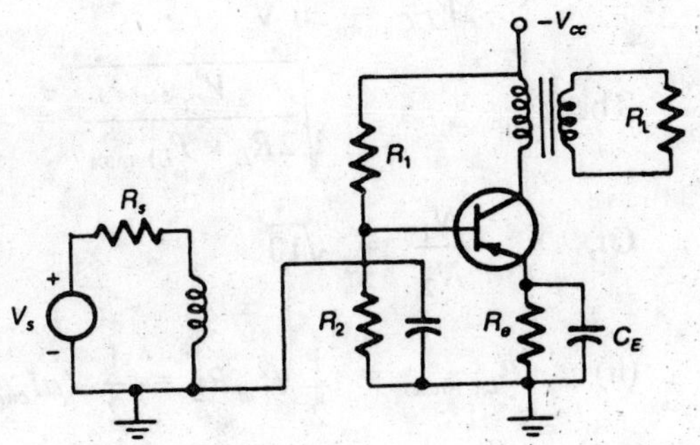

Fig. P. 13.1 (a)

$$P_L = \frac{1}{T}\int_0^T i_L{}^2 R_L dt$$

$$= \frac{1}{T}\int_0^T I_{Lm}^2 \sin^2 \omega t \; dt = \frac{R_L T_{Lm^2}}{2T}\int_0^T (1 - \cos 2\omega t)\, dt$$

$$= \frac{I_{Lm^2}}{2T} R_L \times T = \frac{I_{Lm^2} R_L}{2} = \frac{I_{Lm^2} R_L}{2}$$

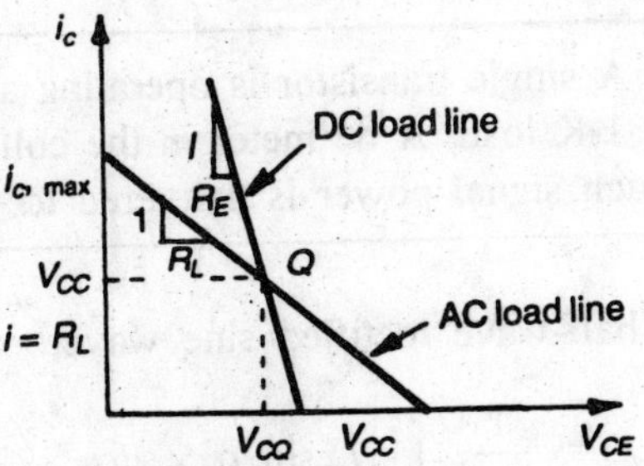

Fig. P. 13.1 (b)

From Fig. 13.1 (d) for maximum symmetrical swing

$$I_{cm} \approx I_{cQ} = \frac{V_{CC}}{R'_L}$$

hence, $P_{L1\,max} \approx \dfrac{I_{cQ^2} R'_L}{2} \approx \dfrac{V_{CC^2}}{2R'_L} = \dfrac{V_{CC^2}}{2a^2 R_L}$

Given, $P_{L1\,max} = 5$ W,

$$R_L = 4\ \Omega,$$
$$V_{CC} = 20\ \text{V}$$

Then, $$a = \sqrt{\frac{V_{CC^2}}{2R_L \times P_{L1\,max}}} = \frac{20}{\sqrt{2\times4\times5}} = \sqrt{10}$$

Or, $$\frac{N_1}{N_2} = \sqrt{10}$$

(ii) $$P_{L1\,max.} = \frac{1}{2} I_{Lm}^2 R_L = \frac{1}{2}(aI_{cm})^2 R_L = \frac{I_{cm^2}}{2} R'_L$$

Or, $$I_{cm} = \sqrt{\frac{2P_{L1max}}{R_{L'}}} = \sqrt{\frac{2\times5}{4\times10}} = 0.5\ \text{A}$$

(iii) From the load line

$$I_C = I_{cm} = 0.5\ \text{A};\ V_{CE} = V_{CC} = 20\ \text{V}$$

(iv) $$\eta = \frac{\text{Load power}}{\text{Supplied power}} \times 100\ \%$$

$$= \frac{5}{20\times0.5} \times 100 = 50\ \%$$

Problem 13.2: A single transistor is operating as an ideal class B amplifier With a 1-K load. A dc meter in the collector circuit reads 10 mA. How much signal power is delivered to the load?

Solution: For a half-wave rectified sine wave,

$$I_{dc} = \frac{1}{2\pi}\int_0^{\pi} I_m \sin\alpha \times d\alpha$$

$$= \frac{1}{2\pi} \times I_m - \cos\alpha \Big|_0^{\pi} = \frac{I_m}{\pi} \qquad \ldots\text{(i)}$$

Power delivered to the load,

$$P_L = \frac{R_L}{T}\int_0^{T/2} I_m^2 \sin^2\omega t dt$$

$$= R_L \times \frac{I_m^2}{T}\int_0^{T/2} \frac{[1-\cos 2\omega t]}{2} dt$$

$$= \frac{I_m^2}{2T} \times R_L \times \left[t - \frac{\sin 2\omega t}{2\omega} \right]_0^{T/2}$$

$$= \frac{I_m^2 R_L}{4} \qquad \text{...(ii)}$$

Using expression for I_m from eq (i) in eq. (ii), we get,

$$P_L = \frac{(\pi I_{dc})^2 R_L}{4}$$

$$= \frac{\pi^2 \times \left(10 \times 10^{-3}\right)^2 \times 10^3}{4} = 0.246$$

Problem 13.3: Give an ideal class B transistor amplifier whose characteristics are as in Fig. 13.3. The collector supply voltage V_{CC} and the effective load resistance $R'_L = (N_1/N_2)^2 R_L$ are fixed as the base-current excitation is varied. Show that the collector dissipation P_C is zero at no signal ($V_m = 0$), rises as V_m increases, and passes through a maximum given by E q at $V_m = 2 V_{CC}/\pi$.

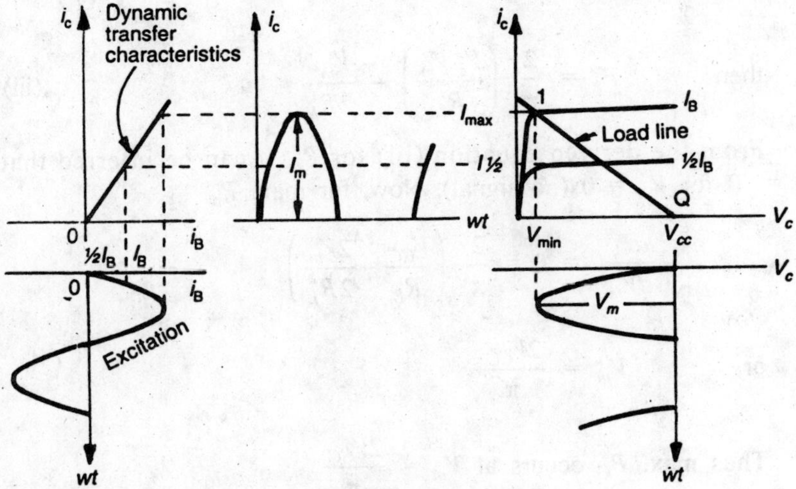

Fig. P. 13.3

Solution: From the given wave forms, it can be inferred that for a sinusoidal excitation the output is sinusoidal during one half of each period and in zero during the second half cycle. Also the wave forms represent one transistor Q only. Then the output of Q_2 will be series of sine pulse that are 180° out of phase with those of Q_1.

Hence, the load current will be perfect sine wave. The power output is,

$$P = \frac{V_m I_m}{2} = \frac{I_m(V_{CC} - V_{min})}{2} \quad \text{...(i)}$$

and the dc power input is,

$$p_i = \frac{2 I_m V_{CC}}{\pi} \quad \text{...(ii)}$$

[2 used in expression for two transistors]

Then the collector dissipation,

$$P_c = P_i - P$$

$$= \frac{2 I_m V_{CC}}{\pi} - \frac{I_m V_m}{\pi}$$

since, $I_m = \frac{V_m}{R_{L'}}$. (where $R_L' = \left(\frac{N_1}{N_2}\right)^2 R_L$)

then $$P_c = \frac{2}{\pi}\left(\frac{V_{cc} V_m}{R_L'}\right) - \frac{V_m^2}{2R_L'} \quad \text{...(iii)}$$

From the derived equation (iii) for P_c, it can be inferred that $P_c = 0$ for $V_m = 0$ (no signal). Now, for max. P_c

$$\frac{dP_c}{dV_m} = 0 = \frac{2}{\pi}\left(\frac{V_{CC}}{R_L'} - \frac{2V_m}{2R_L'}\right)$$

or, $$V_m = \frac{2V_{CC}}{\pi}$$

Thus max. P_C occurs at $V_m = \frac{2V_{CC}}{\pi}$

Problem 13.4: The idealized push-pull class B power amplifier shown has $R_2 = 0$, $V_{CC} = 20$ V, $N_2 = 2$ N, and $R_L = 20\ \Omega$, and the transistor have $h_{FE} = 20$. The input is a sinusoid. For the maximum output signal at $V_m = V_{CC}$ determine

(a) the output signal power

(b) the collector dissipation in each transistor

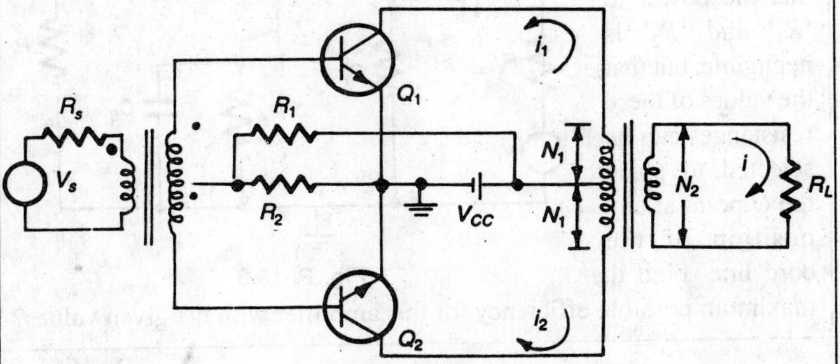

Fig. P. 13.4

Solution:

(i) Output signal power,

$$P = \frac{V_m I_m}{2} = \frac{V_m \times V}{2R_L'} \qquad \left[\text{Where } R_L' = R_L\left(\frac{N_1}{N_2}\right)^2\right]$$

$$= \frac{V_m^2}{2R_L'} = \frac{V_{CC}^2}{2R_L}\left(\frac{N_2}{N_1}\right)^2 \qquad [\because \text{ Given } V_m = V_{CC}]$$

$$= \frac{(20)^2}{2 \times 20}\left(\frac{4}{1}\right) = 40 \text{ W}$$

(ii) Now, power input,

$$P_i = \frac{2I_m}{\pi} \times V_{CC} = \frac{2V_{CC}^2}{\pi R_L}$$

Then, collector dissipating in each transistor

$$= \frac{1}{2}(P_i - P) = \frac{1}{2} \times \left(\frac{2V_{CC}^2}{\pi R_L'} - 40\right)$$

$$= \frac{1}{2} \times (50.8 - 40) = 5.47 \text{ Watt.}$$

Problem 13.5: In the amplifier of given Fig. let $R_E = 470\ \Omega$, $R_C = 800\ \Omega$, and $R_L = 1.2\ k\Omega$. Assume that the power to 'R_1' and 'R_2' is negligible, but that the values of these resistances can be selected to place the Q point at any position of the cord line. Find the maximum possible efficiency for this amplifier with the given values?

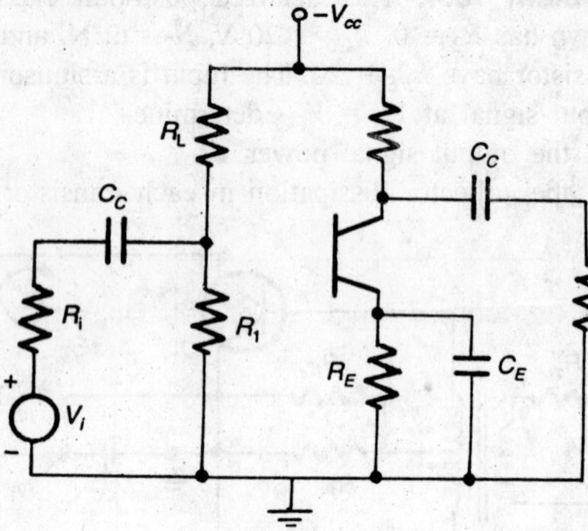

Fig. P. 13.5

Solution: Max. efficiency occurs with max. ac load current, so we assume R_1 and R_2 are selected for max. symmetrical swing. Which occurs at mid-point of ac load line,

$$I_{LQ} = \frac{1}{2}\, I_{(\max)} = \frac{1}{2}\left(\frac{V_{CEQ}}{R_{ac}} + I_{LQ}\right)$$

since, $\quad V_{CEQ} \approx V_{CC} - I_{CQ}R_{dc}$

$$\therefore \quad I_{LQ} = \frac{V_{CC}}{R_{ac} + R_{dc}} = \frac{V_{CC}}{R_c \| R_L + R_E + R_C}$$

Then, Delivered power,

$$p_{cc} = V_{cc} I_{LQ} = \frac{V_{CC}^2}{R_c \| R_L + R_c}$$

Typically $V_{CE\,sat} \approx 0$ and we assume $I_{CBO} = 0$

$$I_{cm} = I_{LQ} = \frac{V_{CC}}{R_C \| R_L + R_E + R_C}$$

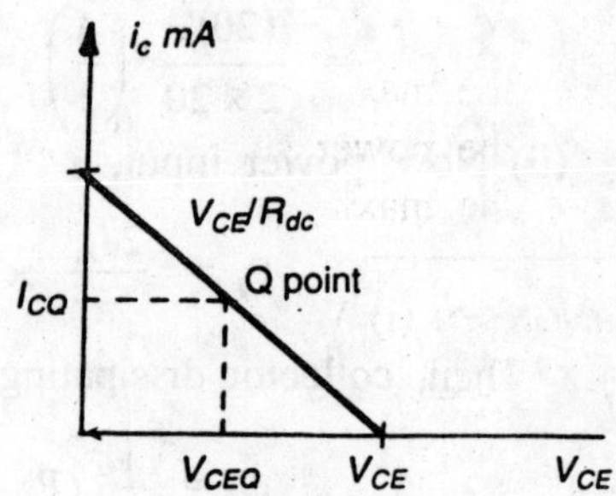

Fig. P. 13.5 (a)

Current division gives

$$I_{Lm} = \frac{R_C}{R_L + R_C} I_{cm}$$

Problem 13.6 : If the maximum collector power dissipation $P_{c(max)}$ is known, show that, for the case of maximum symmetrical swing, the Q point is described by

$$I_{CQ} = \sqrt{P_{c(max)} / P_{ac}}$$

and $$V_{CEQ} = \sqrt{R_{ac} P_{c(max)}}$$

Solution: Since, the Q point lies on the maximum dissipation hyperbola, thus,

$$I_{CQ} = \frac{P_{c(max)}}{V_{CEQ}} \quad \text{...(i)}$$

At the Q point,

$$V_{CEQ} = I_{CQ} R_{ac} \quad \text{...(ii)}$$

Substituting equation (ii) into equation (i) and rearranging yield.

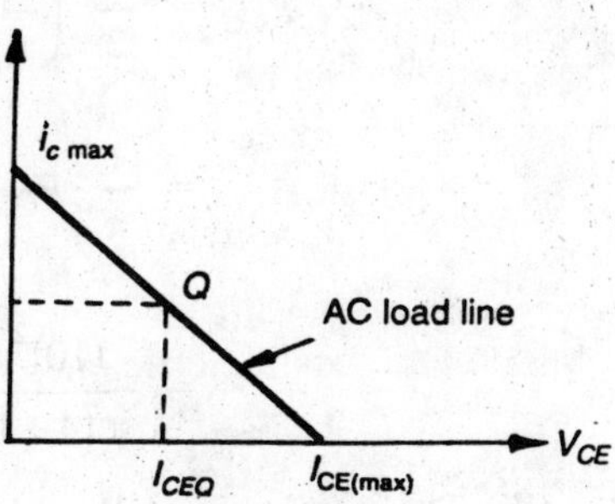

Fig. P. 13.6

$$I_{CQ} = \frac{P_{c\,max}}{I_{ca} R_{ac}} = \sqrt{\frac{P_{c\,max}}{R_{ac}}} \quad \text{...(iii)}$$

$$V_{CEQ} = \frac{P_{c\,max}}{I_{CQ}} = \frac{P_{c\,max}}{\sqrt{P_{c\,max} / R_{ac}}}$$

$$= \sqrt{R_{ac} P_{c\,max}} \quad \textbf{Proved}$$

Problem 13.7 : In the complementary-symmetry amplifier of $R_L = 4\ \Omega$, $R_C = 1\ \Omega$ and $V_{CC} = 10$ V. Find,

(i) the maximum power delivered to the load R_L.
(ii) the power supplied by each dc bias source, and
(iii) the maximum power dissipated by each transistor.

Solution: (i) When $V_s > 0$, Q_2 conducts and Q_1 is cut off; when $V_s < 0$, Q_1 conducts and Q_2 cut off.

If $V_{CE\,sat} \approx 0$ then

$$I_{cm1} = I_{cm2} = \frac{V_{CC}}{(R_L + R_C)}$$

and, $$P_{L\max} = \frac{1}{2} I_{cm}^2 R_L = \frac{1}{2}\left(\frac{V_{CC}}{R_L + R_C}\right)^2 R_L$$

$$= \frac{1}{2}\left(\frac{10}{4+1}\right)^2 (4) = 8 \text{ W}$$

(ii) Since each source must supply half the total power P_{CC}, we have,

$$\frac{1}{2} P_{CC} = \frac{1}{2}\left(\frac{2}{\pi}\right) V_{CC} I_m$$

$$= \frac{1}{\pi} V_{CC} \frac{V_{CC}}{R_L + R_C} = \frac{V_{CC}^2}{\pi(R_L + R_C)}$$

$$= \frac{(10)^2}{\pi(4+1)} = 6.370 \text{ W}$$

(iii) The maximum collector power dissipation per transistor follows with $a = 1$ and R_L replaced with $R_L + R_C$:

$$P_{C\max} = \frac{1}{R_L + R_C}\left(\frac{V_{CC}}{\pi}\right)^2$$

$$= \frac{1}{4+1}\left(\frac{10}{\pi}\right)^2 = 2.06 \text{ W}$$

Problem 13.8: Consider a BJT transistor to be ideal in complete range from cut-off to saturation (at $v_{CE} = V_{\min}$). Show that the power conversion efficiency is given by

$$\eta = \frac{25(V_{CC} - V_{\min})}{V_{CC}} \times 100 \text{ percent}$$

What is the maximum possible efficiency and under what conditions in this maximum achieved?

Solution: From Fig. 13.8

$$V_{CQ} = \frac{V_{CC} - V_{min}}{2}$$

$$\therefore \quad I_{CQ} = \frac{V_{CC} - V_{min}}{2R_L'}$$

$$P_{dc} = V_{CC} I_{CQ} = \frac{V_{CC}(V_{CC} - V_{min})}{2R_L}$$

$$\therefore \quad v_{CE}\text{ (p-p)} = (V_{CC} - V_{min})$$

$$\therefore \quad i_c\text{ (p-p)} = \left(\frac{V_{CC} - V_{min}}{R_L}\right)$$

$$\therefore \quad P_{ac} = \frac{v_{CE}(\text{p-p})\, i_c(\text{p-p})}{8R_L} = \frac{(V_{CC} - V_{min})^2}{8R_L}$$

We know $\eta = \dfrac{P_{ac}}{P_{dc}}$

$$= \frac{(V_{CC} - V_{min})^2}{8R_L} \times \frac{2R_L}{(V_{CC} - V_{min})} = \frac{V_{CC} - V_{min}}{4V_{cc}}$$

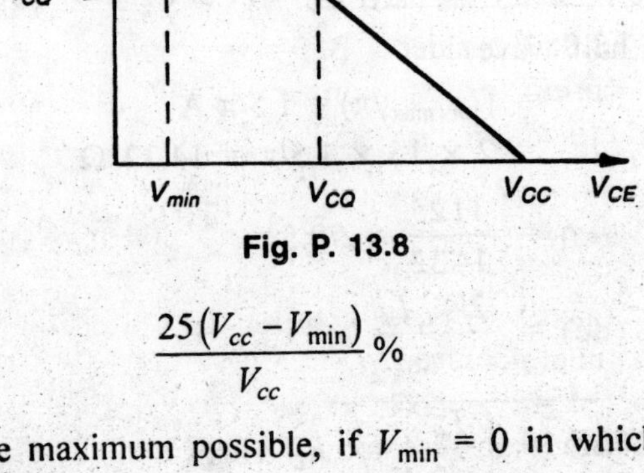

Fig. P. 13.8

or $$\frac{25(V_{cc} - V_{min})}{V_{cc}}\,\%$$

η will be maximum possible, if $V_{min} = 0$ in which case n (max possible) = 25%.

Problem 13.9: In the ideal class-B amplifier with complimentary symmetry shown in Fig. 13.9 $V_{CC} = 15$ V, $R_L = 10\ \Omega$. The input is sinusoidal. Determine :

(a) The minimum signal output power, the corresponding collector dissipation and conversion efficiency.

(b) The maximum dissipation of each transistor and the corresponding conversion efficiency.

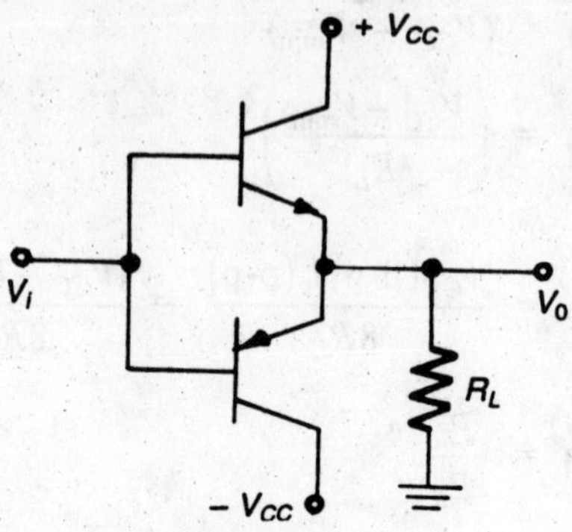

Fig. P. 13.9

Solution:

$$V_{CC} = 15 \text{ V},$$

$$R_L = 10\ \Omega$$

(a)
$$I_{m\,(\max)} = 15/10 = 1.5 \text{ A}$$

$$I_{\text{rms}\,(\max)} = 1.5/\sqrt{2} \text{ A; full-wave}$$

$$P_o\,(\text{ac}) = I^2_{\text{rms}\,(\max)}\, R_L = (1.5/\sqrt{2}\,)^2 \times 10 = 11.25 \text{ W}$$

I_{dc} (each half wave)

$$= (I_{m\,(\max)}/\pi) = 1.5/\pi \text{ A}$$

$$P_i\,(\text{dc}) = 2 \times 15 \times 1.5/\pi = 14.32\ \Omega$$

$$\eta = \frac{11.25}{14.32} = 78.5\%$$

(b)
$$P_i\,(\text{dc}) = \frac{2V_{CC}I_m}{\pi}$$

$$P_o\,(\text{ac}) = \left(\frac{I_m}{\sqrt{2}}\right)^2 \times R_L$$

$$P_D = \frac{2V_{CC}I_m}{\pi} - \frac{I_m^2 R_L}{2}$$

which is maximum at

$$I_m = \frac{2V_{CC}}{\pi R_L} = \frac{2 \times 15}{\pi \times 10} = 0.955 \text{ A}$$

$$P_{D\,(max)} = \frac{2 \times 15 \times 0.955}{\pi} - \frac{(0.955)^2 \times 10}{2} = 4.56 \text{ W}$$

At $P_{D\,(max)}$ i.e. at

$$I_m = 0.955 \text{ A}$$

$$\eta = \frac{P_o(ac)}{P_i(dc)} = \frac{(0.955)^2 \times 10 \times \pi}{2 \times 2 \times 15 \times 0.955} = 0.5 \text{ or } 50\ \%$$

Problem 13.10: Consider the complimentary symmetry class-B amplifier of Fig. 13.10 wherein battery biasing is provided in place of potential divider biasing with value $+kV_\gamma$ for *n-p-n* transistor and $-kV_\gamma$ for the *p-n-p*; V_γ being the cut-in voltage of each transistor. *Given* : k = constant such that $0 < k < 1$, $V_\gamma = 0.6$ V. v_i (input signal) $= V_s \sin \omega t$; $V_s = 1$ V.

(a) Sketch the output v_o for $k = 0$, 0.5 and 1. Calculate the cut-in angle for each value of k.
(b) What happens to distortion as V_s increases?
(c) What happens if k exceeds unity?
(d) If a resistor R is added between each emitter and output is taken from the junction of the two resistors.
(e) Is the push-pull operation class *A*, *B*, *AB* or *C* in (a) and (d)?

Solution : The circuit is drawn below

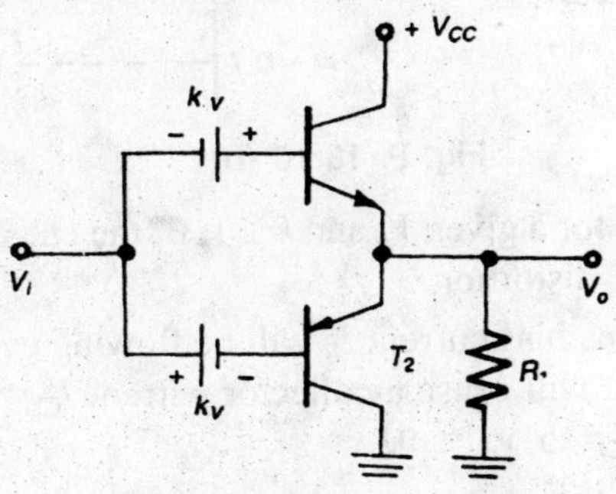

Fig. P. 13.10 (a)

(a) As $kV_\gamma \le V_\gamma$, no base-emitter bias current flow when there is no signal input i.e., $v_i = 0$. With signal applied the base emitter loop begin conducting via load when

$$V_s \sin \omega t - (1-k)V_\gamma \ge 0 \qquad \text{...(i)}$$

or $\sin \omega t - 0.6\,(1-k) \ge 0$; V_s = IV

The conduction begins when (cut-in angle)

$$-0.6\,(1-k) + \sin \theta_1 = 0$$

or $\theta_1 = \sin^{-1} 0.6\,(1-k)$

and ends at $(\pi - \theta_1)$

For $k = 0 \rightarrow \theta_1 = 37°$

$k = 0.5 \rightarrow \theta_1 = 17.5°$

$k = 1 \rightarrow \theta_1 = 0$

Voltage appearing across R_L

$$v_o = -0.6 + \sin \omega t; \quad k = 0; \text{ Fig. 13.10 (a)}$$

$$= -0.3 + \sin \omega t; \quad k = 0.5; \text{ Fig. 13.10 (b)}$$

$$= \sin \omega t; \quad k = 1; \text{ Fig. 13.10 (c)}$$

(b) In general cut-in angle

$$\theta_1 = \sin^{-1} \frac{V_s}{V_\gamma(1-k)} \qquad \text{...(ii)}$$

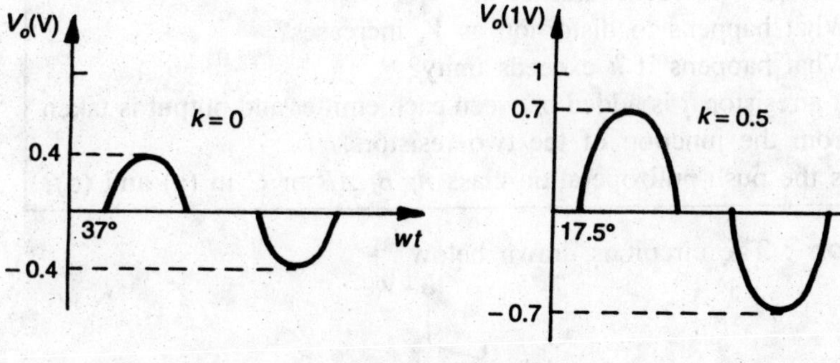

Fig. P. 13.10 (b)

As V_s increases for a given V_γ and $k \le 1$, θ_1 the cut-in angle reduces. This means reduced distortion.

(c) If $k > 1$, it means bias current I_B will be flowing even before signal is being applied. This will cause a collector current $I_C = \beta I_B$ which does not flow through R_L, so $v_o = 0$.

When v_i is applied, i_c the collector current caused by signal flows in the appropriate transistor. Output voltage is now given by

$$v_o = \sin \omega t$$

with each half contributed by one transistor at a time.

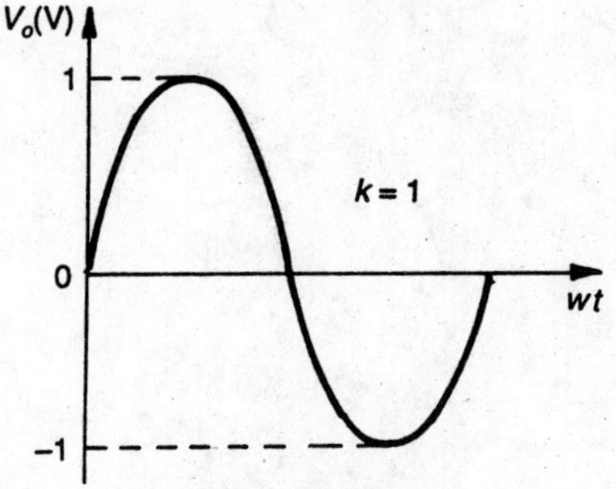

Fig. P. 13.10 (c)

(d) $k\ V_\gamma \leq V_\gamma$ = conduction will begin only when signal is applied.

$$V_s \sin \omega t - (1 - k)\ V_\gamma \geq 0 \qquad \text{...(iii)}$$

So the cut-in angles remain the same as calculated in part (a), When the above condition is met

$$i_b = \frac{V_s \sin \omega t - (1-k) V_\gamma}{(1+\beta) R} \qquad \text{...(iv)}$$

because current through R in the emitter circuit current is $(1 + \beta)i_b$. Because of symmetry i_b does not flow through R_L.

$$i_c = \beta i_b \qquad \text{...(v)}$$

Using Eqs. (iv) and (v), we can express the output voltage as

$$v_o = i_c R_L = \left(\frac{\beta}{1+\beta}\right)\left(\frac{R_L}{R}\right) [V_s \sin \omega t - (1 - k)V_\gamma]$$

$$\approx \left(\frac{R_L}{R}\right) [V_s \sin \omega t - (1 - k)\ V_\gamma];\ (1 + \beta) \approx \beta$$

Thus v_o will have the same wave form as in Figs. 13.10 (a), (b) and (c) except being scaled by (R_L/R).

(e) Class of operation

Part (a)

(i) $k = 0$ → push-pull class-C

(ii) $k = 0.5$ → push-pull class-C

(iii) $k = 1$ → push-pull class-B

IES Examination Papers with Solutions

Problem 1 : Give the circuit of a two level clipper with clipping levels at – 3 V and 5 V. Draw the output wavefrom when the input is sinusoidal of 5 V_{rms} value. Assume the device to be ideal. IES-92

Solution :

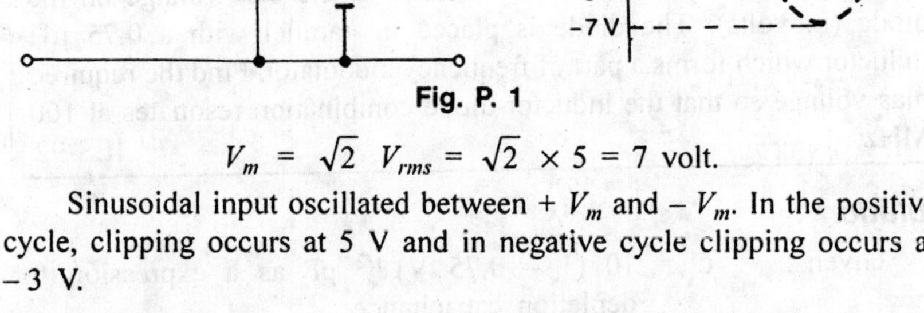

Fig. P. 1

$$V_m = \sqrt{2}\ V_{rms} = \sqrt{2} \times 5 = 7 \text{ volt.}$$

Sinusoidal input oscillated between + V_m and – V_m. In the positive cycle, clipping occurs at 5 V and in negative cycle clipping occurs at – 3 V.

Problem 2 : The leakage currents of transister, with usual notations, are I_{CEO} = 410 mA and I_{CO} = 5 μA, the base current I_B being 30 μA. Calculate the collector current. IES-92

Solution :

In the transistor,

$$I_E + I_C + I_B = 0$$

and $\quad I_C = I_{CO} - \alpha I_E$

or $\quad I_C = I_{CO} + \alpha (I_C + I_B)$

or, $\quad I_C = \dfrac{I_{CO}}{1-\alpha} + \dfrac{\alpha I_B}{1-\alpha}$

For $\quad I_B = 0,$

$$I_C = -I_E = \frac{I_{CO}}{1-\alpha} = I_{CEO}$$

or $\quad \dfrac{I_{CEO}}{I_{CO}} = \beta + 1$

or, $\quad \beta = 81$

Then collector unit,

$$\begin{aligned} I_C &= I_{CEO} + \beta I_B \\ &= (\beta + 1)\, I_{CO} + \beta I_B \\ &= (410 + 81 \times 30)\ \mu\text{A} \\ &= 2840\ \mu\text{A} \end{aligned}$$

Problem 3 : A varactor diode has a depletion capacitance C given by $C = 10\,(1 - 0.75\ V)^{-1/2}$ pF where V is the bias voltage on the diode (in volts). The diode is placed in parallel with a 0.75 μH inductor which forms a part of frequency modulator. Find the required bias voltage so that the inductor-diode combination resonates at 100 MHz.

IES-92

Solution :

Given : $\quad C = 10\,(1 - 0.75\ V)^{-1/2}$ pF as a expression for depletion capacitance.

For resonance at 100 MHz, turning capacitor would be,

$$C = \frac{1}{4\pi^2 \times 0.75 \times 10^{-6} \times \left(10^8\right)^2}\ \text{F}$$

$$= \frac{1}{4\pi^2 \times 0.75 \times 10^{10}} \text{ F} = 3.37 \text{ pF}$$

Substituting the value in the expression given for depletion capacitance

$$3.337 = \frac{10}{\sqrt{(1 - 0.75\text{ V})}}$$

$$\text{or } 1 - 0.75\,V = \left(\frac{10}{3.377}\right)^2$$

or Thus $V = -10.35$ volts is required bias voltage.

Problem 4 : Analyse the effect of the emitter bypass capacitor on the low-frequency of an *R-C* coupled amplifier. State the assumption you make. Use hybrid parameters. Also given an asymptotic Bode plot of the magnitude of response giving important values when the emitter resistor is 1 K, emitter bypass capacitor is 200 μF, collector load resistor is 2 K and the transistor parameters are $h_{ie} = 1.1$ K and $h_{fe} = 50$. Neglect the effect of other parameters. IES-92

Solution :

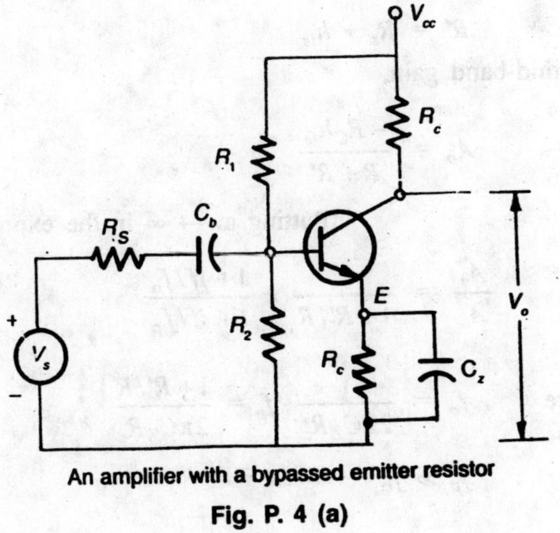

An amplifier with a bypassed emitter resistor

Fig. P. 4 (a)

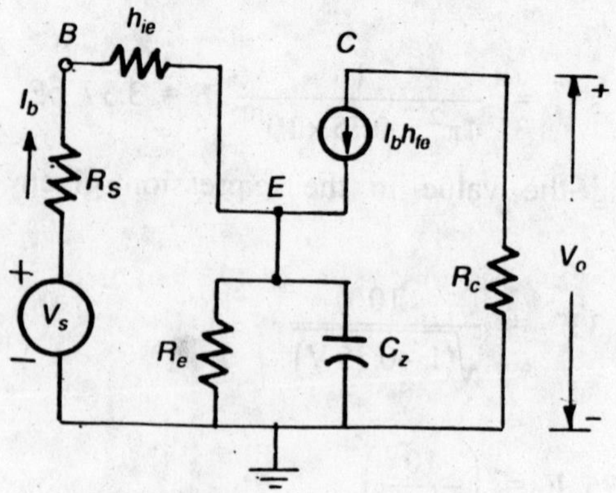

Simplified *h*-parameter mould

Fig. P. 4 (b)

From Fig. P. 4 (b),

$$V_o = I_b \times h_{fe} R_C = \frac{-V_s h_{fe} R_C}{R_s + h_{ie} + Z'_e}$$

where, $$Z'_e = (1 + h_{fe}) \frac{R_e}{1 + j\omega C_Z R_e}$$

then, $$A_v = \frac{V_o}{V_s} = \frac{-R_C h_{fe}}{R + R'} \times \frac{1 + j\omega C_Z R_e}{1 + j\omega C_Z \left[R_e R / (R + R')\right]}$$

$$R' = R_s + h_{ie}$$

The mid-band gain,

$$A_o = \frac{-R_C h_{fe}}{R + R'}$$

[Putting $\omega \to \infty$ in the expression of A_v]

Now, $$\frac{A_v}{A_o} = \frac{1}{1 + R'/R} \times \frac{1 + jf/f_o}{1 + jf/f_p}$$

Where $$f_o = \frac{1}{2\pi C_Z R_e},\ f_p = \frac{1 + R'/R}{2\pi C_Z R_e}$$

For $f_p \gg f_o$

and at $\qquad f = f_p,$

$$\left|\frac{A_v}{A_o}\right| = \frac{1}{1+R'/R} \times \frac{f_p/f_o}{\sqrt{1+1}} = \frac{1}{\sqrt{2}}$$

Thus low 3-*dB* frequency f_2 is approximately equal to f_p. If $f_p \neq f_o$ then $f_L \neq f_p$ and 3-*dB* frequency may not exist. In this way bypass capacitor affects low frequency response of the amplifier.

> **Problem 5 :** Draw the circuit diagram of a Schmitt Trigger using transistors and explain its working with reference to the transfer characteristics for various values of loop gain. IES-93

Solution :

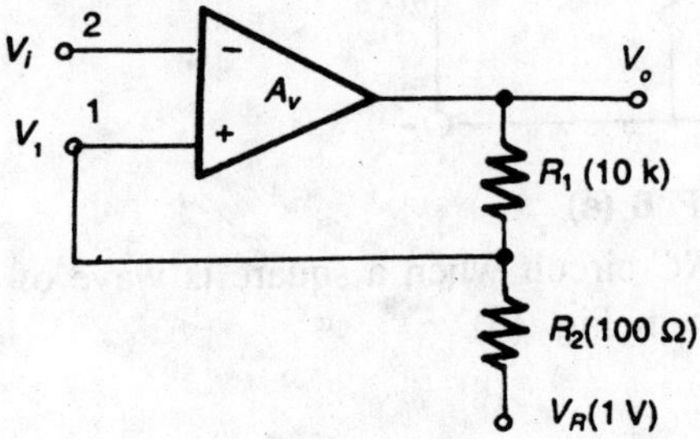

Fig. P. 5 (a)

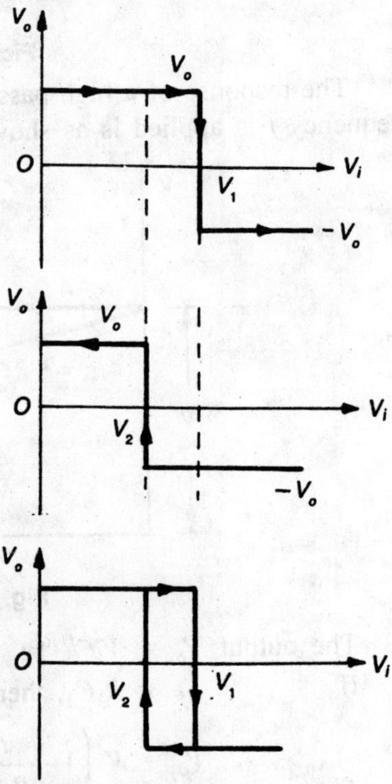

Fig. P. 5 (b)

From Fig. P. 5 (a),

$$\frac{V_1 - V_R}{R_2} = \frac{V_1 - V_o}{R_1}$$

or, $$V_1 = \frac{R_1 V_R}{R_1 + R_2} + \frac{R_2 V_o}{R_1 + R_2}$$

$$\equiv V_1$$

For, $V_i < V_1$, $V_o = + V_o$ (+ 5 V).

For $V_i > V_1$,

The voltage at the non-inverting terminal is,

$$V_1 = \frac{R_1 V_R}{R_1 + R_2} - \frac{R_2 V_o}{R_1 + R_2} = V_2$$

The hysteresis value

$$V_H = V_1 - V_2 = \frac{2R_2V_o}{R_1 + R_2} = 0.10 \text{ Volt.}$$

Because of the hysterisis, the circuit triggers at a higher voltage for increasing than for decreasing signals.

Problem 6 : Sketch the response of a high pass *RC* circuit when a square wave of frequency *f* is applied at its input. Derive an expression for the percentage tilt. IES-93

Solution :

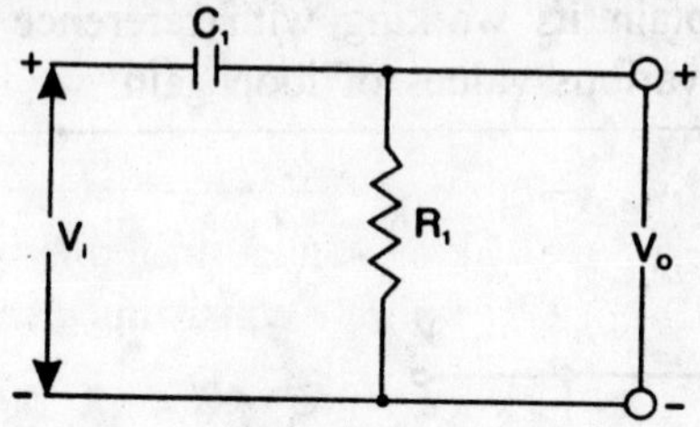

Fig. P. 6 (a)

The response of a high pass *RC* circuit when a square is wave of frequency *f* is applied is as shown in Fig. (b).

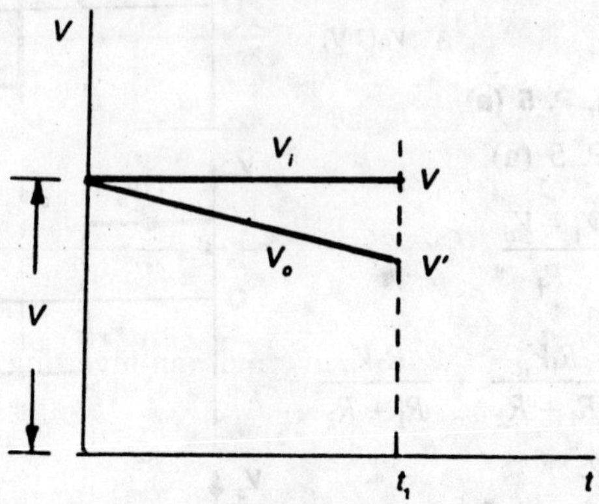

Fig. P. 6 (b)

The output $V_o = Ve^{-t/R_1C_1}$

If $t \ll R_1C_1$, then,

$$V_o = V\left(1 - \frac{t}{R_1C_1}\right)$$

Then, % tilt or sag $= \dfrac{V - V'}{V} \times 100 = \dfrac{t_1}{R_1C_1} \times 100\%$

If $t_1 = \dfrac{T}{2}$ then the wave input would be symmetrical, and

$$P = \frac{T}{2R_1C_1} \times 100$$

$$= \frac{1}{2fR_1C_1} \times 100 = \frac{\pi f_L}{f} \times 100\%$$

It is important to be noted that $P \alpha f_L$ (lower 3-dB frequency).

Problem 7 (a) : Find the output voltage in terms of V_1, V_2, R_1 and R_2 for the differential input amplifier consisting of a base amplifier of infinte gain as shown in the figure. IES-93

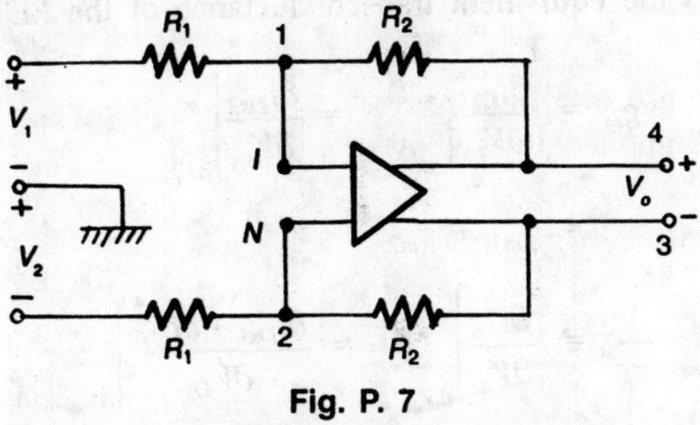

Fig. P. 7

Solution :

Let the voltages at the inverting and non-inverting terminal are V_I and V_N. Then,

$$V_I = (V_4 - V_1)\frac{R_1}{R_1 + R_2} + V_1 \quad \text{...(A)}$$

and $$V_N = (V_3 - V_2)\frac{R_2}{R_1 + R_2} + V_2 \quad \text{...(B)}$$

As $I_i = 0$,

$V_I = V_N$, then equating (A) and (B), we get

$$V_4 - V_3 = R_2/R_1\,(V_2 - V_1)$$

or, $$V_o = \frac{R_2}{R_1}(V_2 - V_1),$$

As $$V_o = V_4 - V_3$$

Problem 7 (b) : Show that if two identical FETs are connected in parallel, then g_m is double and r_d is half of individual FETs. If two FETs are not identical, show that :

$$\frac{1}{r_d} = \frac{1}{r_{d1}} + \frac{1}{r_{d2}}$$

and $$\mu = \frac{\mu_1 r_{d2} + \mu_2 r_{d1}}{r_{d1} + r_{d2}}$$ IES-93

Solution :

Let g_m is the equivalent transconductance of the FET's.

Then $$g_m = \left.\frac{\partial i_{Ds1}}{\partial V_{gs}}\right|_{\Delta V_{Ds=0}} = \left.\frac{\partial i_{Ds2}}{\partial V_{gs}}\right|_{\Delta V_{Ds=0}}$$

$$= g_{m1} + g_{m2}$$

Also, $$\frac{1}{r_d} = \left.\frac{\partial i_{Ds}}{\partial V_{gs}}\right|_{\Delta V_{gs=0}} = \left.\frac{\partial i_{Ds1} + \partial i_{Ds2}}{\partial V_{Ds}}\right|_{\Delta V_{gs=0}}$$

$$= \left.\frac{\partial i_{Ds1}}{\partial V_{Ds}}\right|_{\Delta V_{gs=0}} + \left.\frac{\partial i_{D2}}{\partial V_{Ds}}\right|_{\Delta V_{gs=0}}$$

$$= \frac{1}{r_{d1}} + \frac{1}{r_{d2}}$$

Now, $$\mu = g_m r_d$$

$$= (g_{m1} + g_{m2}) \times \frac{r_{d1} r_{d2}}{r_{d1} + r_{d2}}$$

$$\mu = \frac{\mu_1 r_{d2} + \mu_2 r_{d1}}{r_{d1} + r_{d2}}$$

Problem 8 : A transistor's maximum average dissipation capability is 2.2 W. When used in a class A direct coupled power amplifier, find (i) the maximum a.c. power deliverable to the load, and (ii) the maximum d.c. power needed from the power supply to provide the power in part (i). IES-94

Solution:

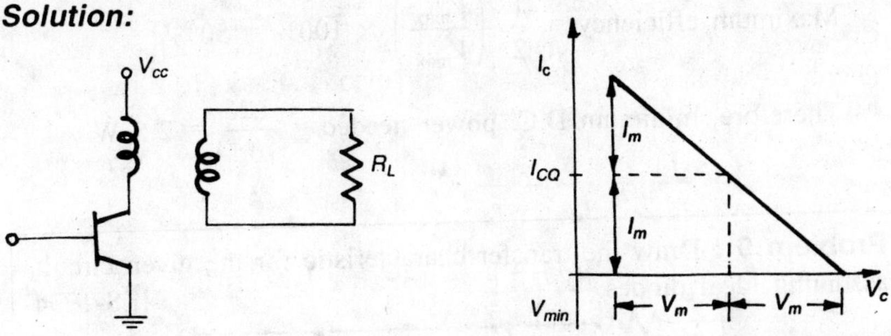

Fig. P. 8 (a) & (b)

Maximum ac power deliverable to the load,

$$P_{D\ max} = V_c I_c = (V_{CC})\ (I_{CE}) = 2.5\ \text{W}$$

(i) $$(P_{ac})_{max} = \frac{1}{2} V_m I_m = \frac{1}{2}\ (V_{CE})\ (I_{CE})$$

$$= \frac{1}{2}\ (2.5) = 1.25\ \text{W}$$

$$\eta = \frac{\text{Signal power supplied to the load}}{\text{D.C. power supplied (input)}} = V_o I_c$$

$$= \frac{\frac{1}{2} V_m I_m}{V_{cc} I_c} \times 100\%$$

$$= \frac{1}{2}\left(\frac{V_m}{V_{cc}}\right) \times 100\%\ (\text{Since, } I_m = I_c) \qquad \ldots(A)$$

Also $$V_{cc} = \frac{(V_{max} + V_{min})}{2} \qquad \ldots(B)$$

Substituting equation (B) in (A) we get,

$$\eta = \frac{\dfrac{(V_{max} - V_{min})}{2}}{\dfrac{(V_{max} + V_{min})}{2}} \times 100\%$$

Assuming $V_{min} = 0$

$$\text{Maximum efficiency} = \frac{1}{2}\left(\frac{V_{max}}{V_{max}}\right) \times 100\% = 50\%$$

$$\text{Therefore, minimum D.C. power needed} = \frac{1.25}{0.5} = 2.5 \text{ W.}$$

Problem 9 : Draw the transfer characteristic for the given circuit, assuming ideal diodes.

IES-1994

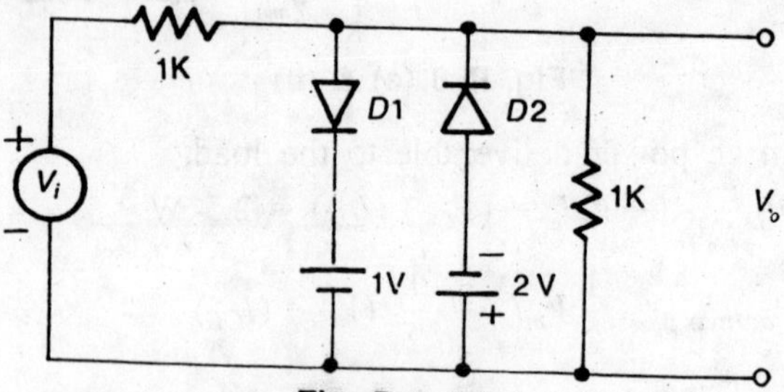

Fig. P. 9 (a)

Solution :

Case I :

When $V_i > 1$ V, D_1 ON D_2 OFF

and, $V_o = 1$ Volt

Case II :

When $V_i < 2$V, D_1 OFF D_2 ON

and, $V_o = 2$ Volt

Case III :

For $-2V < V_i < 1$ V

$$V_o = \frac{V_i}{1+1} \times 1$$

$$V_o = \frac{V_i}{2}$$

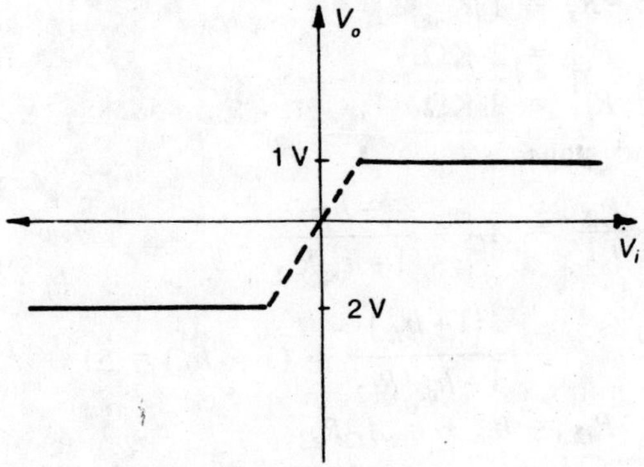

Fig. P. 9 (b)

Problem 10 : For the casacaded amplifier shown, find the input impedance, output impedance, current gain and the voltage gain. Make any reasonable assumption. The h-parameters are :

$$h_{ie} = 1.1 \text{ k}\Omega$$
$$h_{fe} = 50 \text{ (Neglect } h_{re}, h_{oe})$$

IES-94

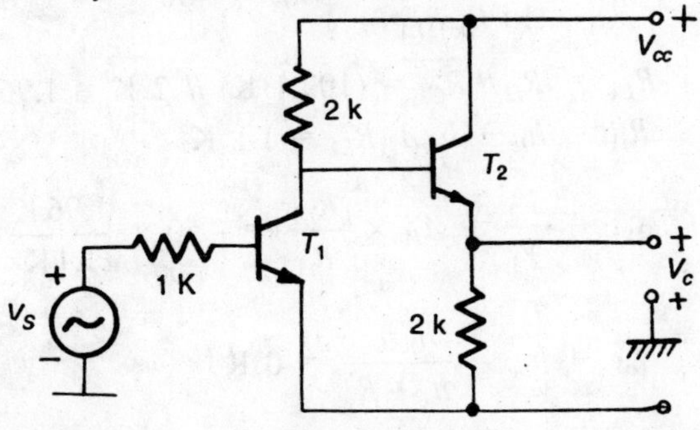

Fig. P. 10

Solution : Given :

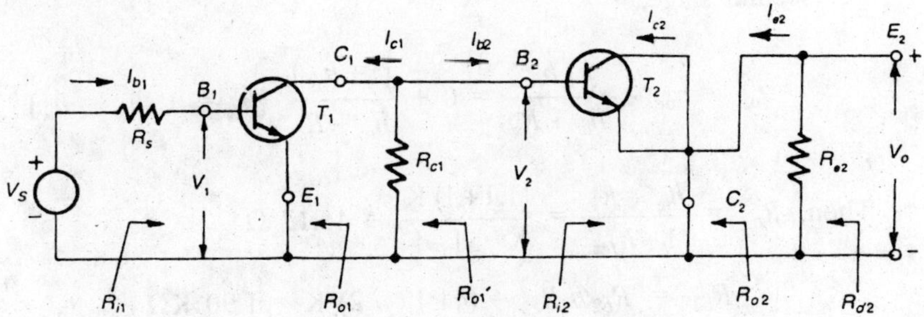

Fig. P. 10 (a)

$$R_s = 1\ \text{K}$$
$$R_{e2} = 2\ \text{K}\Omega$$
$$R_{c1} = 2\ \text{K}\Omega$$

The second stage:

$$-\frac{I_{e2}}{I_{b2}} = A_{I2} = \frac{-h_{fc}}{1 + h_{oc} R_{e2}}$$

$$= \frac{(1 + h_{fe})}{1 + h_{oe}\ R_{e2}} \approx (1 + h_{fe}) = 51 \qquad [h_{oe} \approx 0]$$

and, $$R_{i2} = h_{ic} + h_{rc} A_{I2} R_{e2}$$
$$= h_{ie} + A_{I2} R_{e2}\ (h_{ic} = h_{ie} \text{ and } h_{rc} = 1)$$
$$= 1.1 + 51 \times 2 = 103.1\ \text{K}$$

$$A_{v2} = \frac{V_o}{V_2} = A_{I2} \times \frac{R_{e2}}{R_{i2}} = 51 \times \frac{2\ \text{K}}{(103.1\ \text{K})} = 0.989$$

The first stage:

$$A_{I1} = \frac{-h_{fe}}{1 + h_{oe} R_{L1}} = -h_{fe} = -50$$

Here, $$R_{L1} = R_{i2} // R_{C1} = (103.1\ \text{K}) // 2\ \text{K} = 1.96\ \text{K}$$
$$R_{i1} = h_{ie} + h_{re} A_{I1} R_{L1} = 1.1\ \text{K}$$

$$A_{v1} = \frac{V_2}{V_1} = A_{I1} \times \frac{R_{L1}}{R_{i1}} = -50 \times \frac{1.96\ \text{K}}{1.1\ \text{K}} = -89.09$$

$$\gamma_{o1} = h_{oe} - \frac{h_{fe} h_{re}}{h_{ie} + R_s} = 0\ \text{K}$$

and, $$R_{o1} = \infty\ \text{K}$$

Then, $$R_{o1'} = R_{C1} = 2\ \text{K}$$

For second stage,

$$\gamma_{o2} = h_{oe} - \frac{h_{fc} h_{rc}}{h_{ic} + R_s} = 0 + \frac{(1 + h_{fe})}{h_{ie} + R_s} \quad [h_{rc} = 1,\ h_{ic} = h_{ie}]$$

Then, $$R_{o2} = \frac{h_{ie} + R_s}{1 + h_{fe}} = \frac{(1.1 + 1)\ \text{K}}{51} = 41.17\ \Omega$$

$$R_{o2'} = R_{o2} // R_{e2} = (41.17 // 2)\ \text{K} = 1.90\ \text{K}\Omega$$

Also, $$I_{b2} = -I_{c1} \times \frac{R_{c1}}{R_{c1} + R_{i2}}$$

$$-\frac{I_{b2}}{I_{c1}} = \frac{2}{2+103.1} = \frac{2}{105.1}$$

Now, $$A_v = A_{v1} \times A_{v2}$$
$$= (+0.989)(-89.09) = -85.43$$

and $$A_I = A_{I2} \times A_{I2} \times \frac{2}{(105.1)}$$

$$= (-50)(51)\left(\frac{2}{105.1}\right) = -48.52$$

Also, $$A_{vs} = A_v \frac{R_{i1}}{R_{i1} + R_s} = (-85.43) \times \frac{1.1}{2.1} = -44.74$$

Problem 11 : Draw the circuit of an *RC* phase shift oscillator using an FET. Derive :

(*i*) An expression for the frequency of oscillations and

(*ii*) the condition for sustained oscillastions IES-1994

Solution :

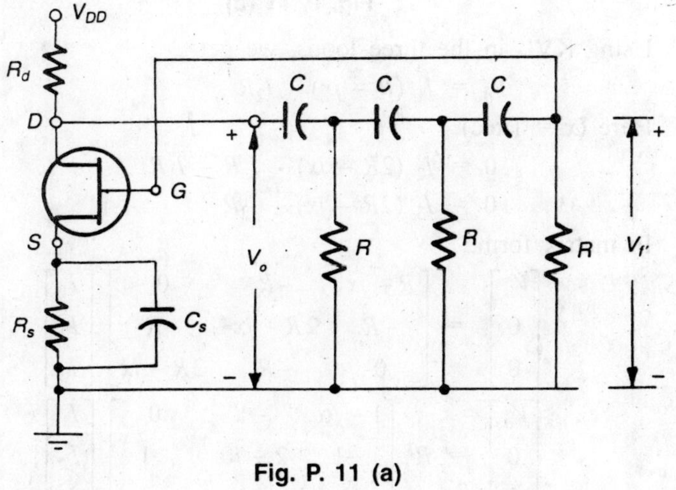

Fig. P. 11 (a)

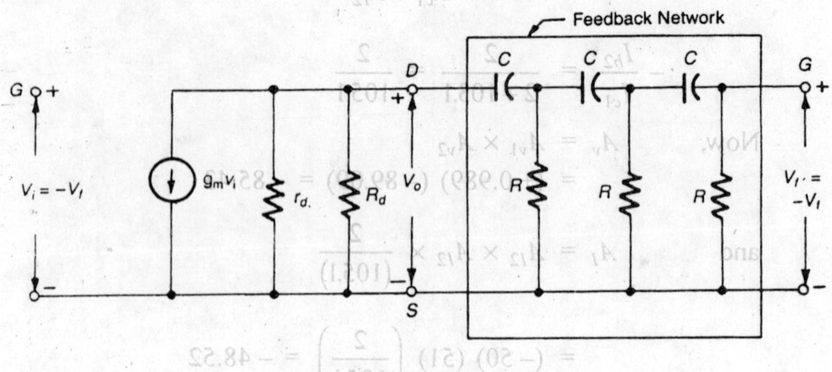

Fig. P. 11 (b)

Considering only the feedback network

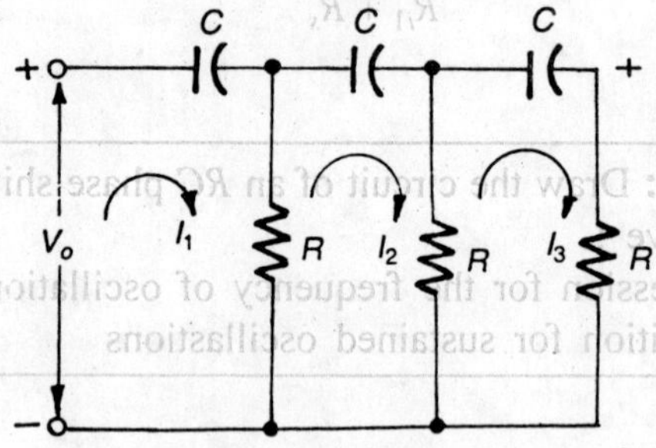

Fig. P. 11 (c)

Using KVL in the three loops, we get

$$V_o = I_1 (R - jx) - I_2 R$$

Here ($x = 1/\omega c$)

$$0 = I_2 (2R - jx) - I_3 R - I_1 R$$

$$0 = I_3 (2R - jx) - I_2 R$$

In matrix form,

$$\begin{bmatrix} V_o \\ 0 \\ 0 \end{bmatrix} = \begin{bmatrix} R - jx & -R & 0 \\ -R & 2R - jx & -R \\ 0 & -R & 2R - jx \end{bmatrix} \begin{bmatrix} I_1 \\ I_2 \\ I_3 \end{bmatrix}$$

$$\begin{bmatrix} V_o \\ 0 \\ 0 \end{bmatrix} = R^3 \begin{bmatrix} 1 - j\alpha & -1 & 0 \\ -1 & 2 - j\alpha & -1 \\ 0 & -1 & 2 - 1\alpha \end{bmatrix} \begin{bmatrix} I_1 \\ I_2 \\ I_3 \end{bmatrix},$$

$$\alpha = x/R = \frac{1}{\omega CR}$$

Now, $$-\beta = \frac{V_f'}{V_o} = \frac{I_3 R}{V_o}$$

$$I_3 = \frac{\Delta_3}{\Delta} = \left(\frac{1}{R}\right) \frac{\begin{bmatrix} 1-j\alpha & -1 & V_o \\ -1 & 2-j\alpha & 0 \\ 0 & -1 & 0 \end{bmatrix}}{\begin{bmatrix} 1-j\alpha & -1 & 0 \\ -1 & 2-j\alpha & -1 \\ 0 & -1 & 2-j\alpha \end{bmatrix}}$$

then, $$-\beta = \frac{1}{1-\alpha^2 + j(\alpha^3 - 6\alpha)}$$

For 180° phase-shift,

$$\alpha^3 - 6\,\alpha = 0$$

$$\alpha = \sqrt{6}$$

or, $$\frac{1}{\omega CR} = \sqrt{6}$$

or, $$f = \frac{1}{2\pi RC\sqrt{6}}$$

(ii) At this frequency

$$\beta = +\frac{1}{29}$$

For oscillation $|\beta A| > |$

or, $A > 29$ is the required condition

Problem 12 : Draw the circuit of an stable multivibrat for using *OP AMP*(s) and explain its working with the help of waveforms. Derive an expression for frequency of oscillations. IES-94

Solution :

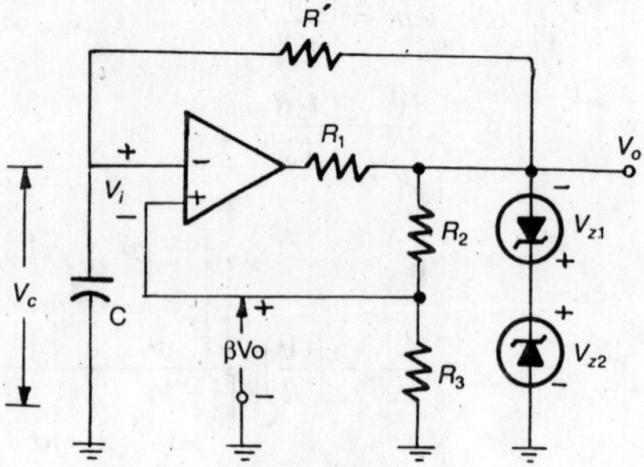

A stable generator

Fig. P. 12 (a)

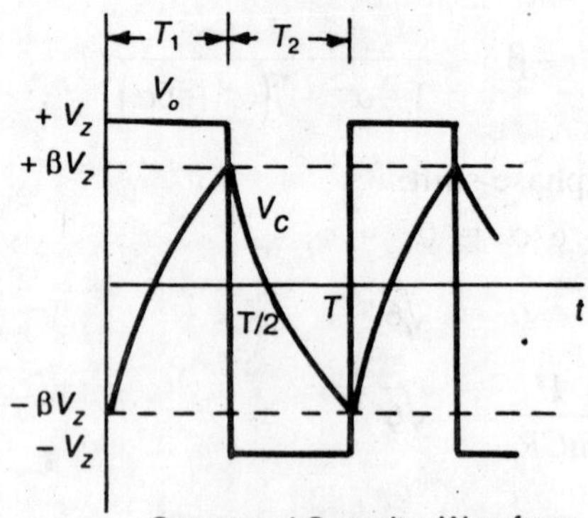

Output and Capacitor Waveform

Fig. P. 12 (b)

A fraction β, of the V_o given by $\beta = \dfrac{R_3}{R_2 + R_3}$ is fed back to the non-inverting terminal of the *OP–AMP*. The differential input

$$V_i = V_c - \beta V_o.$$

If $\quad V_i > 0$ Then $V_o = -V_{21}$

and if $\quad V_i < 0,$ then $V_o = +V_{22}$

Consider when $V_i < 0$ or, $V_c < \beta V_0 = \beta V_{Z2}$, then capacitor C charges exponentially toward $+V_{Z2}$ through $R'C$ combination. The output remains constant at $+V_{Z2}$ untill V_c equal βV_{Z2} at which time the comparator output reverses to $-V_{Z1}$. Now V_c changes towards $-V_{Z1}$. The capacitor voltage, is given by

$$V_c(t) = V_Z[1 + (1 + \beta)\, e^{-t/R'C}]$$

at $\quad t = T/2,$

$$V_c(t) = +\beta V_Z$$

Then, $\quad \beta V_Z = V_Z[1 - (1 + \beta)\, e^{-t/R'C}]$

or, $\quad e^{-t/R'C} = \dfrac{1-\beta}{1+\beta}$

or, $\quad e^{-T/2R'C} = \dfrac{1-\beta}{1+\beta}$

or, $\quad T = 2\, R'C \ln \dfrac{1+\beta}{1-\beta}$

and, $\quad f = \dfrac{1}{T} = \dfrac{1}{2R'C\ln\left(\dfrac{1+\beta}{1-\beta}\right)}$

Problem 13 : Draw the circuit of a When Bridge Oscillator using a single operational amplifier. Derive the condition for sustained oscillations. IES-95

Solution :

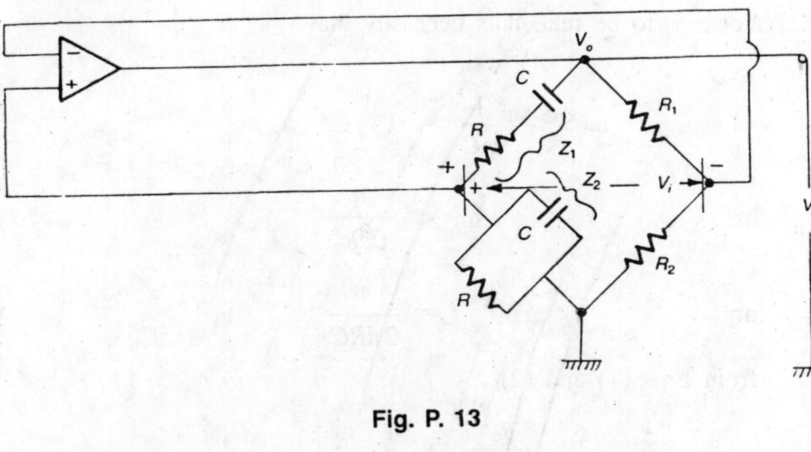

Fig. P. 13

In the circuit of Fig. (a)

$$V_i = \frac{V_o}{Z_1 + Z_2} \times Z_2 - \frac{V_o}{R_1 + R_2} R_2 \qquad \ldots(1)$$

As, $\quad Z_1 = R + 1/(j\omega C)$

and, $$Z_2 = \frac{R \times 1/(j\omega C)}{R + 1/(j\omega C)}$$

$$= \frac{R}{1 + j\omega CR}$$

then, $$\frac{Z_2}{Z_1 + Z_2} = \frac{R/(1 + j\omega CR)}{R + 1/(j\omega C) + R/(1 + j\omega CR)}$$

$$= \frac{R}{\left(R + \frac{1}{j\omega C}\right)(1 + j\omega CR) + R}$$

$$= \frac{R}{R + \frac{1}{j\omega C} + j\omega CR^2 + R + R}$$

$$= \frac{R}{3R + j\left[\omega CR^2 - 1/(\omega C)\right]} \quad \text{...(ii)}$$

For $\frac{V_o}{V_i}$ to be real, it is necessity that

$$\omega CR^2 - \frac{1}{\omega C} = 0$$

or, $$\omega^2 = \frac{1}{C^2 R^2}$$

or $$f = \frac{1}{2\pi RC}$$

from Eqs. (1) and (2)

$$\frac{R_2}{R_1 + R_2} = \frac{1}{3}$$

Hence $$R_1 = 2R_2$$

This is the necessary condition for sustained oscillations.

Problem 14 : Why is desirable to incorporate current limiting circuitry in a voltage regulator? Draw the circuit of a typical series pass regulator with built-in overload current limiting feature and describe its functioning. IES-95

Solution :

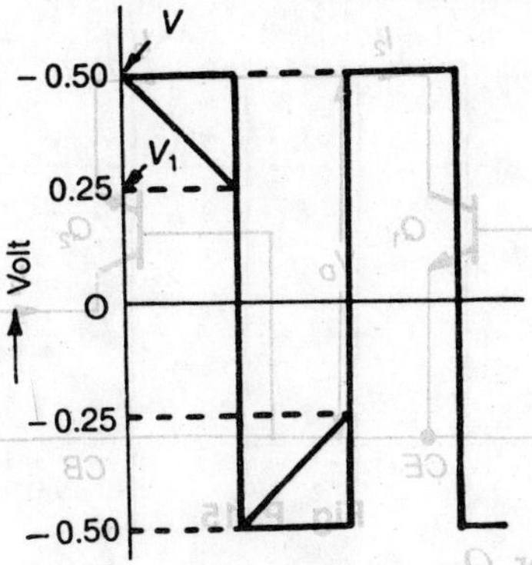

Fig. P. 14

(a) Continuous variation of frequency is accomplished by varying simultaneously the two capacitors.

(b) Changes in frequency range are accompanied by switching in different values for the two identical resistors R.

(c) Thermistors having negative temperature coefficient should be used in place of R_1 and R_2 for amplitude stabilization.

Tilt for the square wave is expressed as

$$P = \frac{V - V'}{V} \times 100 = \frac{T}{2RC} \times 100$$

$$= \frac{0.1}{0.2} \times 100 = 50\%$$

or, $$\frac{V - V'}{V} = \frac{1}{2}$$

or $$2V = 2V' = V$$

or, $$V = 2V'$$

Thus, $$V' = \frac{V}{2} = \frac{0.5}{2} = 0.25 \text{ volt}$$

Now, maximum steady state value = 0.5 – 0.25 = 0.25 V

and minimum steady state value = – 0.5 + 0.25 = – 0.25 V

Problem 15 : Draw the circuit of transistorized cascade amplifier and determine its *h*-parameters in terms of those of the component transistors. What are the special advantages of this configuration? IES-95

Solution :

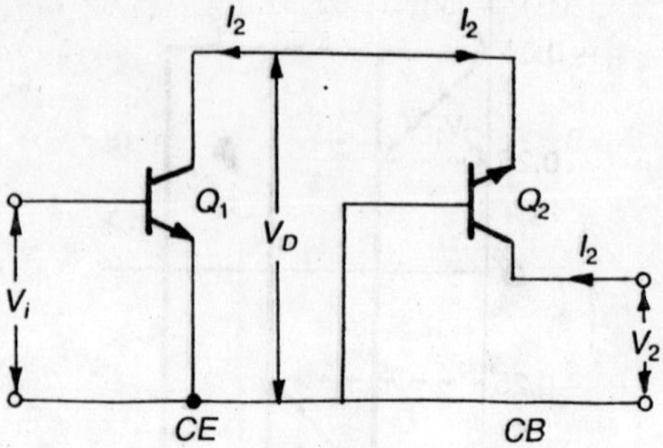

Fig. P. 15

For *CE* transistor Q_1,

$$V_i = h_{11}I_1 + h_{12} \times V_o \qquad ...(1)$$

$$I_2 = h_{21} \times I_1 + h_{22}V_o \qquad ...(2)$$

For *CB* transistor Q_2,

$$V_o = h'_{11}I'_2 + h'_{12} \times V_2 \qquad ...(3)$$

$$I''_2 = h'_{21} \times I'_2 + h'_{22} \times V_2 \qquad ...(4)$$

But $\quad I'_2 = -I_2$

Hence $\quad V_o = -h'_{11}I_2 + h'_{12} \times V_2 \qquad ...(5)$

$$I''_2 = -h_{21} \times I_2 + h'_{22} \times V_2 \qquad ...(6)$$

Form Eqs. (2) and (5),

$$I_2 = h_{21} \times I_1 + h_{22} \times (-h'_{11} \times I_2 + h'_{12} \times V_2)$$

$$= h_{21} \times I_1 - h'_{11} \times h_{22} \times I_2 + h'_{12} \times h_{22} \times V_2$$

Hence $\quad I_2 = \dfrac{h_{21} \times I_1 + h'_{12} \times h_{22} \times V_2}{1 + h'_{11} \times h_{22}} \qquad ...(7)$

Form Eqs. (1), (5) and (7),

$$V_i = h_{11} \times I_1 - h'_{11} \times h_{12} \times I_2 + h_{12} \times h'_{12} \times V_2$$

$$= h_{11} \times I_1 - h_{12} \times h'_{11}\left[\frac{h_{21} \times I_1 + h'_{12} \times h_{22} \times V_2}{1 + h'_{11} \times h_{22}}\right] + h_{12} \times h'_{12} \times V_2$$

$$= h_{11}\left[I_1 + \frac{h_{12} \times h'_{11} \times h_{21}}{1 + h'_{11} \times h_{22}}\right] + V_2\left[\frac{h'_{12} \times h_{22}}{1 + h'_{11} \times h_{22}} + h_{12} \times h'_{12}\right] \qquad ...(8)$$

From Eqs. (6) and (7),

$$I_2'' = -\frac{h_{21}' \times h_{21}}{1+h_{11}' \times h_{22}} I_1 + \left[\frac{h_{12}' \times h_{22}}{1+h_{11}' \times h_{22}} + h_{22}'\right] V_2 \qquad \ldots(9)$$

From Eqs. (8) and (9), we get the values of the h parameters of the cascade amplifier as below :

$$h_{11}'' = h_{11}\left[I_1 - \frac{h_{12} \times h_{11}' h_{21}}{1+h_{11}' \times h_{22}}\right]$$

$$h_{12}'' = \frac{h_{12}' \times h_{22}}{1+h_{11}' \times h_{22}} + h_{12} \times h_{12}'$$

$$h_{21}'' = \frac{h_{21}' \times h_{21}}{1+h_{11}' \times h_{22}}$$

and

$$h_{22}'' = \frac{h_{12}' \times h_{22}}{1+h_{11}' \times h_{22}} + h_{22}'$$

Problem 16 : In the transistor circuit shown below $I_{CBO} = 2\mu$ Amp at 25°C and doubles for every 10°C increase in temperature.

(i) Find maximum allowable value of R_B if the transistor is to remain cut off at 75°C. Assume $V_{BE\,(\text{cut off})} = -0.1$ V.

(ii) If $V_{BB} = 1.0$ V and $R_B = 50$ KΩ, how high may the temperature increase before the transistor comes out of cut off? IES-97

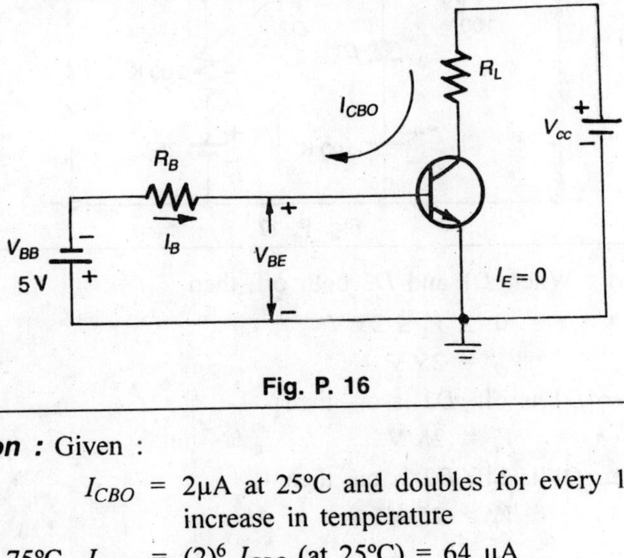

Fig. P. 16

Solution : Given :

$$I_{CBO} = 2\mu\text{A at 25°C and doubles for every 10°C increase in temperature}$$

At 75°C $\quad I_{CBO} = (2)^6\, I_{CBO}\text{ (at 25°C)} = 64\ \mu\text{A}$

At cut off $I_B = -I_{CBO}$ and $V_{BE} = -0.1$ V

Then $V_{BE} = -V_{BB} + R_B I_{CBO} \le 0.1$ V (For transistor to be cutoff)

or $-0.10 = -5 + R_B \times 64 \times 10^{-6}$

$$\therefore \quad R_B = \frac{4.9}{64 \times 10^{-6}} = 76.5 \text{ V}$$

Again $V_{BE} = -V_{BB} + R_B \times I_{CBO}$

(using KVL in B–E loop, we have)

or $-0.1 = -1 + 50 \times 10^3 \times I_{CBO}$

or $0.9 = 50 \times 10^3 \times I_{CBO}$

$$\therefore \text{ Then, } I_{CBO} = \frac{0.9}{50 \times 10^3} = 1.8 \times 10^{-5} = 18 \ \mu\text{A}$$

For temperature rise we have,

$$18 = 2 \times 2^{(T-25)/10}$$

$$\text{or} \quad \frac{T-25}{20} = \log_2 9 \approx 3.1$$

$$\therefore \quad T = 87°\text{C}$$

Problem 17 : The input voltage V_i to the two-level clipper shown below varies linearly from 0 to 150 V. Sketch the output voltage to the same scale as the input voltage. Assume ideal diodes. IES-97

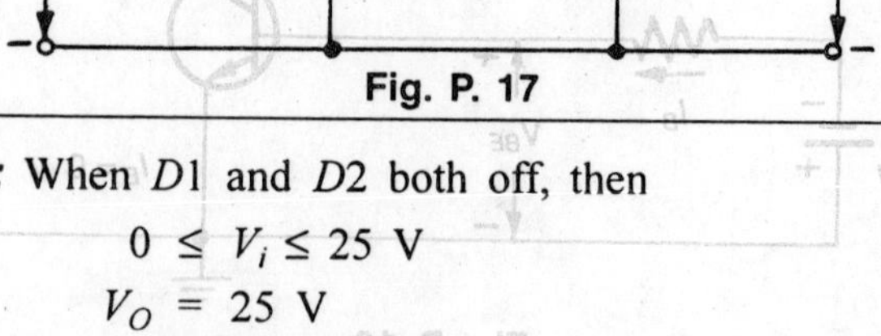

Fig. P. 17

***Solution* :** When $D1$ and $D2$ both off, then

$$0 \le V_i \le 25 \text{ V}$$

$$\therefore \quad V_O = 25 \text{ V}$$

When $D1$ is off, $D2$ is on, then

$$V_i = 25 \text{ V}$$

When $D1$ is off, $D2$ is on, then

$$V_i > 25 \text{ V}$$

$$\therefore \quad V_o = \frac{200}{300}(V_i - 25) + 25$$

For $D1$ to be to go ON, we have

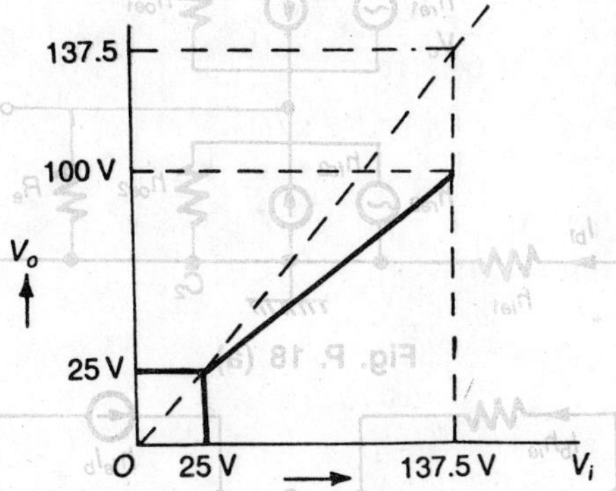

Fig. P. 17 (a)

$$V_o = 100 \text{ V}$$

$$\therefore \text{ and,} \quad 100 = \frac{2}{3}(V_i - 25) + 25$$

$$\text{Thus,} \quad V_i = \frac{75 \times 3}{2} + 25 = 137.5 \text{ V}$$

Now, $D1$ is ON and $D2$ is also ON, for $V_i > 137.5$

$$\therefore \Rightarrow \quad V_o = 100 \text{ V}$$

Problem 18 : Derive the expression for the voltage gain A_v and the resistance R_{in} of the amplifier shown. FInd the values of A_v and R_{in} for the following values of h parameters for the transistors :

$h_{ie} = 1000$ W;

$h_{fe} = 100$;

$h_{re} = h_{oe} = 0$

IES-97

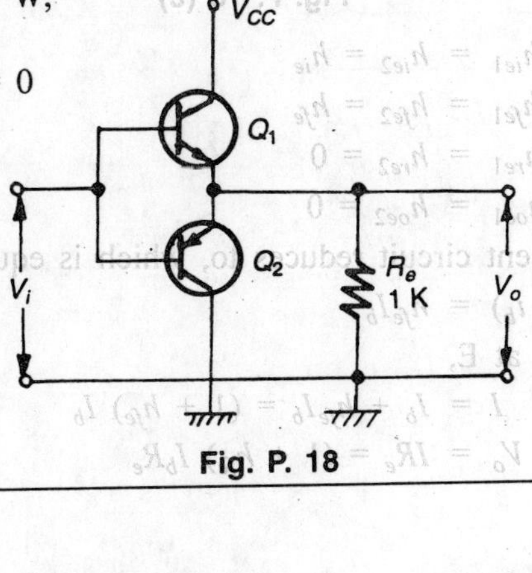

Fig. P. 18

Solution : Equivalent circuit

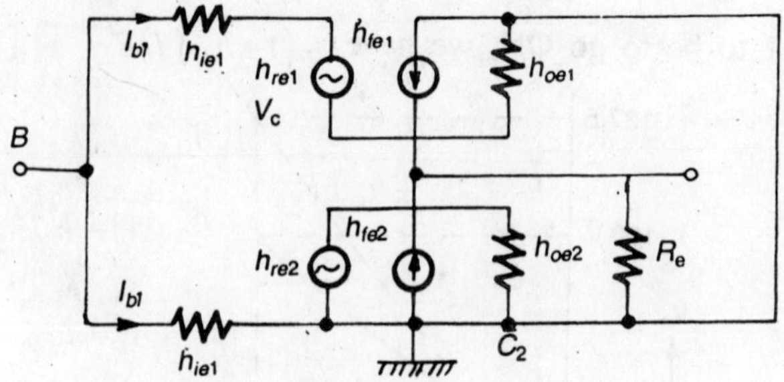

Fig. P. 18 (a)

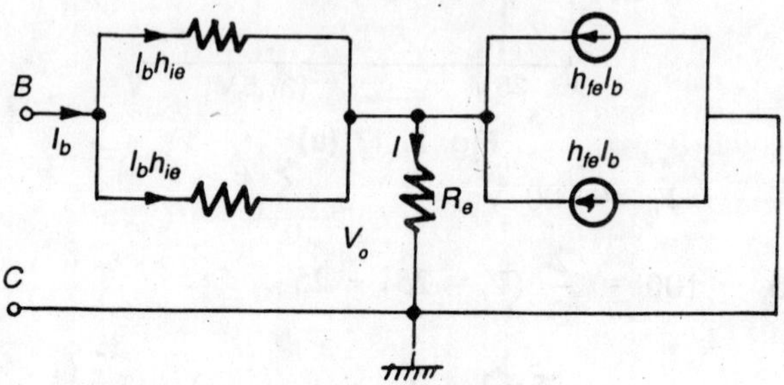

Fig. P. 18 (b)

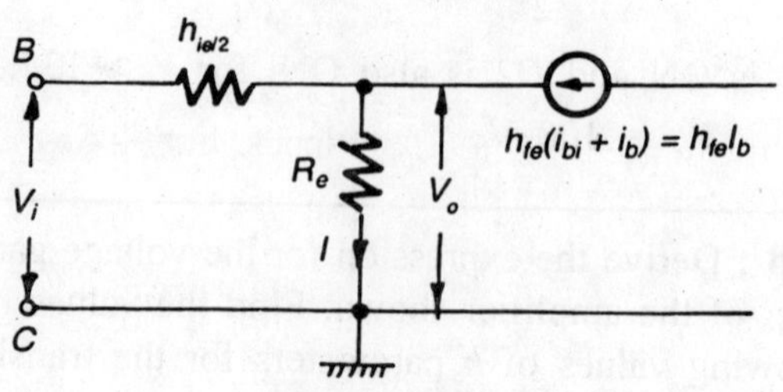

Fig. P. 18 (c)

For $\quad h_{ie1} = h_{ie2} = h_{ie}$

$\quad h_{fe1} = h_{fe2} = h_{fe}$

and $\quad h_{re1} = h_{re2} = 0$

$\quad h_{oe1} = h_{oe2} = 0$

The equivalent circuit reduces to, which is equivalent to

$h_{fe}\ (i_{b1} + i_b) = h_{fe} I_b$

Using KCL at E,

$$I = I_b + h_{fe} I_b = (1 + h_{fe})\ I_b$$

$$V_o = IR_e = (1 + h_{fe})\ I_b R_e$$

Voltage gain,

$$A_V = \frac{V_o}{V_i} = \frac{(1+h_{fe})\,I_b R_e}{I_b h_{ie}/2 + (1+h_{fe})\,I_b R_e}$$

$$= \frac{2(1+h_{fe})\,R_e}{h_{ie} + 2(1+h_{fe})\,R_e} = \frac{1}{1+\dfrac{h_{ie}}{2(1+h_{fe})\,R_e}}$$

Input resistance

$$R_{in} = \frac{V_i}{I_b} = \frac{I_b h_{ie}/2 + (1+h_{fe})\,I_b R_e}{I_b}$$

$$= \frac{h_{ie}}{2} + (1 + h_{fe})\,R_e$$

For given values,

$$A_V = \frac{1}{1+\dfrac{1000}{2\,(1+100)\,1000}} = 0.995$$

$$R_{in} = \frac{1000}{2} + (1 + 100)\,1000 = 101.5\ \text{K}\Omega$$

Problem 19 : For the feedback amplifier shown

(i) draw the circuit without feedback but taking the loading of feedback network into account.

(ii) find the feedback factor β

(iii) assuming loop gain to be much larger than unity, find the voltage gain with feedback. IES-97

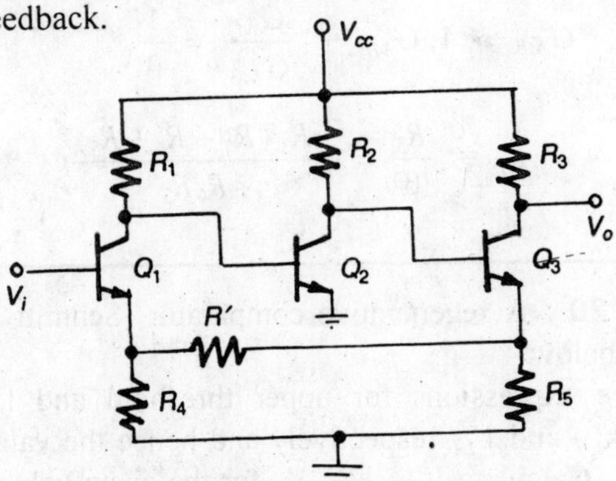

Fig. P. 19

Solution : (i) The required circuit without feedback for Q_3 to Q_1 but taking mutual loading effect of resistance is

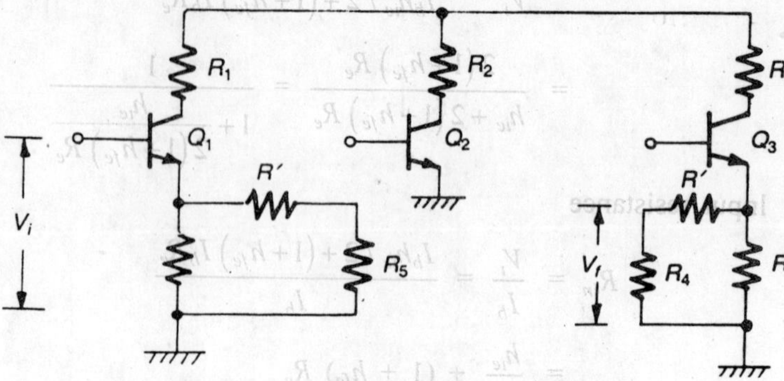

Fig. P. 19 (a)

(ii) The circuit uses current series feedback and for this circuit and, feedback factor is

$$\beta = -\frac{V_f}{I_o} = -\frac{R_5 \times R_4}{R_4 + R' + R_5}$$

then, $$V_f = -I_o \frac{R_5 R_4}{R_4 + R' + R_5}$$

(iii) $$A_{vf} = G_{MF} R_3$$

Gain with feedback,

$$G_{MF} = \left(\frac{G_M}{1 + G_{M\beta}}\right) R_3$$

If $$G_{Mb} \gg 1,\ G_{MF} = \frac{G_M}{G_{M\beta}} = \frac{1}{\beta}$$

$$A_{vf} = \frac{R_3}{\beta} = -\frac{R_3 (R_4 + R' + R_5)}{R_4 R_3}$$

Problem 20 : A regenerative comparator (Schmitt Trigger) circuit is shown below.

(i) Derive expressions for upper threshold and lower threshold voltages, V_{UT} and V_{LT} respectively and hence the value of hysteresis voltage V_H. Calculate V_{UT}, V_{LT}, V_H for the given values of R_1 and R_2.

(ii) A sine wave with 2 V peak-to-peak amplitude and 1 kHz frequency is applied at the input of the circuit. Plot the input and output waveforms. IES-97

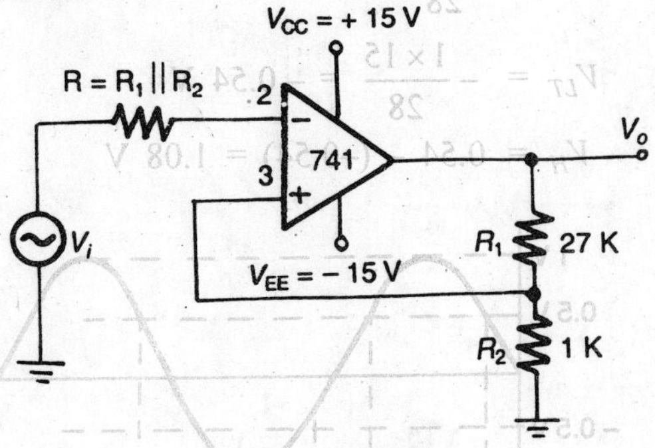

Fig. P. 20

Solution : Let $V_i < V_1$

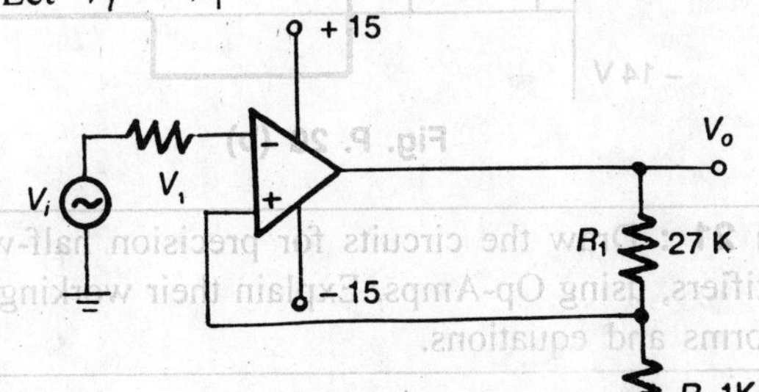

Fig. P. 20 (a)

$$\therefore \qquad V_o = + V_o = 15 \text{ V}$$

$$V_1 = \frac{V_o}{R_1 + R_2} \times R_2$$

If $\qquad V_i > V_1$

then $\qquad V_o = -15$ volt

Upper threshold voltage,

$$V_H = \frac{V_o R_2}{R_2 + R_3}$$

Lower threshold voltage,

$$V_{LT} = -\frac{V_o R_2}{R_1 + R_2}$$

and hysteresis, $V_H = V_{HT} - V_{LT} = \dfrac{2R_2V}{R_1 R_2}$

Hence $V_{HT} = \dfrac{1\times 5}{28} = 0.54$ V

$V_{LT} = -\dfrac{1\times 15}{28} = -0.54$ V

and $V_H = 0.54 - (-0.54) = 1.08$ V

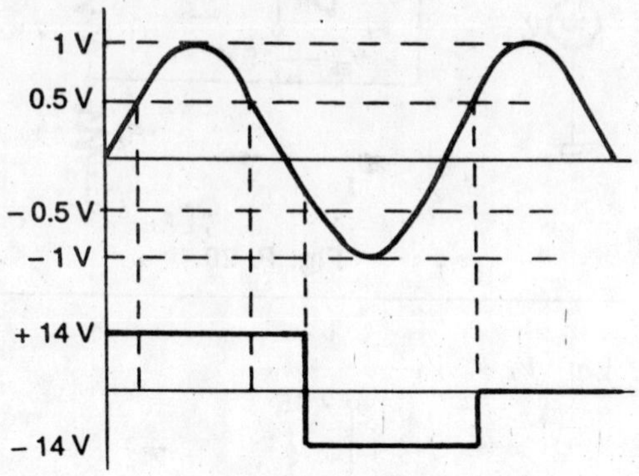

Fig. P. 20 (b)

Problem 21 : Draw the circuits for precision half-wave and full-wave rectifiers, using Op-Amps. Explain their working with the help of waveforms and equations. IES-97

Solution : Precision Half Wave Rectifier

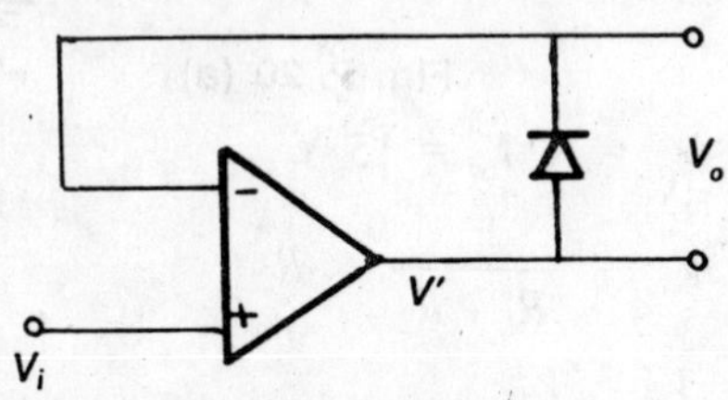

Fig. P. 21

In the figure shown when V_i goes positive then V' exceeds V_γ and D conducts. Due to the virtual connection between the non–inverting and inverting inputs (due to feedback with D ON) $V_0 \approx V_T = V_i$.

Hence, the circuit acts as a voltage follower for positive signals. When V_i swings negatively D is OFF and no current is delivered to the external load and therefore $V_0 = 0$, thus acting as a half–wave rectified.

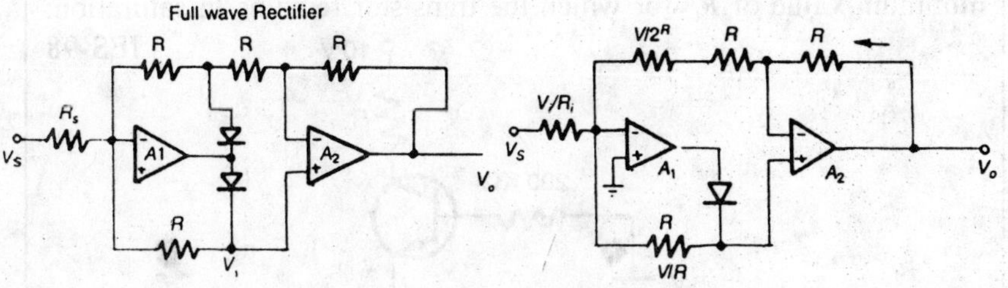

Fig. P. 21 (a) & (b)

Consider first the half cycle when V_s is positive. Then D_1 is non-conducting and there is no current in R which is connected to the non-inverting input to A_2, it follows that $V_1 = 0$. Therefore the system consists of two op-amps in cascade.

With the gain of $A_1 = -\frac{R}{R_1}$, and gain of $A_2 = -\frac{R}{R_1} = -1$, then the output is

$$V_0 = +\frac{R}{R_1} \text{ for } V_s > 0$$

Now consider the cycle where V_s is negative. Then D_1 is OFF and D_2 is ON because of the virtual ground at the input to A_2.

$V_2 = V_1 = V$, since the input terminals of A_1 are at the same (ground) potential.

During the half cycle when V_s is negative, D_1 is OFF and D_2 is ON as indicated.

$$V_1 = V_2 = V \text{ and } I = \frac{V}{2} R$$

$$\frac{V_S}{R_1} + \frac{V}{2R} + \frac{V}{R} = 0$$

or
$$V = -\frac{2}{3} \times \frac{R}{R_1} V_S$$

Output voltage,

$$V_o = iR + V$$

$$= \frac{V}{2R} R + V = \frac{3}{2} V = -\frac{R}{R_1}, \text{ for } V_S < 0$$

Hence full rectification can be obtained.

Problem 22 : A silicon transistor with V_{BE} (sat) = 0.8 V, b = h_{FE} = 100, V_{CE} (sat) = 0.2 V is used in the circuit shown. Find the minimum value of R_C for which the transistor remains in saturation. IES-98

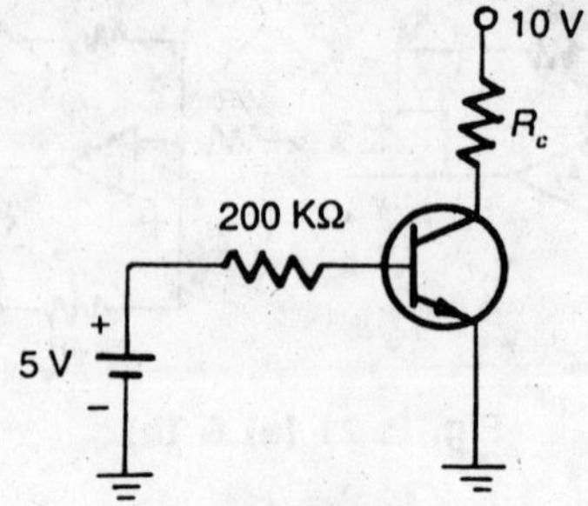

Fig. P. 22

Solution : Given :

$$V_{BE} = 0.7 \text{ V for Si transistor}$$

$$I_\beta = \frac{5-0.7}{200\times10^{-3}} = 21.5 \ \mu\text{A}$$

Condition for the transister to go in saturation is

$$I_\beta \geq \frac{I_c}{h_{FE}} \quad \text{...(A)}$$

where $$I_c = \frac{V_{CC}-V_{CE\,(sat)}}{R_C} \quad \text{...(B)}$$

By eqn. (A) and eqn (B), we have

$$I_\beta \geq \frac{V_{CC}-V_{CE\,(sat)}}{R_C h_{FE}}$$

Thus $$R_C \geq \frac{V_{CC}-V_{CE}\,(\text{sat})}{h_{FE} I_\beta} = \frac{10-0.2}{100\times 21.5\times10^{-6}}$$

$$\Rightarrow \quad R_C \geq 4558.14\ \Omega$$

$$\Rightarrow \quad R_{min} = 4.558\ \text{K}\Omega$$

Problem 23 : A silicon single phase full wave bridge rectifier circuit is shown. Explain what happens if the transformer and the load positions are interchanged. IES-98

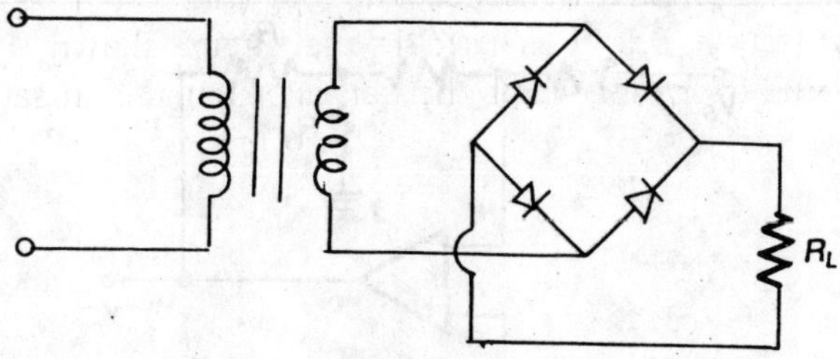

Fig. P. 23

Solution : During the cycle, all the diodes are forward-biased and hence closed. Then

$$V_{AB} = V_{D1} + V_{D2}$$
$$= 0.7 + 0.7 = 1.4 \text{ volt}$$

and it may cause excessive current that would flow through the diodes thus damaging it.

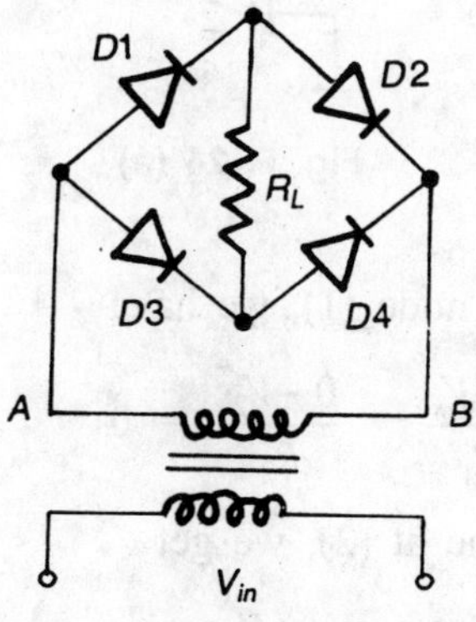

Fig. P. 23 (a)

In the negative cycle, all the diodes are reversed–biased, and the PIV across the diode is double to that in the former position.

Problem 24 : Develop the voltage transfer funciton $\frac{V_o}{V_s}$ for the amplifier shown. IES-98

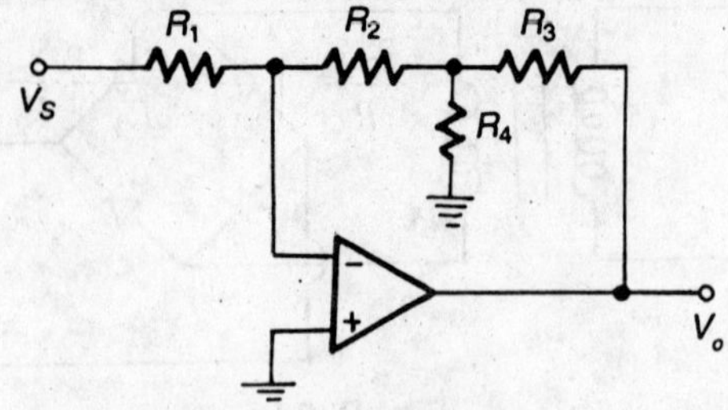

Fig. P. 24

Solution :

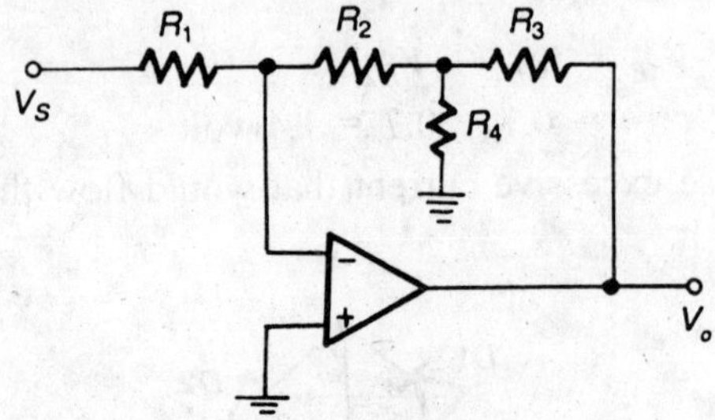

Fig. P. 24 (a)

V_1 Using KCL at node (1), we have

$$\frac{0 - V_s}{R_1} + \frac{0 - V_2}{R_2} = 0 \qquad \text{...(A)}$$

and using the same at (2), we get

$$\frac{V_2 - V_o}{R_3} + \frac{V_2}{R_4} + \frac{V_2 - V_1}{R_2} = 0$$

$$\Rightarrow \frac{V_2 - V_o}{R_3} + \frac{V_2}{R_4} + \frac{V_2}{R_2} = 0$$

$$\Rightarrow V_2 \left(\frac{1}{R_3} + \frac{1}{R_4} + \frac{1}{R_2} \right) = \frac{V_o}{R_3} \qquad \text{...(B)}$$

From eqn (A) and (B),

$$-\frac{R_2}{R_1} V_s \left(\frac{1}{R_3} + \frac{1}{R_4} + \frac{1}{R_2} \right) = \frac{V_o}{R_3}$$

Hence, $T(s) = \frac{V_o}{V_s} = -\frac{R_2 R_3}{R_1}\left(\frac{1}{R_3}+\frac{1}{R_4}+\frac{1}{R_2}\right)$

$$\frac{V_o}{V_s} = -\frac{R_2+R_3}{R_1} - \frac{R_2\,R_3}{R_1\,R_4}$$

Problem 25 : An amplifier with open loop voltage gain $A_v = 1000 \pm 100$ is available. It is required to have an amplifier whose gain varies by no more than ± 0.2%. Find.
(i) reverse transmission factor β of the feedback network IES-98

Solution : Output voltage,

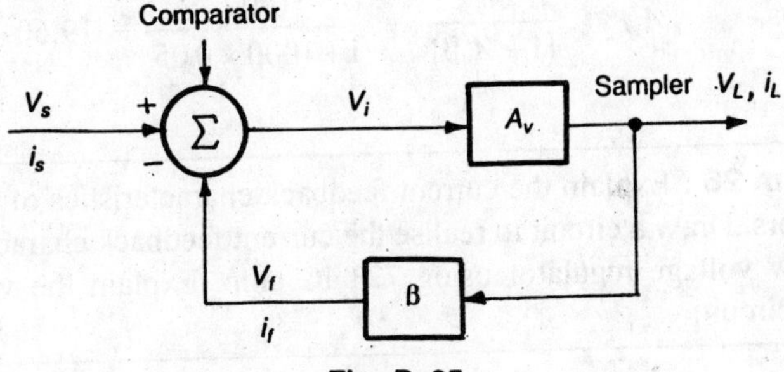

Fig. P. 25

$$V_L = A_V V_1$$

and $V_1 = V_s - V_f$

$$= V_s - \beta V_L - \beta A_v V_1$$

$$A_{vf} = \frac{V_L}{V_s} = \frac{A_v V_1}{(1+A_v\beta)\,V_1} = \frac{AV}{1+A_v\beta} \qquad \text{(A)}$$

Where A_v, $A_{vf}\beta$ are open loop, closed loop gain and feedback transfer ratio respectively.

Difererntiating eqn (A) with respect to A_{vf}, we have

$$\left|\frac{dA_{vf}}{A_{vf}}\right| = \frac{1}{|1+A_v\beta|} \times \left|\frac{dA_v}{A_v}\right| \qquad \text{...(B)}$$

As given, $A_v = 1000 \pm 100$

$$\Rightarrow \quad \frac{dA_v}{A_v} = \frac{100}{1000} \times 100$$

$$= 10\%, \text{ it is required } \frac{dA_{vf}}{A_{vf}} = 0.2\%$$

Then
$$\beta = \frac{\left|\frac{dA_v}{A_v}\right|}{A_v \left|\frac{dA_{vf}}{A_{vf}}\right|}$$

$$= \frac{10}{1000 \times 0.2} = 0.05 = 5\% \text{ (Assuming } A_v\beta \gg 1)$$

Thus,
$$A_{vf} = \frac{A_v}{(1 + A_v\beta)} = \frac{1000}{1 + 1000 \times 0.05} = 19.60$$

Problem 26 : Explain the current feedback characteristics of voltage regulators. Draw a circuit to realise the current feedback characteristic in a low voltage regulator using 723 IC chip. Explain the working of the circuit.

IES-98

Solution :

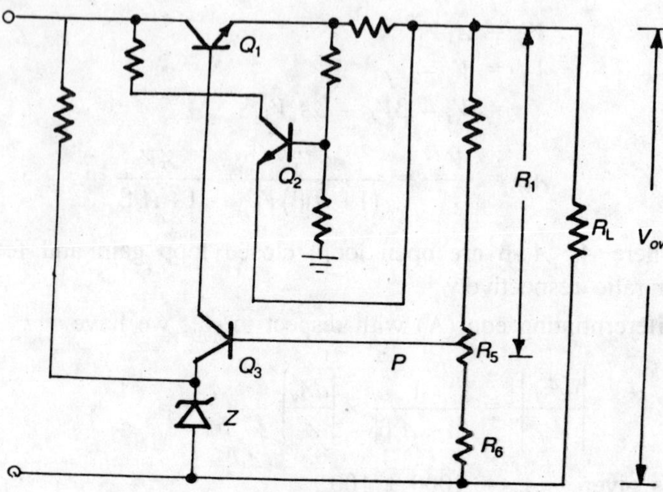

Fig. P. 26 (a)

Fold back limiting implies that shorted load currents is less than the maximum load current which can flow through the circuit with the decrease in R_L.

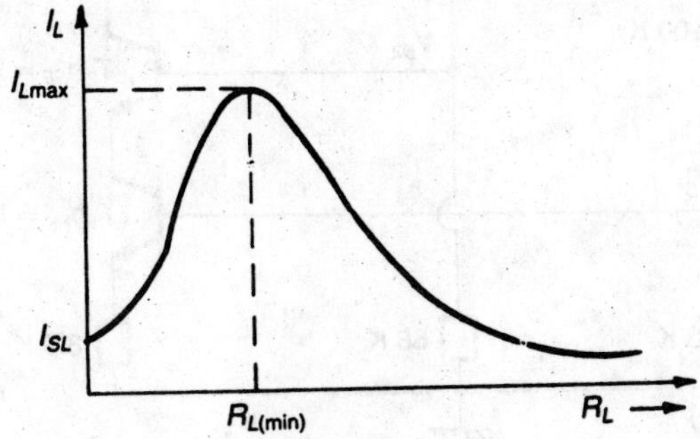

Fig. P. 26 (b)

Q_3 is biased by a voltage divider network consisting of R_5 and R_6. Load current I_L flows through R_L dropping a voltage of app. $I_L R_4$ across it at base of Q_3.

Problem 27 : For the circuit shown below, calculate the Q point for both transistors. Assume silicon transistors, $V_{BE} = 0.7$ V.

IES-99

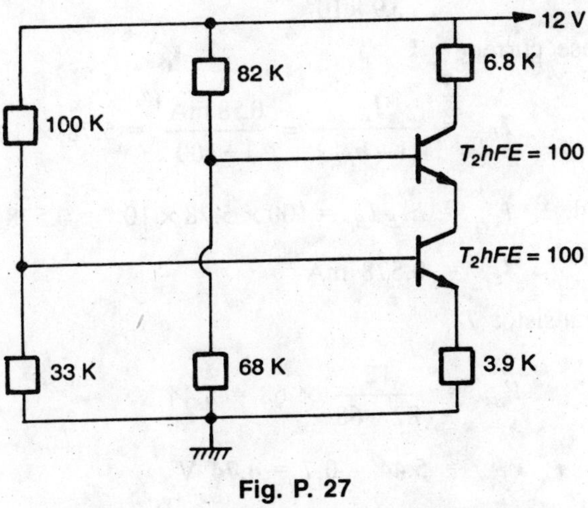

Fig. P. 27

Solution : As given,

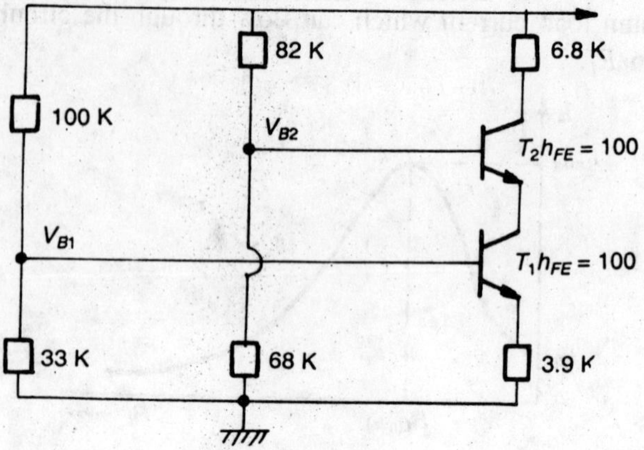

Fig. P. 27 (a)

$$V_{BE} = 0.7 \text{ V}$$

By voltage division,

$$V_{B_1} = \frac{12 \times 33}{(100+33)} = 2.977 \text{ volts}$$

Then, $V_{E_1} = V_{B_1} - V_{BE} = I_{E_1} R_{E_1}$

Now, $I_{E_1} = \dfrac{2.977 - 0.7}{3.9 \times 10^3} = 0.58 \times 10^{-3}$ Amp $= 0.58$ mA

∴ Base current,

$$I_{B_1} = \frac{I_{E_1}}{(1+h_{FE_1})} = \frac{0.58 \text{ mA}}{1+100} = 5.78 \times 10^{-6} \text{ Amp}$$

∴ and, $I_{C_1} = h_{FE} I_{B_1} = 100 \times 5.78 \times 10^{-6} = 0.578 \times 10^{-3}$ Amp

$$I_{C_1} = 0.578 \text{ mA}$$

For transistor T_2,

$$V_{B_2} = \frac{12}{82+68} \times 68 = 5.44 \text{ V}$$

$$V_{E_2} = 5.44 - 0.7 = 4.74 \text{ V}$$

$$I_{E_2} = I_{C_1} = 0.578 \times 10^{-3} \text{ A}$$

$$I_{B_2} = \frac{I_{E_2}}{1+h_{FE}} = \frac{0.578 \times 10^{-3}}{1+100} = 5.72 \times 10^{-6} \text{ A}$$

$$\therefore \quad I_{C_2} = h_{FE} I_{B_2} = 0.572 \times 10^{-3} \text{ Amp}$$

$$V_{C_2} = 12 - 6.8 \times 10^3 \times 0.572 \times 10^{-3} = 8.108 \text{ V}$$

$$V_{CE_2} = V_{C_2} - V_{E_2} = 8.108 - 4.74 \text{ V} = 3.368 \text{ V}$$

$$V_{CE_1} = V_{C_1} - V_{E_1} = V_{E_2} - (V_{B_1} - 0.7)$$

$$= 4.74 - 2.977 + 0.7 = 2.463 \text{ V}$$

Hence, Q-points are

$V_{CE1} = Q = 2.463$ V,	$V_{CE2} = Q = 3.368$ V,
$I_{C1}, Q = 0.578$ mA,	$I_{C2} = Q = 0.572$ mA,
$I_{B1}, Q = 5.8\ \mu$A	$I_{B2}, Q = 5.72\ \mu$A

Problem 28 : A power supply using half-wave rectifier is to have an output dc voltage of 30 V, with a load resistance of 500 Ω. The ripple factor should not exceed 0.01. Find a suitable value for C. Determine the peak diode current. Assume 50 Hz supply frequency.

IES-99

Solution : Half-wave rectifier with capaciter filter :

$$C = \frac{1}{f R_L \gamma} = \frac{1}{50 \times 500 \times 0.01}$$

$$= 4 \times 10^{-3} \text{ F} = 4000\ \mu\text{F}$$

Diode current

$$i = i_L + i_C$$

$$= \frac{v_o}{R_L} + C\frac{dv_o}{dt}$$

$$= \frac{V_m}{R_L} \sin \omega t + \omega c\, V_m \cos\omega t$$

$$\Rightarrow \quad i = I_m \sin(\omega t + \phi)$$

where $$I_m = V_m \sqrt{\frac{1}{R_L^2} + \omega^2 c^2}$$

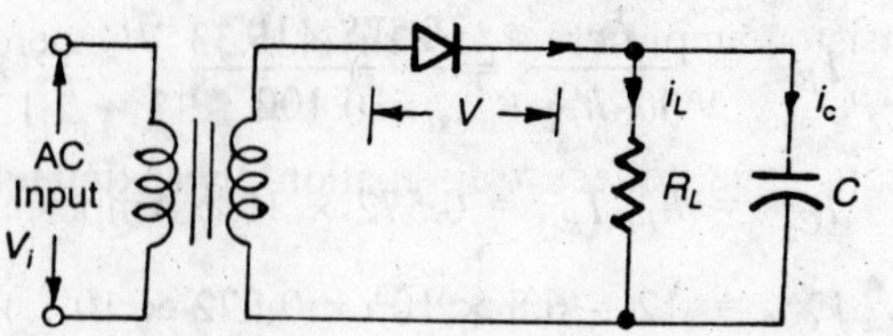

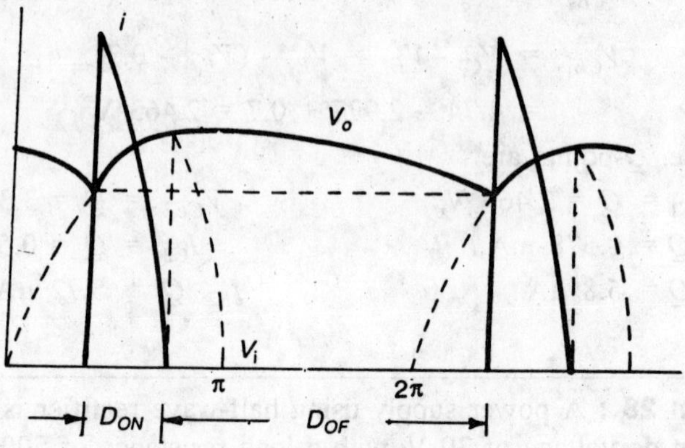

Fig. P. 28

and $\phi = \tan^{-1} R_L wc.$

Thus, peak diode current would be

$$I_m = \sqrt{\frac{1}{R_L^2} + \omega^2 c^2} = 30 \times \sqrt{\frac{1}{(500)^2} + \left(2\pi \times 50 \times 4000 \times 10^{-6}\right)^2}$$

$$= 37.699 \text{ Amp.}$$

Problem 29 : A class B transformer coupled amplifier is to supply 4 W to a 10 Ω load. Available supply voltage V_{CC} = 30 V. The transformer efficiency is 75%. Specify the output transformer and the output transistors. IES-99

Solution : Given,

$$P_o = 4 \text{ W}$$
$$R_L = 10\ \Omega$$
$$V_{CC} = 30 \text{ V}$$
$$\eta_T = 75\%$$

Then, power input to transformer $= \frac{4}{0.75} = 5.33$ W

As, transistor amplifier is to deliver 5.33 W, therefore

$$P_{c\,max.} = 0.4 \times P_{max} = 0.4 \times 5.33 = 2.132 \text{ W}$$

Transistors have collector dissipation approximately 2.132/2, i.e., 1.066 W.

The maximum power which can be delivered if $V_{min} = 0$, is,

$$P_{max} = \frac{V_{CC}^2}{2R_L'}$$

$$\therefore \text{ then,} \quad R_L' = \frac{V_{CC}^2}{2\,P_{max}} = \frac{(30)^2}{2 \times 5.33} = 84.427\ \Omega$$

$$\therefore \quad n^2 = \frac{R_L}{R_L} = \frac{4\ \Omega}{84.427\ \Omega} = 0.0473$$

Turns ratio $n = 0.2176 = \dfrac{1}{4.6}$

Hence, transformer's turns ratio

$$\frac{N_2}{N_1} = \frac{1}{4.6}$$

Problem 30 : Determine the frequency of oscillation for the following circuit and the value of R_1 needed to maintain oscillations. Name the circuit. IES-99

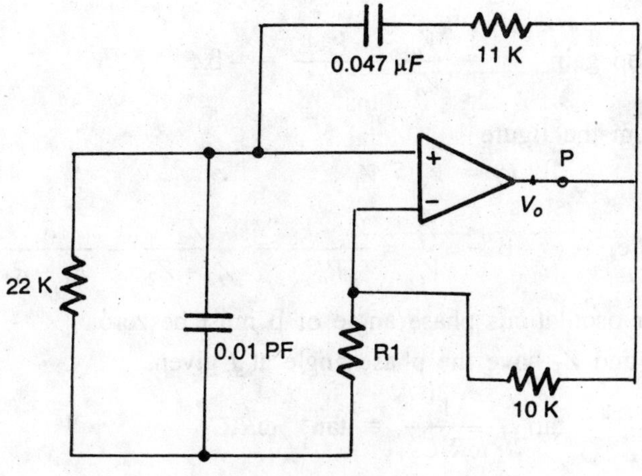

Fig. P. 30

Solution :

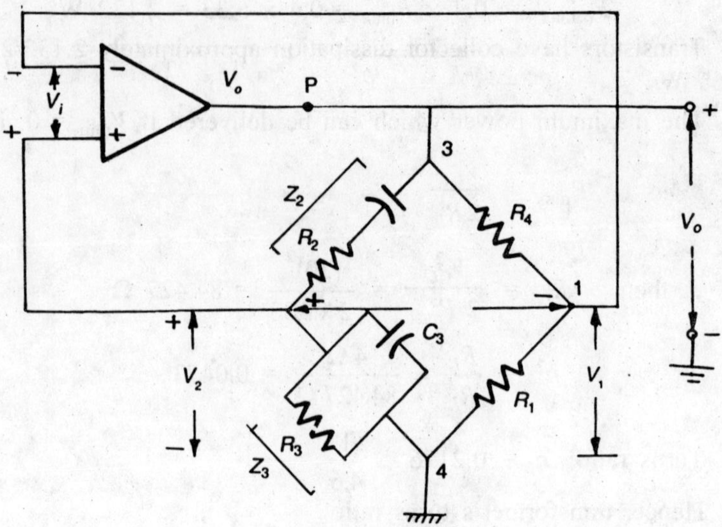

Fig. P. 30 (a)

$$R_3 = 22\ \text{K}\Omega$$
$$R_4 = 10\ \text{K}$$
$$R_2 = 11\ \text{K}\Omega$$
$$C_3 = 0.01\ \mu\text{F}$$
$$C_2 = 0.047\ \mu\text{F}$$
$$R_1 = ?$$

$$\text{Loop gain} = \frac{V_o}{V_o'} = \frac{V_i A_v}{V_o'} = -\beta A$$

From the figure

$$V_i = V_2 - V_1$$

Here, $$-\beta = \frac{V_i}{V_o'} = \frac{V_2 - V_1}{V_o'} = \frac{Z_2}{Z_3 + Z_2} - \frac{R_1}{R_1 + R_4}$$

For oscillations phase angle of β must be zero.

Z_2 and Z_3 have the phase angle at f given,

$$\Rightarrow \quad \tan^{-1} \frac{1}{\omega R_2 C_2} = \tan^{-1} \omega R_3 C_3$$

$$\Rightarrow \quad \tan^{-1}\frac{1}{2\pi f R_2 C_2} = \tan^{-1} 2\pi f R_3 C_3$$

or, $$f = \frac{1}{2\pi\sqrt{R_2 R_3 C_2 C_3}} = \frac{10^3}{2\pi\sqrt{11 \times 22 \times 0.01 \times 0.047}}$$

$$= 472 \text{ Hz.}$$

For null desired,

$$\frac{R_1}{R_1 + R_4} = \frac{1}{3} \text{ as, at frequency } f, \frac{Z_2}{Z_2 + Z_3} = \frac{1}{3}$$

then,

$$\Rightarrow \quad 3R_1 = R_1 + R_4$$

$$\Rightarrow \quad R_1 = R_{4/2} = 5 \text{ K}$$

Problem 31 : For the common emitter amplifier shown in Fig. P. 31, draw the simplified high frequency equivalent circuit and derive an approximate expression for the voltage gain and 3 dB frequency.

IES-2000

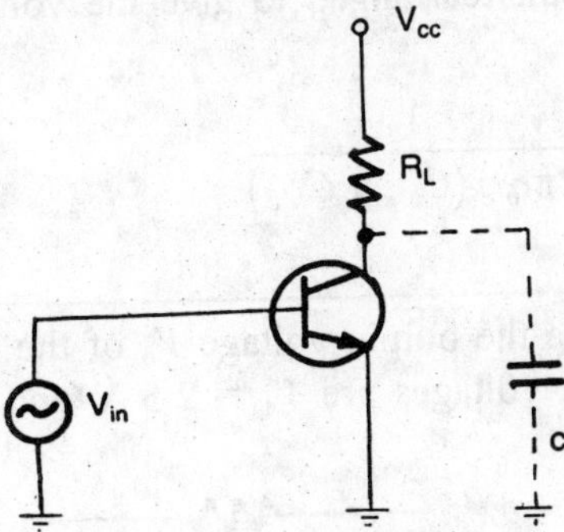

Fig. P. 31

Solution : Equivalent-high frequency model is shown below,

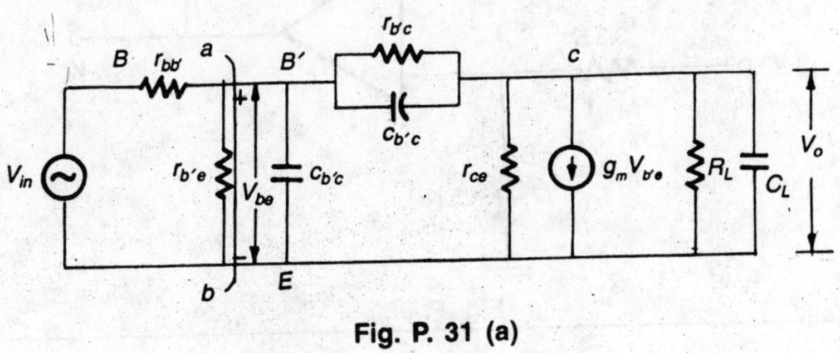

Fig. P. 31 (a)

Using Thevenin's equivalent to the left of determinal ab,

$$V_{Th} = \frac{V_{in}}{r_{bb} + r_{b'e}} r_{b'e},$$

$$R_{Th} = \frac{r_{bb'} \times r_{b'e}}{r_{bb} + r_{b'e}}$$

Now

$$\frac{V_{b'e} - V_{Th}}{R_{Th}} + \frac{V_{b'e}}{\frac{1}{SC_{b'c}}} + \frac{V_{b'e} - V_o}{r_{b'c}} + \frac{V_{b'e} - V_o}{\frac{1}{SC_L}} = 0 \quad \text{...(A)}$$

[Using KCL at node *B*]

or,

$$\frac{V_o - V_{b'e}}{r_{b'c}} + \frac{V_o - V_{b'e}}{\frac{1}{SC_{b'c}}} + \frac{V_o}{r_{ce}} + \frac{V_o}{R_L} + \frac{V_o}{\frac{1}{SC_L}} + g_m V_{b'e} = 0 \quad \text{...(B)}$$

Solving equation (B) for $V_{b'e}$ and substituting value of $V_{b'e}$ in quation (A) and the result rearranged to give the voltage ratio $\frac{V_{Th}}{V_0}$ f_β is given as

$$f_\beta = \frac{1}{2\pi r_{b'e}\left(C_{b'e} + C_{b'c}\right)}$$

Problem 32 : Calculate the output voltage V_o of the circuit shown in Fig. P. 32. The input voltages are V_1 = 2.5 V and V_2 = 1 V.

IES-2000

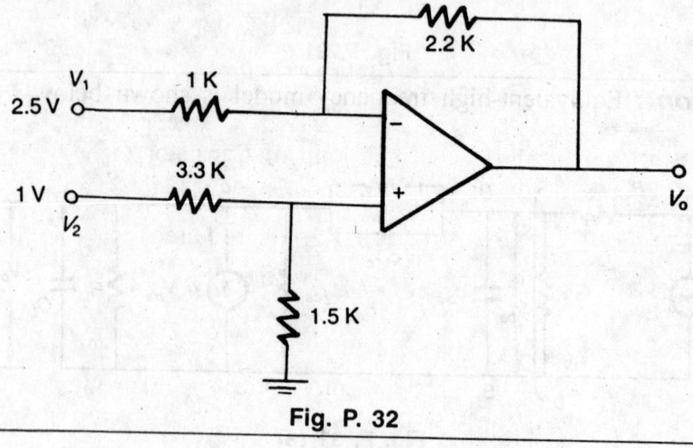

Fig. P. 32

Solution : By voltage division

$$V_T = \frac{V_2 \times 1.5}{1.5 + 3.3} = 0.3125$$

Using KCL at inverting terminal, we get

$$\frac{2.5 - V}{1} = \frac{V_- - V_o}{2.2}$$

Since $V_- = V_+$, then

or, $(2.5 - V_+)\,2.2 = V_+ - V_o$

or, $V_o = -4.5$ volt

Problem 33 : In the level shifter circuit shown in Fig. P. 33, calculate the value of R for V_o to become zero. Assume $V_D = V_{BE} = 0.7$ V and h_{FE} is very large. IES-2000

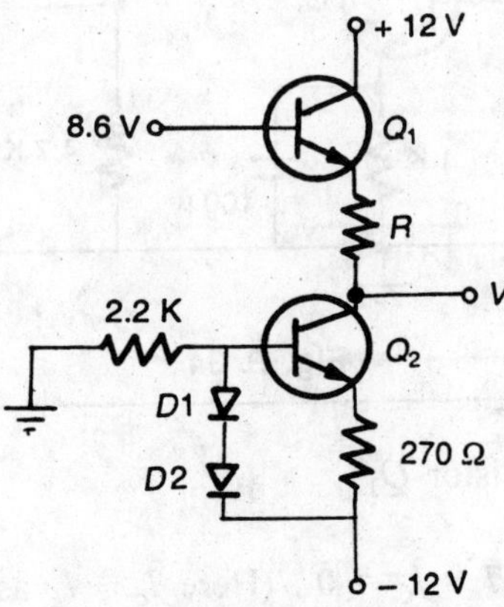

Fig. P. 33

Solution : Using KVL in the B-E loop of transistor Q_1,

$8.6 - 0.7 - IR = V_o$

or $IR = 7.9$ V (Since $V_o = 0$, required) ...(A)

Using KVL for the transistor Q_2, we have,

$2.2\, I_1 + 1.4 = 12$

$$I_1 = \frac{10.6}{2.2} = 4.82 \text{ mA}$$

and $2.2 \times 4.82 - 0.7 - 0.270\ I = 12$

or $I = 2.578$ mA ...(B)

From equations (A) and (B), we obtain,

$$R = \frac{7.9}{2.578} = 3.06\ \text{k}\Omega$$

Problem 34 : Calculate V_o in the circuit shown in Fig. P. 34. Assume h_{FE} to be large and neglect the base current. $V_{BE} = 0.7$ V.

IES-2000

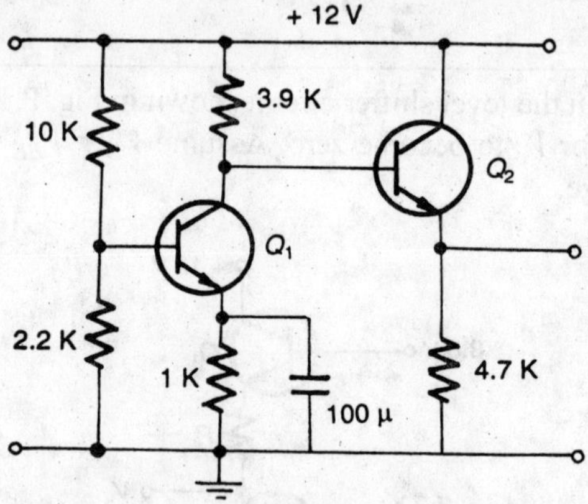

Fig. P. 34

***Solution* :** For transistor Q_1

$$\frac{12}{12.2} \times 2.2 - 0.7 - I_C = 0 \quad \text{(Here } I_c \approx I_E \text{ as } h_{FE} \text{ is too large).}$$

or collector current

$$I_C = 1.464 \text{ mA}$$

Now, for transistor Q_2, using KVL in the B-E loop, we have

$$12 - 3.9 \times 1.464 - 0.7 = V_o$$

$$V_o = 5.59 \text{ V}$$

Problem 35 : Derive an expression for the stability factor of the circuit shown in Fig. P. 35.

IES-2000

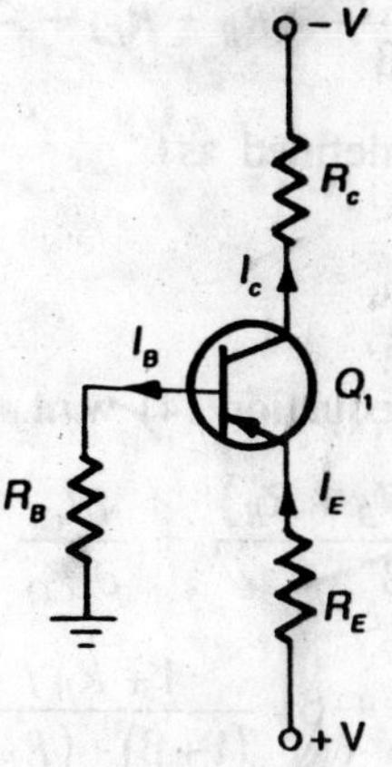

Fig. P. 35

Solution : From the circuit given, using KVL in the B-E loop, we get

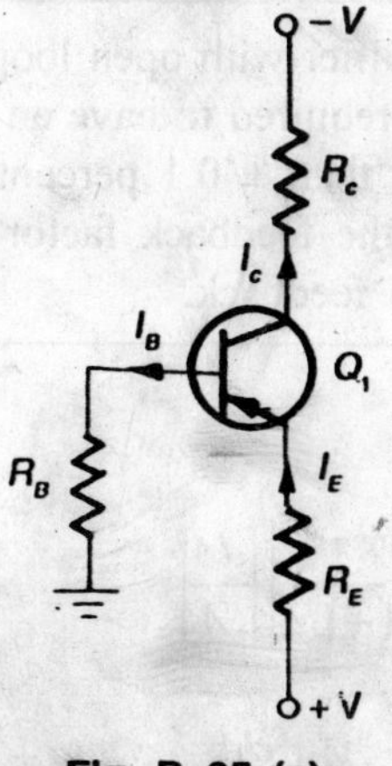

Fig. P. 35 (a)

$$V_{BE} = I_B (R_E + R_B) + R_E I_C - V \qquad \ldots(1)$$

Collector current

$$I_C = (1 + \beta) I_{CO} + \beta I_B \qquad \ldots(2)$$

or

$$I_B = \frac{I_C - (1+\beta) I_{CO}}{\beta} \qquad \ldots(3)$$

Substituting the expression of I_B from equation (2), in equation (1) we have

$$V_{BE} = -V - I_{CO}\frac{(1+\beta)}{\beta}(R_B + R_E) + \frac{I_C}{\beta} R_B + R_E(1+\beta) \quad ...(4)$$

Stability factor, S is defined as

$$S = \frac{\partial I_C}{\partial I_{CO}} \approx \frac{\Delta I_C}{\Delta I_{CO}}$$

Now, differentiating equation (4) w.r.t. I_{co}, we get

$$0 = 0 - \frac{(1+\beta)(R_B + R_E)}{\beta} + \frac{\partial I_C}{\partial I_{CO}} \quad \frac{R_B + R_E(1+\beta)}{\beta}$$

or $$\frac{\partial I_C}{\partial I_{CO}} = (1+\beta)\frac{1 + R_B / R_E}{(1+\beta)+(R_B / R_E)}$$

or Thus, $$S = (1+\beta)\frac{1 + R_B / R_E}{(1+\beta)+R_B / R_E}$$

Problem 36 : An amplifier with open loop voltage gain A_o = 1000 ± 100 is available. It is required to have an amplified whose voltage gain varies by no more than ± 0.1 percent.
(i) Find the value of the feedback factor required.
(ii) Find the gain with feedback. IES-2000

Solution :

As, $$\left|\frac{dA_f}{A_f}\right| = \left|\frac{1}{1+\beta A}\right|\left|\frac{dA}{A}\right|$$

$$\frac{dA_f}{A_f} = 0.1\%, \quad \frac{dA}{A} = 10\%, \; A = 1000$$

where A_f, A and β are respectively gain with feedback, open loo gain and feedback factor.

Now, $$0.1 = \frac{1}{1+\beta A} \times 10$$

or $\beta = 0.099$ or 0.1

(ii) Gain with feedback,

$$A_f = \frac{A}{1+\beta A} = \frac{1}{1+1000 \times 0.999} = 10$$

Problem 37 : In the circuit shown in Fig. calculate the value of the output current I_o for $V_{in} = 5$ V. If $V_{CE\,sat}$ of Q_2 is 0.3 V, calculate the maximum value of R_L that can be used in the circuit.

IES-2000

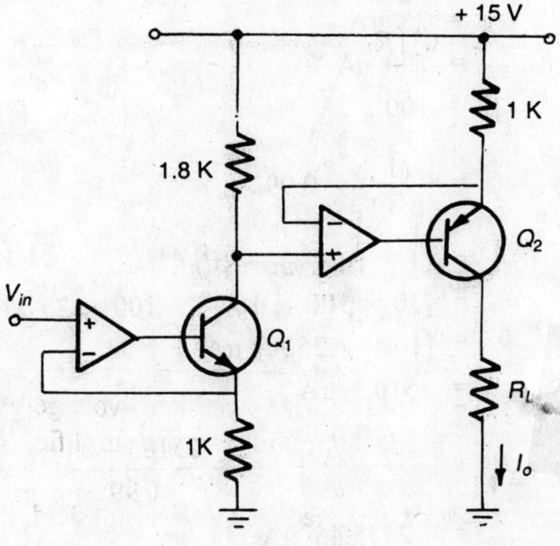

Fig. P. 37

Solution :

$$V_{E,Q_1} = 5 \text{ V}$$

(Since OPAMP is ideal, $V_{in} = V_{E,Q_1} = V_+ = V_-$)

then, $$I_{E1} = \frac{5 \text{ V}}{1 \text{ K}} = 5 \text{ mA}$$

Now, Voltage across 1.8 K $= 1.8 \times 5 = 9$ V ($I_{E1} \approx I_{C1}$)

Thus, Input voltage to OP-AMP $= 15 - 9 = 6$ V

and, Voltage across 1 K resistor $= 15 - 6 = 9$ V

Hence, $$I_{E2} = \frac{9}{1} = 9 \text{ mA}$$

$$I_o = I_{E2} = 9 \text{ mA}$$

Problem 38 : A transistor is connected in common emitter configuration as an amplifier. The parameters of the transistor specified are :

$$I_B = 25\ \mu A,$$
$$I_{CBO} = 100\ nA,$$
$$\beta = 100$$

Find its I_C, I_E, I_{CEO} and alpha, where symbols have their usual meaning.

IES-01

Solution : Given :

$$I_B = 25\ \mu A$$
$$I_{CBO} = 100\ nA$$
$$\beta = 100$$

Now, $\alpha = \dfrac{\beta}{\beta+1} = 0.99$

and $I_C = (1 + \beta)\ I_{CBO} + \beta I_B$

$$= [10 \times 100 \times 10^{-19} + 100 \times 25 \times 10^{-6}]\ A$$
$$= [10.1 + 2500]\ \mu A$$
$$= 2510.1\ \mu A$$

$$I_E = -\frac{I_C - I_{CBO}}{\alpha} = -\frac{2510.1 - 0.1}{0.99}$$
$$= -2535.45\ \mu A$$
$$= -2.535\ mA$$

Now, if $I_B = 0$

$$I_E = -I_C$$

$$I_C = -I_E = \frac{I_{CBO}}{1-\alpha} = I_{CEO}$$

or, $\dfrac{I_{CBO}}{I - \alpha} = I_{CEO}$

$\Rightarrow$ $I_{CEO} = 101\ nA$

Problem 39 : A simple full wave bridge rectifier circuit has an input voltage of 240 V a.c. r.m.s. Assume the diodes to be ideal. Find the output d.c. current, d.c. voltage, rms values of output currents and voltages and the peak inverse voltage that appears across the non-conducting diode. Assume load resistance to be 10 KΩ. IES-01

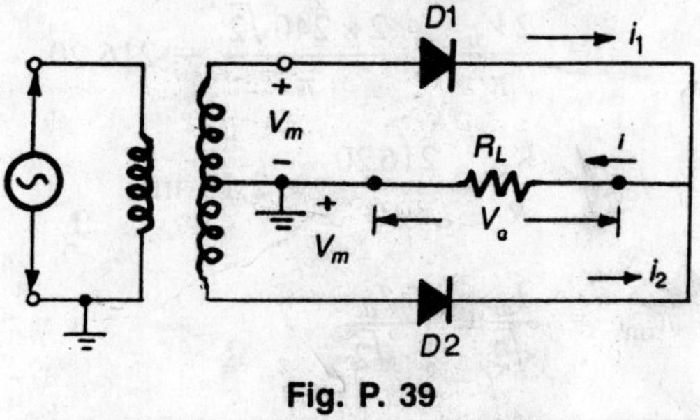

Fig. P. 39

Solution :

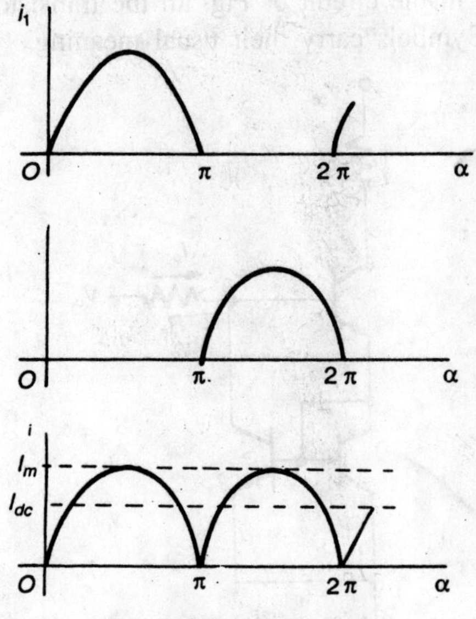

Fig. P. 39 (a), (b) & (c)

Direct current

$$I_{dc} = \frac{1}{2\pi}\left[\int_0^{\pi} I_m \sin\alpha\, d\alpha + \int_{\pi}^{2\pi} (-I_m \sin\alpha)\, d\alpha\right]$$

$$= \frac{I_m}{2\pi}\left[-\cos\alpha\Big|_0^{\pi} + \cos\alpha\Big|_{\pi}^{2\pi}\right] = \frac{2\,I_m}{\pi}$$

Then, $V_{dc} = \dfrac{2\, I_m R_L}{\pi}$

$$= \frac{2\, V_m}{\pi} = \frac{2 \times 240\sqrt{2}}{\pi} = 216.20$$

Thus, $I_{dc} = \dfrac{V_{dc}}{R_L} = \dfrac{216.20}{10} = 2.16 \text{ mA}$

and $\pi_{rms} = \dfrac{I_m}{\sqrt{2}} = \dfrac{\pi \cdot I_{dc}}{2 \cdot \sqrt{2}}$

Problem 40 : In the circuit of Fig. all the transistors are matched Si transistors. Symbols carry their usual meaning.

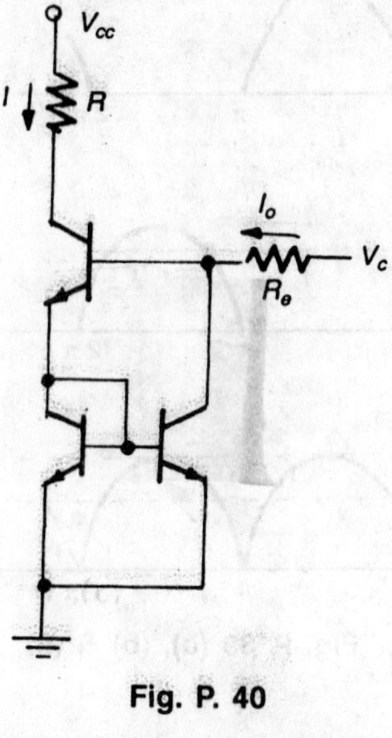

Fig. P. 40

(i) Find the expression of the current I in terms of I_o.
(ii) Calculate the value of I if $V_C = 5$ V, $R_e = 2$ KΩ, $V_{CC} = 6$ V and $R = 2.2$ KΩ. Take $V_{BE} = 0.7$ V and $\beta = 100$.
(iii) What can be the maximum value of I? If R_C is varied between 1 KΩ and 4 KΩ give the range of the variation of I. IES-01

Solution :

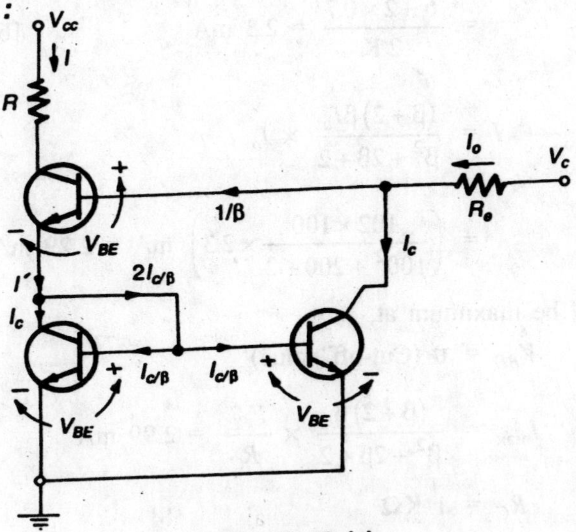

Fig. P. 40 (a)

(i) From the circuit shown here, we have

$$I' = I_C + \frac{2I_C}{\beta} = \frac{(\beta+2)}{\beta} I_C \quad \ldots(1)$$

and,
$$I' = I + I/\beta = \frac{(\beta+1)}{\beta} I \quad \ldots(2)$$

and
$$I_o = I_C + I/\beta \quad \ldots(3)$$

From eqn (1) and eqn (2)

$$I_C = \frac{(\beta+1)}{\beta+2} I$$

putting the expression of I_C in eqn (3), we get

$$I_o = \frac{(\beta+1)}{\beta+2} I + I/\beta$$

or
$$I = \frac{(\beta+2)\beta I_o}{\beta^2+2\beta+2}$$

(ii) As,
$$I_o = \frac{V_{CC}-2V_{BE}}{R_C}$$

$$= \frac{6 - 2 \times 0.7}{2\text{ K}} = 2.3 \text{ mA} \qquad \text{[by KVL]}$$

then,

$$I = \frac{(\beta + 2)\beta I_o}{\beta^2 + 2\beta + 2} \times I_o$$

$$= \left(\frac{102 \times 100}{100^2 + 200 + 2} \times 2.3\right) \text{mA} = 2.29 \text{ mA}$$

(iii) I will be maximum at,

$$V_{BE} = 0 \text{ (Cut-off value)}$$

or

$$I_{max} = \frac{(\beta + 2)\beta}{\beta^2 + 2\beta + 2} \times \frac{V_{CC}}{R_C} = 2.99 \text{ mA}$$

when $R_C = 1\text{ K}\Omega$

$$I_1 = \frac{(\beta + 2)\beta}{\beta^2 + 2\beta + 2} \times \frac{(6 - 1.4)}{1} = 4.55 \text{ mA}$$

at $R_C = 4\text{ K}\Omega$

$$I_2 = I_1/4 = 1.13 \text{ mA}$$

Hence, the value of I varies between 1.13 mA to 4.55 mA

Problem 41 : Derive the transfer function of the circuit shown below and identify the function of the circuit IES-02

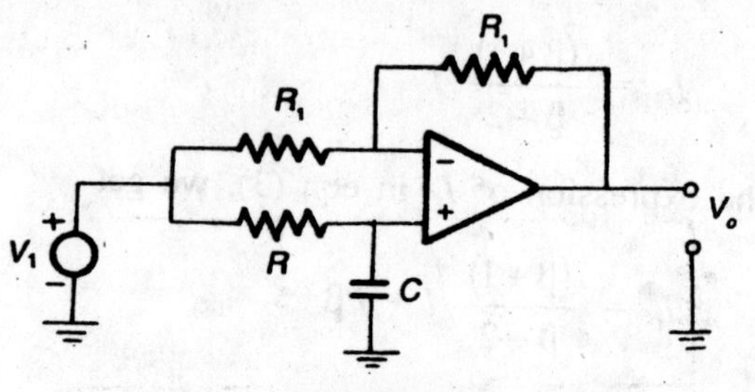

Fig. P. 41

Solution : Using KCL at the inverting terminal, we have

$$\frac{V_1 - V}{R_1} = \frac{V_1 - V_o}{R_1}$$

$$\left[\begin{array}{r} V \text{ be the potential at inverting terminal} \\ \text{and } V_- = V_+ = V \end{array}\right]$$

or, $$\frac{V_1 + V_o}{2} = V \quad \ldots(A)$$

Now, applying KCL at the non–inverting terminal, we have

$$\frac{V_1 - V_o}{R} = V.Cs$$

or $$V_1 = V\,(RCs + 1) \quad \ldots(B)$$

Equations (A) and (B) simultaneously gives,

$$H(s) = \frac{V_O}{V_1} = \frac{1 - RCs}{1 + RCs}$$

or $$H(j\omega) = \frac{1 - j\omega RC}{1 + j\omega RC} = 1\ \angle - \tan^{-1}\ \omega RC$$

When $\omega = 0$, $H(j\omega) = 1$. As frequency increases $j\omega RC \gg 1$ so that $H(j\omega)$ continues to be unity (–1). Therefore it is an all-pass filter.

Problem 42 : Draw the circuit of an stable multi-vibrator using p-n-p transisotrs, whose output is a squatre wave with steep edges. Also draw the labelled waveforms at the collectors and bases of the two transistors. Determine the frequency of the output waveform. Discuss its aplication as voltage to frequency converter. IES-02

Solution : Determination of frequency of the output waveform.

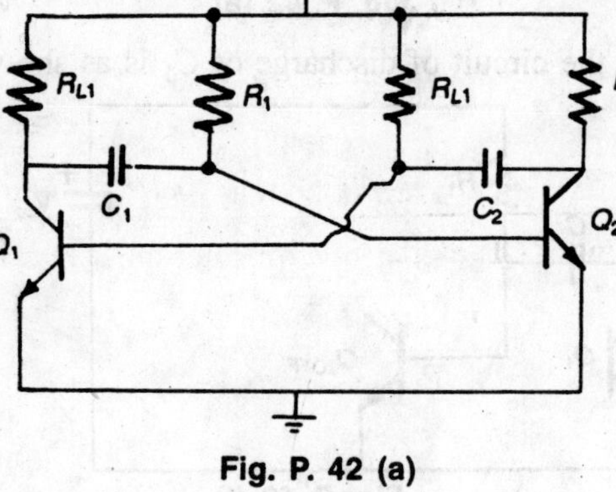

Fig. P. 42 (a)

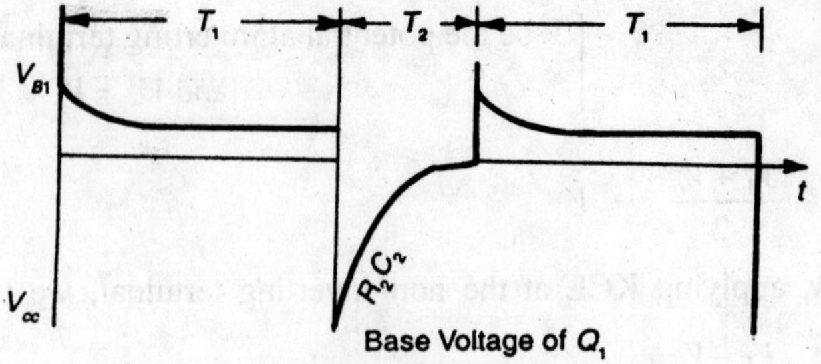

Fig. P. 42 (b)

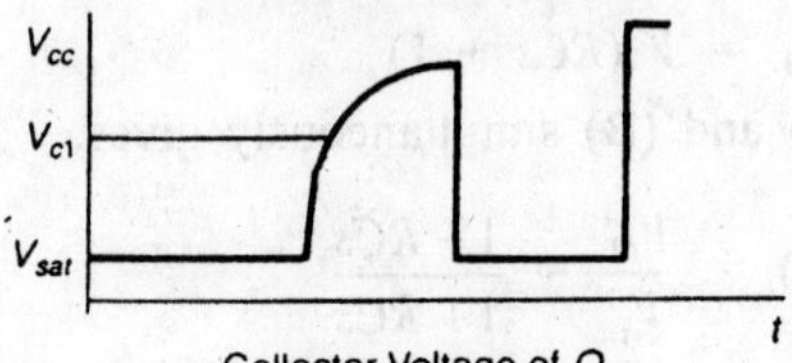

Fig. P. 42 (c)

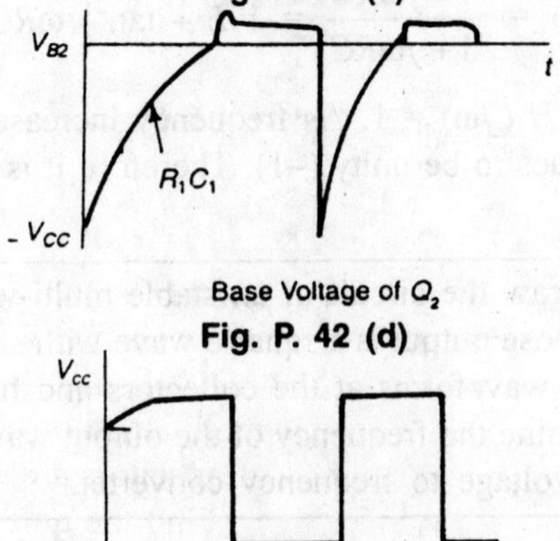

Fig. P. 42 (d)

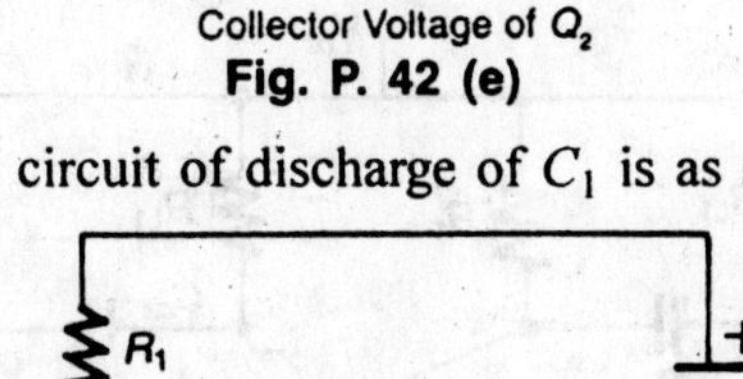

Fig. P. 42 (e)

During T_1, the circuit of discharge of C_1 is as shown in the figure given below

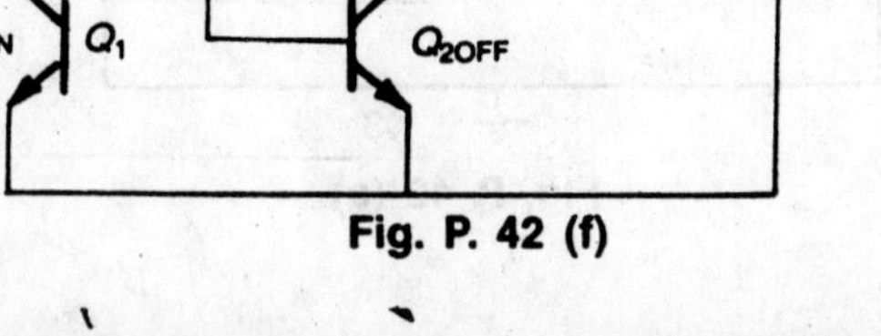

Fig. P. 42 (f)

Now, recalling that C_1 was chareged to V_{CC} when Q_1 was OFF, at the beginning of time T_1, as Q_1 turns ON, the collector end of C_1 goes to zero. Hence the base end of C_1 is at $-V_{CC}$. This is also the voltage of V_{B2}.

i.e., $V_{B2} = -V_{CC}$

Since, in the figure above initially $V_o = -V_{CC}$ and the time constant is R_1C_1 and the final potential should be $+V_{CC} = V_1$

Now, V_{B2} can be expressed as,

$$V(t) = V_f - (V_f - V_o)\, e^{-t/RC_1}$$

From the values given,

$$V_{B2}(t) = V_{CC} - (V_{CC} + V_{CC})\, e^{-t/R_1C_1}$$

This operation will continue till $V_{B2}(T_1) = 0$

Putting $t = T_1$ and $V_{B2}(T_1) = 0$, we get

$$T_1 = R_1C_1 \ln 2$$

Similarly, $T_2 = R_2C_2 \ln 2$

Hence, frequency of the output waveform $= \dfrac{1}{T_1 + T_2}$

$$f = \frac{1}{(R_1C_1 + R_2C_2)\ln 2}$$

Problem 43 : In an *R-C* coupled ampllifier, shown below, the BJT has $h_{fe} = 50$. All bypass and coupling capacitors are assumed to have zero reactance at the signal frequency. Find the quiescent conditions and draw the small-signal equivalent circuit, neglecting h_{oe} and h_{re}.

IES-02

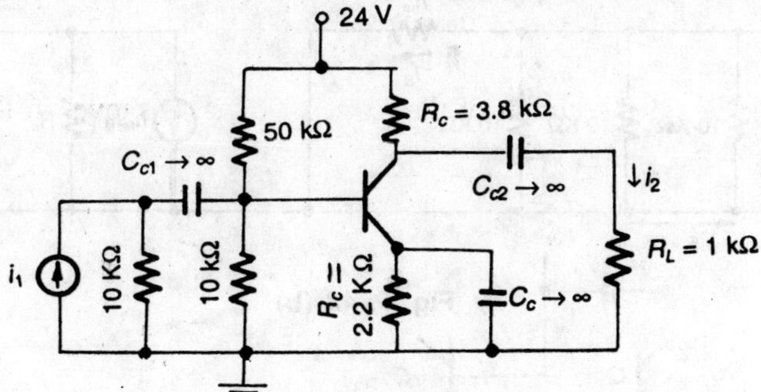

Fig. P. 43

Solution :

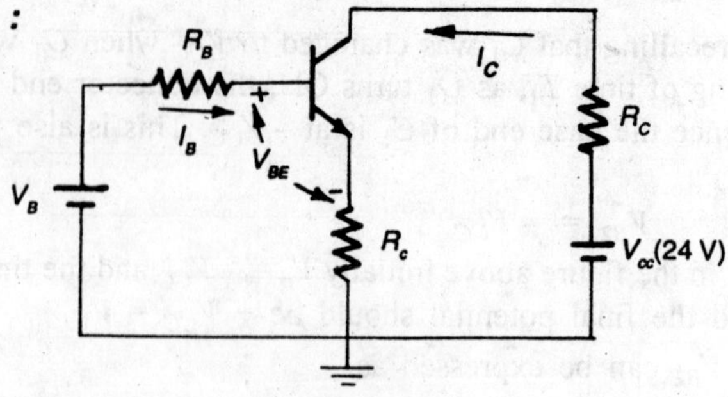

Fig. P. 43 (a)

From, Thevenin's theorem,

$$V_B = 24 \times \frac{10}{60} = 4 \text{ volt}$$

and, $\quad R_B = R_1 \parallel R_2 = 10 \text{ K}\Omega \parallel 5 \text{ K}\Omega = 8.33 \text{ K}\Omega$

Now, applying KVL in the base-emitter loop, we have

$V_B - I_B R_B - V_{BE} - I_E R_e = 0$

or $\quad I_B = \dfrac{V_B - V_{BE}}{R_B + R_e(1+\beta)} = \dfrac{4-0.7}{8.33+51\times 2.2} = 0.027 \text{ mA}$

then $\quad I_C = \beta\, I_B = 1.36 \text{ mA}$

Thus $\quad V_{CE} = V_{CC} - I_C R_C - I_E R_e$

$= 24 - (1.36 \times 3.8) - (1.387 \times 2.2)$

$= 24 - 5.168 - 3.0514 = 15.78 \text{ volts}$

Thus, Q-point is 11.36 mA, 15.78 volt.

Small-signal equivalent circuit shown below :

As given, h_{oe} and h_{re} are neglected.

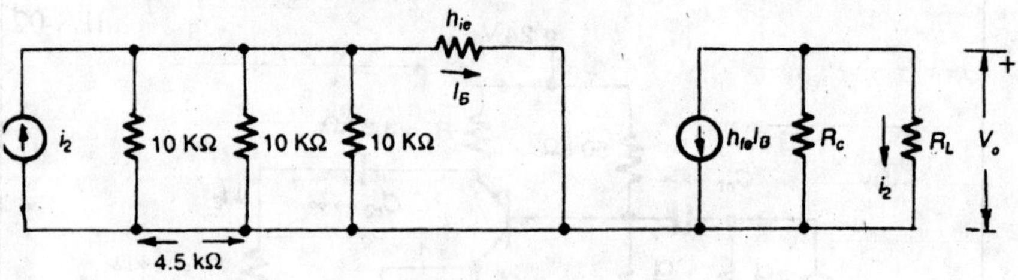

Fig. P. 43 (b)

Reader's Notes

Reader's Notes

Reader's Notes

Reader's Notes